Barbary Mural
Roc

D0926789

Petrologic Phase Equilibria

A SERIES OF BOOKS IN GEOLOGY

EDITOR: James Gilluly

Petrologic Phase Equilibria

W. G. ERNST

UNIVERSITY OF CALIFORNIA
LOS ANGELES

W. H. FREEMAN AND COMPANY
San Francisco

Library of Congress Cataloging in Publication Data

Ernst, Wallace Gary, 1931–
   Petrologic phase equilibria.

   Bibliography: p.
   Includes index.
   1. Phase rule and equilibrium. 2 Petrology. I. Title.
QE431.5.E76   552   76–3699
ISBN 0–7167–0279–7

Copyright © 1976 by W. H. Freeman and Company

No part of this book may be reproduced by any mechanical,
photographic, or electronic process, or in the form of
a phonographic recording, nor may it be stored in a retrieval
system, transmitted, or otherwise copied for public or private
use, without written permission from the publisher.

Printed in the United States of America

1 2 3 4 5 6 7 8 9

Contents

Preface

The principles of phase equilibrium are central to an understanding of the disciplines of mineralogy, petrology, and geochemistry. The mineralogic and bulk-chemical diversity of igneous and metamorphic rocks results in large part from the operation of well-known physical-chemical processes. Although statistical mechanics and statistical thermodynamics provide the atomistic basis for all such interactions, mineral equilibria are as readily comprehended at the macroscopic level. Indeed, classical thermodynamics, which is the foundation of the subject of phase equilibrium, had its origin in observational, readily measurable, macroscopic phenomena. Therefore, although at times in this text we will inquire into the atomistic nature of certain processes, it will be sufficient for our purposes to describe most interactions involving mineralogic and petrologic systems in terms of macroscopic classical thermodynamics.

This book is the outgrowth of a course I have given for the past 14 years at UCLA. It is aimed at advanced undergraduates and beginning graduate students, and assumes a bachelor's level training in earth sciences, including fundamental courses in calculus and chemistry, as well as geology. Mineralogic, petrologic, and geochemical problems will be attacked at appropriate places, and we will assume a basic appreciation for the problems. However, the emphasis here will be on the elucidation of the principles of mineral equilibria as applied to petrology rather than on broad attempts to systematically cover all processes and rock types. The chemical principles and applications presented will be confined to those pertaining to igneous and metamorphic rocks under crustal—and, to a lesser extent, mantle—conditions. Many sedimentary processes are more a function of inherently physical, mechanical properties of the material undergoing erosion, transportation, and deposition, for instance, and fall outside the prescribed area of concern. Chemical interactions are important for a good many sedimentary rocks, but they have already been adequately covered by many other texts, and so will not be discussed here. Finally, although the principles certainly are applicable, the subjects of lunar petrogenesis and trace-element geochemistry will not be addressed.

I have made use of many experimental petrology, geochemistry, and thermodynamics texts and research works in the course given at UCLA. Among those found to be most helpful, I would like to call attention especially to the following: Bowen (1928); Glasstone (1947); Ricci (1951); Tuttle and Bowen (1958); Fyfe, Turner, and Verhoogen (1958); Korzhinskii (1959); Lewis and Randall (1961); Abelson (1959, 1967); Kern and Weisbrod (1967); Turner (1968); Wyllie (1971); Broecker and Oversby (1971); Moore (1972); Miyashiro (1973); Winkler (1974); Carmichael, Turner, and Verhoogen (1974); and Ringwood (1975).

This book has greatly benefited from critical reviews by A. L. Albee, California Institute of Technology, H. J. Greenwood, University of British Columbia, and P. M. Orville, Yale University. The lectures on which the text is based received helpful feedback in one form or another from many UCLA students. The illustrations were drafted by Julie Guenther and Vicki Jones, technical typing was done by Lenore Aagaard and Julie Knaack, photography was done by Lowell Weymouth. Finally, I would like to express my appreciation to the UCLA Department of Geology and the Institute of Geophysics and Planetary Physics for continued support.

December 1975   *W. G. Ernst*

Petrologic Phase Equilibria

CHAPTER 1

Elements of Classical Thermodynamics

The principles of classical thermodynamics will be useful to us in understanding mineral equilibria, but before we begin discussing them, we need to have a few working definitions. The significance of these defined terms may not seem obvious at this stage, but we must have them in order to introduce precision into our discussions and, by clearly spelling out the concepts embodied in these definitions, to avoid future confusion.

DEFINITIONS AND CONVENTIONS

Phase

A phase is a substance homogeneous throughout and mechanically separable, or at least distinguishable, from other substances—other phases—having different properties. Minerals are examples of phases. Individual minerals may show compositional zoning, which results from chemical disequilibrium. Such inhomogeneity causes some minerals (and other substances as well) to depart from the strict definition of a phase given here, but to avoid an awkward problem in nomenclature, we will allow some chemical heterogeneity in the substances to which we apply the term phase. In principle, each mineral is separable, or at least distinguishable, from other minerals present in the material being examined. Minerals, of course, are not the only examples of phases.

A chemically homogeneous gas is a phase; obviously, its properties are very different from those of minerals. Liquid water is a phase, and its properties contrast strongly with those of other forms of H_2O.

State of Aggregation

The properties of a phase are a function of the attractive forces that bind the atoms, ions, or molecules of the substance together. There are three readily distinguished conventional states: solid, liquid, and gas. A solid is characterized by strong bonding and a high degree of three-dimensional atomic periodicity; it has very high viscosity, hence possesses a coherent shape which is modified only slowly (if at all) with the passage of time. Gas molecules, on the other hand, are only weakly attracted to one another; so viscosity in a gas is extremely low—that is, there is no long-range ordering of the particles—and hence a gas spontaneously fills any volume to which it has access. A liquid has properties, such as bonding and atomic periodicity, that are intermediate between those of solids and gases. It is condensed like a solid and has a specific molar volume, yet has a low-enough viscosity to flow relatively rapidly and to change shape with time. The term fluid will be employed where properties of the substance are transitional between those of the liquid and gaseous states. Although many condensed phases (generally of differing chemistry) can coexist at equilibrium (see definition further on), we can have only one gas phase within a given container, because gases mix completely and homogeneously in all proportions.

Component

The components of a phase consist of the smallest number of chemically distinct substances needed to specify the bulk composition of the phase, including all of its chemical variations. The number of components depends to some extent on the kind of reaction we wish to consider. For instance, given stoichiometric H_2O, we may define the composition of the phase "water" as one component, namely, H_2O itself; certain equilibria, such as the polymorphic transition of water to ice, are adequately described in terms of this single component. However, for the molecular dissociation of H_2O, we must consider a two-component system, namely, a system involving atomic or molecular hydrogen and oxygen; for ionic dissociation of water, the two-component system consists of ions of hydrogen and hydroxyl. In the first example, we have chosen the elements hydrogen and oxygen as the components for convenience. Instead of using H and O (or H_2 and O_2), we could employ H_2 and H_2O as components; the species O_2 would then be described by the expression $2H_2O - 2H_2$. That is, we need not preclude negative quantities of components.

Usage of Phase and Component

Many petrologists use mineralogic phases and chemical components in a some-what ambiguous and interchangeable way. For instance, the compositional system $Na_2O \cdot Al_2O_3 \cdot 6SiO_2 - K_2O \cdot Al_2O_3 \cdot 6SiO_2 - SiO_2 - H_2O$ is often referred to as "the system albite–orthoclase–quartz–water." Obviously, the chemical compositions of some phases, such as quartz, tridymite, and cristobalite, may be identical to a component, in this case, SiO_2. But the term "quartz" implies P–T conditions below the quartz-tridymite equilibrium (see Figure 4.3). Not only does "orthoclase" describe a structural state (reflecting a certain degree of Si–Al ordering) and hence certain physical conditions, but it also has a restrictive chemical connotation, i.e., $\sim KAlSi_3O_8$; in fact, the actual phase encountered under most petrologically interesting conditions is a $(K,Na)AlSi_3O_8$ solid solution. Finally, "water" refers to liquid H_2O, not to the rather tenuous aqueous phase containing dissolved silica and alkalis. To avoid unwarranted inferences about phase compositions and physical conditions, we will use capitalized abbreviations for components, such as **Ab** for $Na_2O \cdot Al_2O_3 \cdot 6SiO_2$. Where writing out the compositon of a component, we will generally present it in terms of oxide proportions rather than as mineralogic structural formulas. Where mineral names are abbreviated, we will use lowercase letters, e.g., **ms** for muscovite, **ab** for albite.

System

The system is that portion of the universe being considered. It may be, for instance, the contents of a small crucible, a hand-size volume of rock, or a granite batholith.

Open System

In open systems, changes of energy, mass, and bulk composition can occur. Material may be added to or subtracted from that portion of the universe being considered.

Closed System

In a closed system, mass is not transported beyond the boundaries of the system. In other words, the system retains constant mass and constant bulk composition. However, energy transfer beyond the boundaries of the system is possible. Thus heat may be added to or subtracted from the system, work may be done on the system or by the system, and the total energy need not remain constant.

Isolated System

In an isolated system, the total energy and mass of the system are fixed. There is no energy transfer or mass transfer beyond the margins of the system.

Homogeneous Reaction

A homogeneous reaction is one which occurs within a single phase. For instance, we can compare the distribution of ferrous iron and magnesium between two different octahedrally coordinated cation structural sites in orthopyroxene (hypersthene). This distribution will be a function of temperature: at higher temperatures there is a more random distribution of iron and magnesium between the two cation sites; whereas at lower temperatures iron is preferentially ordered in one of the two positions, magnesium in the other. The important point to remember here is that we are dealing with a single, macroscopically homogeneous phase. Another familiar example is the dissociation—or association—of substances in a multicomponent gas phase. We can imagine a variety of reactions taking place in the atmosphere: for instance, the production of eye irritants in smog is due to the formation of nitrous oxides and degradation of hydrocarbons, but prior to the disconcerting precipitation of particulate matter, this is a homogeneous reaction.

Heterogeneous Reaction

In contrast to a homogeneous reaction, which involves only one phase, a heterogeneous reaction involves two or more phases. The melting of ice is a heterogeneous reaction. So is the decomposition of muscovite to form K-feldspar, corundum, and H_2O.

Equilibrium

The equilibrium state is defined as the condition of minimum energy for that portion of the universe being considered; as will be proven later on, if temperature and pressure are specified, the equilibrium configuration possesses the lowest possible Gibbs free energy (denoted G). A consequence of this minimum-energy configuration is that, for an equilibrium situation, the observable properties of the system undergo no change with time, as long as temperature and pressure remain constant. An exception to this proviso occurs where two or more phase assemblages possess identical values for the Gibbs free energy. In such a situation, an infinitesimal perturbation of the system would cause a slight change in the proportions of the phase assemblages, but the equilibrium, or *stable reaction* would take place with $\Delta G = 0$.

Stability

Stability is the condition of equilibrium (minimum G).

Disequilibrium or Instability

Instability is the condition of nonequilibrium, in which the energy of the system is not a minimum; that is, there is another configuration in which the system would have a lesser value (lower G). Where instability occurs, properties of the system do change with time, although not necessarily at a perceptible rate; $\Delta G \neq O$.

Metastability

The condition of metastability is such that there is no apparent change in the observable properties of the system with time; yet there exists another state not present in this system to which the metastable configuration can alter but not conversely. What this really means is that the energy, G, of a metastable state exceeds that of one or more other phase configurations; a prohibitively large activation energy must be supplied to convert the metastable state to a more stable (lower G) configuration, and so the reaction is inhibited.

Metastable Reaction

Where two or more metastable assemblages have the same value of Gibbs free energy, an infinitesimal perturbation of the system will promote the incipient conversion if one phase compatibility to another under conditions in which $\Delta G = O$. This is a special kind of "equilibrium" reaction, but one which takes place in the absence of the minimum-G configuration.

Intensive and Extensive Parameters

An intensive property is one whose value is independent of the mass of the system, whereas the value of an extensive property does depend on the mass. For instance, the intensive parameters temperature, pressure, density, and index of refraction are not functions of the amount of matter under consideration, whereas the extensive parameters length, heat content, volume, and weight are.

Convention for Reactants and Products

For all heterogeneous reactions considered here, the righthand side of an equation of the type "reactants = products" will refer to the chemically equivalent high-temperature assemblage. As far as possible, for homogeneous reactions the more dissociated (or disordered) assemblage will be given on the right. The change during the reaction in any extensive property (except mass, which obviously must remain constant for a stoichiometrically balanced reaction), such as volume, is by convention taken as the sum of the product values minus the sum of the reactant values: thus ΔV for a reaction is simply

$$V_{\text{products}} - V_{\text{reactants}}.$$

Standard-State Conditions

Reference states for substances participating in a particular reaction must be defined somewhat arbitrarily, and the choice is generally based on convenience. All standard-state substances are chosen as pure and, except where expressly stated otherwise, at one atmosphere pressure. With the temperature held constant at some conveniently specified value, the activities of condensed substances and the fugacities of gases are given unit values. (See following section, *Critical Relationships*, for discussions of activity and fugacity.) For any property of interest, we will indicate the standard-state value by a superscript o; thus the volume of phase i as a pure substance is V_i^0.

Now that some basic concepts have been defined and conventions stated, we can turn to a consideration of selected thermodynamic relationships. Additional definitions will be introduced further on as the need arises. Conversion factors which will also be useful to us later are listed in Appendix 1 (see also Clark, 1966).

CRITICAL RELATIONSHIPS

Zeroth Law of Thermodynamics

Two bodies, each of which is in thermal equilibrium with a third body, are in thermal equilibrium with each other—that is, are at the same temperature. This relationship, which is simple enough to require no elaboration, says nothing about chemical equilibrium; the bodies referred to may, for example, be at the same temperature, but if placed in contact would tend to interact because of compositional incompatibility.

First Law of Thermodynamics

The total energy of an isolated system must remain constant, although there may be changes from one form of energy to another.* If we designate the total, or internal, energy of an isolated or closed system as E, heat as Q, and work

---

*We acknowledge, of course, the relationship between mass and energy given by Einstein's equation, which tells us that energy equals mass times the square of the speed of light, but this does not concern us in a treatment of classical thermodynamics.

as W (the latter two being forms of energy), then the circuit integral (displacement from, then returning to, the original state) which involves no change in net internal energy is

$$\oint (dQ - dW) = 0. \tag{1.1}$$

The convention which we will employ throughout is that heat is positive if *added to* the system, and that work is positive if *performed by* the system. Now suppose that the system is closed to import or export of matter, but not to the various forms of energy. Then the first law becomes

$$dE = dQ - dW. \tag{1.2}$$

You may recall from elementary physics that mechanical work is the product of force times distance. In turn, force is simply pressure, P, times surface area. Therefore, mechanical work is simply the product of pressure times surface area times distance, or pressure times volume V. At constant pressure, then,

$$dW = P \, dV. \tag{1.3}$$

Substituting this expression for work in (1.2) yields the most familiar form of the first law of thermodynamics:

$$dE = dQ - P \, dV. \tag{1.4}$$

Ideal Gas Law

The absolute temperature, T, of a gas is proportional to the product of its pressure times its molar volume. For thermochemical computation, the temperature is always given in degrees Kelvin. The behavior of an ideal gas is given by the expression

$$PV = nRT, \tag{1.5}$$

where n is the number of moles, and R is the universal gas constant. Real gases depart from ideal behavior to some extent, especially at low temperatures and high pressures, where interactions between the molecules become important. Later we shall replace pressure by fugacity, in order to adapt the ideal gas law to real gas behavior.

Enthalpy

The enthalpy, or heat content, is defined as

$$H = E + PV. \tag{1.6}$$

It is clear from this relationship that enthalpy is defined as the sum of two energy terms. Differentiating (1.6) at constant pressure,

$$dH = dE + P\,dV; \tag{1.7}$$

since from (1.4) we have $dE = dQ - P\,dV$, (1.7) reduces at constant pressure to

$$dH = dQ. \tag{1.8}$$

Heat Capacity

Heat capacity, C, is defined as the heat added to the system divided by the rise in temperature. The heat capacity of a substance increases as the ways increase in which energy is absorbed kinetically. Various modes of energy absorbance include atomic and molecular translations, rotations, and vibrations. However, as long as a new mode of energy absorbance does not come into play, the heat capacity of a specific substance may be considered to be nearly constant—at least within a limited temperature range. It is convenient to distinguish two different heat-capacity values, appropriate for constant pressure and constant volume,

$$C_P = \left(\frac{\partial Q}{\partial T}\right)_P \tag{1.9}$$

and

$$C_V = \left(\frac{\partial Q}{\partial T}\right)_V, \tag{1.10}$$

respectively.

The heat capacity of a substance at constant pressure exceeds the corresponding heat capacity at constant volume. In other words, the temperature rise within a system to which a certain amount of heat has been added is less for a system in which pressure is fixed than for one where the volume is fixed. This is so because, when pressure is constant, expansion of the substance results in work being performed by the system; hence a portion of the supplied energy is used up, and only the remainder is left to contribute to the elevation of the temperature. When volume is constant, no work is done, and all the heat added is reflected in the thermal energy and temperature increase of the system.

Second Law of Thermodynamics

Two statements of the second law are: it is impossible to transfer heat from a cold body to a hot body; and spontaneous reactions run toward lower-energy states and are not reversible. The second law is an embodiment of the concept of entropy. Entropy, S, is a measure of the disorder of a system, and is a single-valued function of the state of the system. Like the internal energy, it is an extensive variable, dependent on the mass of the system. Under equilibrium conditions, the change of entropy during a reaction is given as

$$dS = \left(\frac{dQ}{T}\right)_{rev} \qquad (1.11)$$

For a spontaneous, disequilibrium process,

$$dS > \left(\frac{dQ}{T}\right)_{irr}. \qquad (1.12)$$

To appreciate this difference between reversible and irreversible processes, let us consider the fusion of solid H_2O. At one atmosphere total pressure and $0°C$ (i.e., $273°K$), where water and ice are in equilibrium, the heat which must be added to melt one gram of ice is about 80 cal. Under equilibrium conditions, then, the entropy increase on fusion is 80/273 or 0.293 cal/gram/deg. In contrast, if we could superheat ice $3C°$ above its equilibrium melting point (assuming as well that the heat capacities of water and ice are the same, hence that the $\Delta H_{melting}$ is constant in that interval), the nonequilibrium fusion, a spontaneous irreversible process, would yield a $(\Delta H/T)_{irr}$ value less than the entropy increase:

$$0.293 \text{ cal/gram/deg} = dS > \left(\frac{dQ}{T}\right)_{irr} = 80/276 = 0.290 \text{ cal/gram/deg.}$$

Or consider the freezing of liquid H_2O at equilibrium in contrast to freezing at a supercooled state of, say, $-3°C$. The equilibrium entropy change is -0.293 cal/gram/deg, whereas the $(dQ/T)_{irr}$ value of the spontaneous reaction is $-80/270$ or -0.296 cal/gram/deg. Again it is seen that the quantity $(dQ/T)_{rev}$, and therefore dS, is greater than $(dQ/T)_{irr}$.

Third Law of Thermodynamics

We have just defined entropy as a measure of the degree of disorder or randomness in the arrangement of constituents within a system or a phase. Provided reaction kinetics do not inhibit rearrangement, individual phases become more ordered—hence have lower entropies—at lower temperatures. For a perfect crystal, one which contains no lattice defects, impurities, etc., the entropy, S_0, becomes zero at $0°K$ (i.e., $-273°C$). Knowing this hypothetical $0°K$ entropy allows us to calculate the isobaric entropy at any temperature as follows. Rewriting equation (1.9), the definition of heat capacity at constant pressure, we have

$$dQ = C_P \, dT, \qquad (1.13)$$

which may be substituted in the expression for entropy change, equation (1.11):

$$dS = \frac{C_P \, dT}{T}. \tag{1.14}$$

The entropy at some fixed pressure and any temperature T therefore can be determined by evaluating the expression

$$S_T = S_0 + \int_0^T \frac{C_P \, dT}{T} + \sum \frac{\Delta H_{trans}}{T}, \tag{1.15}$$

or

$$S_T = S_0 + \int_0^T C_P \, d \ln T + \sum \frac{\Delta H_{trans}}{T}. \tag{1.16}$$

To calculate the heat capacity at constant pressure, we need an analytical (polynomial) expression which is a function of temperature. The ΔH_{trans} values are heats of transitions at discrete temperatures (or within temperature intervals), such as polymorphic reactions, fusion, etc.; these, of course, also contribute to the total entropy of a system or phase, and must be evaluated in order to compute S_T.

Entropy of Mixing

The mixture of components in solid, liquid, and gaseous solutions involves an increase in compositional disorder, and hence an increment in entropy. Evaluation of this ΔS_{mixing} may be derived as follows. From a combination of the first and second laws of thermodynamics, we have

$$dE = T \, dS - P \, dV \tag{1.17}$$

for a reversible process. Rearrangement of this expression gives

$$dS = \frac{dE + P \, dV}{T} \tag{1.18}$$

(see also equation 1.36). For one mole of ideal gas occupying a particular volume, we can write

$$dE = C_V \, dT \tag{1.19}$$

and, of course, $P = RT/V$. Hence substituting these values into (1.18) gives

$$dS = \frac{C_V \, dT + RT \, d \ln V}{T} = C_V \, d \ln T + R \, d \ln V. \tag{1.20}$$

Integration yields

$$S = C_V \ln T + R \ln V + C, \qquad (1.21)$$

where C is a constant of integration. Consider now n different one-component ideal gases, all at the pressure P and the temperature T, and residing within a compartmentalized pressure vessel of total volume V; each pure gas thus occupies a volume v, and, of course, the sum of these partial volumes is V. The total entropy of the system S' is merely the sum of the entropies of the compartmentalized gases, as given by the last equation (1.21):

$$S' = \sum n(C_V \ln T + R \ln v + C). \qquad (1.22)$$

When the partitions are removed, the one-component gases will mix spontaneously and irreversibly. Now each species occupies the entire volume of the system, and hence the total entropy S'' of the system is

$$S'' = \sum n(C_V \ln T + R \ln V + C). \qquad (1.23)$$

For any gas component i, the ratio of its initial volume v_i to the total volume V is equal to its mole fraction X_i (and its partial pressure), because ideal gas behavior allows the total pressure P to remain constant on mixing:

$$\frac{v_i}{V} = \frac{n_i}{n} = X_i. \qquad (1.24)$$

Replacing v_i by VX_i in equation (1.22), we have

$$S' = \sum n(C_V \ln T + R \ln V + R \ln X_i + C). \qquad (1.25)$$

The entropy of mixing for all n species is therefore $S'' - S'$, or equation (1.23) minus equation (1.25):

$$\Delta S_{\text{mixing}} = \sum -n_i R \ln X_i \qquad (1.26)$$

or, for one mole,

$$\Delta S_{\text{mixing}} = \sum -X_i R \ln X_i. \qquad (1.27)$$

Although derived for perfect gases, this equation can also be applied to situations of ideal configurational disorder in condensed phases. Because the mole fractions range individually from 0 to 1, values of $\ln X_i$ will be negative, and hence the entropy of mixing is invariably positive; it achieves a maximum value at compositions in which the several X_i quantities are equal.

Gibbs Free Energy

The energy available to drive chemical reactions is less than the internal energy intrinsic to a closed system, because a portion of the energy is tied up as disorder and another part as a work function (i.e., a PV term). What is available has been termed the Gibbs free energy, and is defined as

$$G = E + PV - TS = H - TS. \tag{1.28}$$

A useful relationship for us to note here is that, considering an equilibrium process proceeding at constant temperature and pressure,

$$\Delta G = \Delta H - T \Delta S. \tag{1.29}$$

However, let us return to the definition of Gibbs free energy presented in equation (1.28), that is, without the isothermal and isobaric constraints, but remembering that the definition holds only for a system closed to the import or export of matter. Differentiation of (1.28) yields:

$$dG = dE + P\,dV + V\,dP - T\,dS - S\,dT. \tag{1.30}$$

Substituting expressions of the first and second laws of thermodynamics—equations (1.4) and (1.11), respectively—into equation (1.30) gives

$$\begin{aligned} dG &= T\,dS - P\,dV + P\,dV + V\,dP - T\,dS - S\,dT \\ &= V\,dP - S\,dT. \end{aligned} \tag{1.31}$$

Finally, holding pressure and temperature individually fixed yields the important relationships derived from equation (1.31):

$$\left(\frac{\partial G}{\partial T}\right)_P = -S \tag{1.32}$$

and

$$\left(\frac{\partial G}{\partial P}\right)_T = V. \tag{1.33}$$

These equations indicate how the equilibrium Gibbs free energy of a phase (or the ΔG of a reaction) under closed-system conditions changes with change in state variables as a function of entropy and volume (or ΔS and ΔV). We will employ equations (1.32) and (1.33) throughout the book in accounting for, and predicting, various phase relationships.

Let us now proceed to a proof of the relationship that, for a closed system at constant pressure and temperature, reversible equilibrium reactions take place with no change in Gibbs free energy, whereas spontaneous reactions involve a decrease in G. Recall from equations (1.11) and (1.12) that

$$dS = \left(\frac{dQ}{T}\right)_{rev} > \left(\frac{dQ}{T}\right)_{irr};\tag{1.34}$$

hence

$$T\,dS \geqq dQ.\tag{1.35}$$

The first law of thermodynamics, rearranged, then becomes

$$T\,dS \geqq dE + P\,dV.\tag{1.36}$$

If we substitute this relationship into the differentiated version of the general expression for Gibbs free energy, equation (1.30), we have

$$\begin{aligned}dE + P\,dV + V\,dP - T\,dS - S\,dT\\ = T\,dS + V\,dP - T\,dS - S\,dT \geqq dG\end{aligned}\tag{1.37}$$

and, finally,

$$V\,dP - S\,dT \geqq dG.\tag{1.38}$$

Here

$$0 = dG \text{ at constant } P \text{ and } T \text{ for a reversible process,}\tag{1.39}$$

and

$$0 > dG \text{ at constant } P \text{ and } T \text{ for an irreversible process,}\tag{1.40}$$

because at constant temperature and pressure equation (1.38) requires that the Gibbs-free-energy change be zero for a reversible, equilibrium reaction where $dS = (dQ/T)_{rev}$, or dG be negative for a spontaneous reaction where $dS > (dQ/T)_{irr}$. A corollary of equation (1.38) is that, since all reactions proceed at either constant or decreasing Gibbs free energy, G must be a minimum at equilibrium.

Chemical Potential

The Gibbs free energy is a function of temperature, pressure, and the amounts of the constituent components. G is a quantity readily employed to describe the over-all behavior of a system or of a specific phase. However, in phases showing compositional variation, it is useful to establish another, related quantity, the chemical potential. This term, designated by μ, is the partial molar free energy of a component, and is defined by the expression

$$\mu_i = \left(\frac{\partial G}{\partial n_i}\right)_{T, P, n_1, n_2, \ldots n_{i-1}}, \tag{1.41}$$

where $n_1, n_2, \ldots n_i$ are the number of moles of components 1, 2, ... *i*. The general relationship of the Gibbs free energy to the physical conditions and the changes in composition of a system or phase are thus:

$$dG = \left(\frac{\partial G}{\partial T}\right)_{P, n_1, n_2, \ldots n_i} dT + \left(\frac{\partial G}{\partial P}\right)_{T, n_1, n_2, \ldots n_i} dP$$

$$+ \left(\frac{\partial G}{\partial n_1}\right)_{P, T, n_2, \ldots n_i} dn_1 + \ldots + \left(\frac{\partial G}{\partial n_i}\right)_{P, T, n_1, n_2, \ldots} dn_i. \tag{1.42}$$

For a system open to the import or export of matter, equation (1.31) must therefore be written in a more general fashion:

$$dG = V\,dP - S\,dT + \mu_1\,dn_1 + \mu_2\,dn_2 + \ldots + \mu_i\,dn_i. \tag{1.43}$$

For a reaction involving a single component, when the system is in a state of equilibrium we have seen that the Gibbs free energy of reactants and products must be equal, because, from equation (1.39), $dG = 0$; in other words, the G per mole of reactant is identical to the G per mole of product. An analogous relationship holds for the partial molar free energy in a multicomponent system: at equilibrium, the chemical potential of a component has the same value in all phases which contain the component. An important proviso here is that a phase must be able to accommodate variable amounts of, say, component *i*, otherwise μ_i is undefined for the composition of the phase itself. Let us consider a system at some arbitrary but constant temperature and pressure consisting of the phases *A*, *B*, and *C*, each of which contains the components 1, 2, and 3. At equilibrium,

$$_A\mu_1 = {}_B\mu_1 = {}_C\mu_1; \quad _A\mu_2 = {}_B\mu_2 = {}_C\mu_2; \text{ and } _A\mu_3 = {}_B\mu_3 = {}_C\mu_3. \tag{1.44}$$

However, in general,

$$_A\mu_1 \neq {}_A\mu_2 \neq {}_A\mu_3. \tag{1.45}$$

Clapeyron Equation

Let us now derive a relationship which, although of great simplicity, will be of considerable value in our later studies involving *P*–*T* diagrams. Consider a general reaction $r = p$ (equilibrium between reactants and products). For reaction under conditions of stability, with no import or export of matter, we have, from equation (1.31),

$$dG_r = V_r dP - S_r dT,$$

and   (1.46)

$$dG_p = V_p dP - S_p dT.$$

Moreover, from equation (1.39), we know that the Gibbs-free-energy changes of the products and reactants at equilibrium must be equal; hence

$$V_p dP - S_p dT = V_r dP - S_r dT. \tag{1.47}$$

By rearrangement, this becomes

$$(V_p - V_r) dP = (S_p - S_r) dT. \tag{1.48}$$

And finally, we have

$$\frac{dP}{dT} = \frac{\Delta S}{\Delta V}. \tag{1.49}$$

This important relationship, the Clapeyron equation, gives the P-T trajectory of equilibrium states for a reaction as a function of the entropy and volume changes. Remembering the definition of enthalpy, it is easily shown that, at any specific pressure,

$$\Delta S = \frac{\Delta H}{T}; \tag{1.50}$$

hence the Clapeyron equation is often given as

$$\boxed{\frac{dP}{dT} = \frac{\Delta H}{T \Delta V}.} \tag{1.51}$$

If the reaction considered involves the evolution or consumption of a gas phase, as a first approximation we may substitute V_{gas} for the ΔV of the total reaction, since the volumes of the condensed phases are characteristically small in contrast to that of the volatile phase, especially at low and moderate pressures. Then equation (1.51) may be approximated by

$$\frac{dP}{dT} = \frac{\Delta H}{T V_{gas}}. \tag{1.52}$$

Assuming ideal-gas-law behavior as another approximation, per mole of volatile, we can substitute the relationship

$$V_{gas} = \frac{RT}{P} \qquad (1.53)$$

into equation (1.52):

$$\frac{dP}{dT} = \frac{P\,\Delta H}{RT^2}. \qquad (1.54)$$

Rearranging variables leads to

$$\left(\frac{dP}{P}\right)\left(\frac{T^2}{dT}\right) = \frac{\Delta H}{R}, \qquad (1.55)$$

or finally

$$\frac{d\ln P}{d\left(\frac{1}{T}\right)} = -\frac{\Delta H}{R} \quad \text{or} \quad \frac{d\log P}{d\left(\frac{1}{T}\right)} = -\frac{\Delta H}{2.303R}. \qquad (1.56)$$

This is the Clausius-Clapeyron equation, and is useful for dealing with heterogeneous reactions which evolve or consume a volatile phase. The heat of a reaction is generally constant, or nearly so, within a finite interval of physical conditions; hence a plot of the loci of equilibrium between products and reactants against log P (or log fugacity, as explained farther on) and $1/T$ results in a virtually straight line (the slope of which equals $-\Delta H/2.303R$); such plots allow extrapolation to P–T values not readily reached by experiment (e.g., see Figure 1.1). Because of the assumptions employed, the Clausius-Clapeyron

FIGURE 1.1.

Plot of log f_{CO_2} against $(T°K)^{-1}$ for the reaction $CaCO_3 + SiO_2 + TiO_2 = CaTiSiO_5 + CO_2$ (Schuiling and Vink, 1967). Experimental data indicated by filled rectangles.

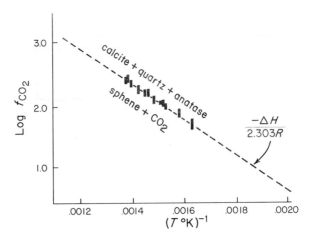

equation is most useful at high temperatures and low pressures, where ideal-gas-law behavior is most closely approached, and where the disparity between the aggregate volume of the condensed assemblage and that of the volatile phase is most marked. The equation may be used for high-pressure equilibria, provided a correction is made for the ΔV_{solids} (Orville and Greenwood, 1965), and provided real gas behavior is adapted to the ideal gas law by substitution of fugacity (see below) for pressure.

Fugacity

For one mole of an ideal gas, the quotient PV/RT is unity, but actual gases depart to some extent from such behavior. For the real pressure (i.e., the partial or total pressure of the volatile component) we therefore substitute the fugacity, f, which allows application of the ideal gas law. The ratio of the fugacity to the real pressure is termed the fugacity coefficient, γ. Thus, for any real gas component, i, we have

$$\gamma_i = \frac{f_i}{P_i}. \tag{1.57}$$

At constant temperature, equation (1.31) reduces to (1.33), or, rearranged,

$$dG = V\,dP. \tag{1.58}$$

Combined with the ideal gas law, this becomes

$$dG = \frac{RT}{P}\,dP = RT\,d\ln P. \tag{1.59}$$

In dealing with a real gas, the pressure must be replaced by the fugacity to be entirely correct:

$$dG = RT\,d\ln f. \tag{1.60}$$

Where a multicomponent gas phase is being treated, we are concerned with the "thermodynamic partial pressure" of an individual volatile species; hence we use an expression analogous to (1.60) involving the partial molar free energy of component i:

$$d\mu_i = RT\,d\ln f_i. \tag{1.61}$$

Integration of this expression results in

$$\mu_i = \mu_i^0 + RT\ln f_i, \tag{1.62}$$

where μ_i^0 is a constant of integration which can be evaluated at unit fugacity (i.e., standard-state conditions); here, because $\ln 1 = 0$, the second term on the righthand side drops out. Of course, care must be taken to specify the particular standard state employed.

Activity

For a single component constituting a gas phase, or for individual species in a homogeneous gas, we employ pressures or partial pressures, or, more correctly, fugacities, in the equations of interest; for equilibria involving condensed phases, a closely analogous term, the activity, a, or "thermodynamic concentration" is used. In this case

$$\gamma_i = \frac{a_i}{X_i}, \tag{1.63}$$

where γ_i is the activity coefficient (similar to the fugacity coefficient defined above), and X_i is the mole fraction of component i in the solid or liquid solution. Because there need be no sharp break in physical properties between a gas and a liquid (e.g., fluid) under some conditions, it is appropriate by analogy with equation (1.60) to write the constant temperature relationship:

$$dG = RT \, d \ln a, \tag{1.64}$$

and

$$d\mu_i = RT \, d \ln a_i. \tag{1.65}$$

Integration yields

$$\mu_i = \mu_i^0 + RT \ln a_i, \tag{1.66}$$

where μ_i^0 is a constant of integration readily evaluated at unit activity. The correspondence of activity and fugacity can be realized by remembering the following: because, at equilibrium, the value of μ_i must be the same in all phases in which i occurs, so too must $d\mu_i$ for any change; hence equations (1.61) and (1.65) are related expressions. Although the ideal gas law was employed in deriving these relationships, they are clearly more generally applicable because, for a liquid or solid as well as a gas phase, the pressure-volume product must be proportional to the absolute temperature.

Van't Hoff Reaction Isotherm

The generalized form of the law of mass action is derived in the following way. Consider any reaction $aA + bB = lL + mM$ for a specific but arbitrarily chosen temperature. Here A and B are reacting species, L and M are product species, and a, b, l, and m are the stoichiometric coefficients, i.e., the number of moles of each species, "species" here meaning participating phases, or components of a homogeneous phase or phases. From equation (1.64) we have for a single participating species,

$$dG_B = bRT\, d \ln a_B = RT\, d \ln a_B{}^b;\qquad(1.67)$$

this is the Gibbs-free-energy change for b moles of species B, i.e., b times equation (1.64). For the entire reaction

$$dG = RT\,(d \ln a_L{}^l + d \ln a_M{}^m - d \ln a_A{}^a - d \ln a_B{}^b)\qquad(1.68)$$

and

$$dG = RT\, d \ln \frac{a_L{}^l\, a_M{}^m}{a_A{}^a\, a_B{}^b}.\qquad(1.69)$$

Integrating this expression yields

$$\Delta G = C + RT \ln \frac{a_L{}^l\, a_M{}^m}{a_A{}^a\, a_B{}^b},\qquad(1.70)$$

where C is a constant of integration. This constant of integration may be evaluated by considering the reaction under standard-state conditions. The standard Gibbs-free-energy change of the reaction is simply the total Gibbs free energy of pure products in their stoichiometric proportions minus the over-all G^0 of the pure reactant assemblage, or ΔG^0. Because the activities of pure solids and liquids, and the fugacities of pure gaseous species, are all of unit value at one atmosphere pressure, the second term in the righthand side of equation (1.70) is zero; hence $C = \Delta G^0$. This relationship could equally well have been derived starting with equation (1.66). Finally, we have:

$$\Delta G = \Delta G^0 + RT \ln \frac{a_L{}^l\, a_M{}^m}{a_A{}^a\, a_B{}^b}.\qquad(1.71)$$

In general, $\Delta G^0 \neq 0$, which means that reactants and products will not be stable together under standard-state conditions. Of course, at equilibrium ΔG (not ΔG^0) must be zero, as has been demonstrated in equation (1.39); so for this condition,

$$\Delta G^0 = -RT \ln \frac{a_L{}^l\, a_M{}^m}{a_A{}^a\, a_B{}^b} = -RT \ln K.\qquad(1.72)$$

This new term, K, is the equilibrium constant, and is seen to define the value of the activity quotient at equilibrium. Only when the activity quotient (a variable) actually achieves the value of K (a constant) does stability occur between reactants and products. Where these two quantities are unequal, ΔG will be either less than or greater than zero; hence the reaction will drive to the right or left, respectively. The generalized form of the law of mass action, or the van't Hoff reaction isotherm, may therefore be written:

$$\Delta G = -RT \ln K + RT \ln \frac{a_L{}^l \, a_M{}^m}{a_A{}^a \, a_B{}^b} \qquad (1.73)$$

for the considered reaction. This is another rather important equation for us, because it shows the critical relationships under equilibrium conditions between "thermodynamic concentrations" (fugacities of gas components may be substituted for activities in this expression, as appropriate) and the standard-state Gibbs-free-energy change or K. To study the influence of the state variables on K, let us recall from equations (1.32) and (1.33) that the Gibbs-free-energy changes of a reaction under standard-state conditions at constant pressure and temperature are

$$\left(\frac{\partial \Delta G^0}{\partial T}\right)_P = -\Delta S^0 \qquad (1.74)$$

and

$$\left(\frac{\partial \Delta G^0}{\partial P}\right)_T = \Delta V^0, \qquad (1.75)$$

respectively. As discussed later on, (1.75) has meaning only where the definition of standard state does not involve the restriction to one atmosphere pressure. Substituting $-RT \ln K$ for ΔG^0 from equation (1.73) in equation (1.74) and differentiating by parts yields, after several steps,

$$R\left(\frac{\partial \ln K}{\partial T}\right)_P = \frac{\Delta H^0}{T^2}. \qquad (1.76)$$

Rearrangement results in:

$$-T^2\left(\frac{\partial \ln K}{\partial T}\right)_P = \left[\frac{\partial \ln K}{\partial\left(\frac{1}{T}\right)}\right]_P = -\frac{\Delta H^0}{R}. \qquad (1.77)$$

This expression shows the effect of changing temperature on $\ln K$. Because typically the righthand side of this equation is nearly constant, increasing temperature results in a decrease in magnitude of the change in $\ln K$ (which of necessity is proportional to the change in $1/T$). In many petrologically interesting cases, values of K approach unity more closely at elevated temperatures, as will be discussed in Chapter 6. The influence of varying pressure on $\ln K$

depends on the definition of standard state. Inasmuch as the equilibrium constant is defined in terms of ΔG^0, if one atmosphere pressure is taken as part of the specification of standard-state conditions, then obviously isothermal increase in pressure will not alter the value of the standard Gibbs-free-energy change—or, therefore, K. On the other hand, some geochemists prefer to define standard state conditions as reactant and product species pure, and at the temperature *and pressure* to be considered. If we use this latter definition, ΔG^0 in general will be a function of pressure, and so will the equilibrium constant, as seen below. We again substitute $-RT \ln K$ for ΔG^0 from equation (1.73), this time in equation (1.75):

$$RT\left(\frac{\partial \ln K}{\partial P}\right)_T = -\Delta V^0. \tag{1.78}$$

Rearrangement results in:

$$\left(\frac{\partial \ln K}{\partial P}\right)_T = -\frac{\Delta V^0}{RT}. \tag{1.79}$$

For many reactions the volume change is negligible; hence as a first approximation in such cases we may consider the equilibrium constant to be virtually independent of pressure, even where we define standard state in such a way that ΔG^0 technically is a function of pressure. Naturally, when we turn our attention to mantle conditions, where pressures are on the order of 50–100 kilobars or more, even a small dependence on ΔV^0 will result in large changes in the equilibrium constant compared to one-atmosphere conditions.

Phase Rule

The variance or degrees of freedom, denoted by F, possessed by a system at equilibrium is the number of independent parameters which must be fixed, or determined, arbitrarily in order to specify accurately the state of the system, including both the physical conditions and the compositions of all the phases. This variability is simply the number of unknown relationships minus the number of known or dependent relations. Let us consider a closed system consisting of p phases and c components; we will regard temperature and pressure as the state variables, and will ignore other influences, such as the presence of gravitational and magnetic fields. Suppose that all c components occur in each phase. Then, to specify the composition of a phase, we must determine $c - 1$ composition variables (i.e., mole fractions), the remaining one simply being the difference. Because there are p phases in the system, the total number of chemical unknowns is $p(c - 1)$. In addition, of course, temperature and pressure must also be specified, so there are a total of $p(c - 1) + 2$ unknowns; obviously, values of T and P must be identical for all phases in the system at equilibrium. However, we do know that there are certain chemical relationships among the coexisting phases; specifically, at equilibrium the chemical potential of a component is the same in all phases in which it occurs. Hence, if we determine μ_i in a single phase, we have its value for all other phases within the system. Therefore, there are $(p - 1)$ chemical restrictions for each component, or a total of $c(p - 1)$ dependent relationships. The total number of independent variables, or the degrees of freedom, may be found by evaluating the expression (unknown minus known relationships)

$$F = p(c - 1) + 2 - c(p - 1), \tag{1.80}$$

or

$$F = pc - p + 2 - pc + c, \tag{1.81}$$

and finally

$$F = c - p + 2. \tag{1.82}$$

This is the general form of the phase rule for systems closed to import and export of matter. Where physical conditions are arbitrarily fixed, the number of independent variables is correspondingly reduced; for instance, for an isobaric or isothermal system, the degrees of freedom decrease to $c - p + 1$. If a component is absent from a particular phase, there will be one less unknown compositional variable, but this situation also decreases the known restrictions by one, because the chemical potential of the lacking component is also not defined in the phase in question; for this reason, the phase rule does not in any way depend on the presence of all components in each phase.

In a closed system, the chemical potentials of all the components are a function of the bulk composition of the system and the state variables P and T. Such components may be termed the determining "inert" components, although strictly speaking, they are not inert at all but simply have their masses fixed within the system. For an open system, however, we may recognize another class of components, the so-called mobile components, m, the chemical potentials of which are controlled outside of, and imposed on, the system. These additional chemical restrictions, like that of constant temperature or of constant pressure, are reflected in the phase rule as additional variance reductions. Although the total number of unknowns (chemical variables plus pressure and temperature) remains the same as in the closed system case, $p(c - 1) + 2$, the number of dependent relationships is increased by the number of mobile components, since these are additional knowns. The total chemical potential restrictions are therefore $(c - m)(p - 1) + mp$. The total number of independent variables for open-system equilibrium is thus

CRITICAL RELATIONSHIPS   23

$$F = p(c - 1) + 2 - (c - m)(p - 1) - mp, \qquad (1.83)$$

or

$$F = pc - p + 2 - (pc - mp - c + m) - mp, \qquad (1.84)$$

and finally

$$F = c - m - p + 2. \qquad (1.85)$$

This open-system treatment provides an alternative form of the phase rule for situations in which mobility is recognized. Note that at least one component must be considered "inert," because the boundaries of the system under inspection may be moved with a specific component; otherwise, it would seem to be possible for a system consisting entirely of mobile components to contain no phases.

Reaction Rate

The velocity with which a particular reaction, of the type $r = p$ (i.e., reactant assemblage = product assemblage), takes place cannot be predicted from the principles of classical thermodynamics. In general, however, the greater the displacement from equilibrium, the faster the reaction proceeds, with conversion from reactant to product assemblage taking place infinitely slowly at equilibrium. The isobaric, isothermal speed with which the concentration of reactants, C_r, decreases as a function of time, t, is decribed by the general form of the rate equation

$$(\partial C_r / \partial t)_{P,T} = -kC_r^n, \qquad (1.86)$$

Where k is the rate constant, and n represents the order of the reaction. This latter quantity reflects the number of species which must interact to transform reactants into products; the order of a reaction is a function of the actual reaction mechanism. For a zeroth-order reaction, no intermediate species are involved, and the proportion of reactants converted to products per unit time is constant at a given P and T. In contrast, for values of $n \geq 1$, the decrease in concentration of reactants progresses logarithmically; predictably, where intermediate species are necessary for reaction, as their concentrations diminish, so does the reaction velocity. Of course, the equilibrium configuration is independent of reaction mechanism.

Perhaps here it is worth pointing out that equations describing the rate of diffusion and the required activation energy for diffusion are similar in form to those presented here and the following discussion of activation energy; quantitative descriptions of these sorts of processes are based on the concepts of statistical mechanics, and will not be elaborated on in this text.

Activation Energy

In spite of the fact that a Gibbs-free-energy difference may exist between less stable reactants and more stable products, input of energy is generally required to transform the initial configuration to a subsequent one. This excess or activation energy, E^*, is represented schematically in Figure 1.2. Clearly, the type

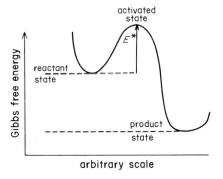

FIGURE 1.2.
Schematic diagram showing activation energy required to transform a reactant assemblage to a product assemblage. E^* is the activation energy.

of conversion illustrated involves an intermediate, highly unstable (i.e., activated) state. Assuming the classical Arrhenius equation applies here, the isobaric rate constant k is related to E^* by the equation

$$k = Ae^{-E^*/RT}, \tag{1.87}$$

where A is a constant of integration (a probability function). Analogous to the Clausius-Clapeyron relationship previously discussed, the slope of the curve relating $\log k$ to $1/T_{(Kelvin)}$ is $-E^*/2.303R$ (see equation 1.56 and Figure 1.1).

CHAPTER 2

Experimental Approach
to Phase Equilibrium

APPARATUS AND TECHNIQUES EMPLOYED

For simplicity we can recognize three different types of equipment, each characterized by a certain pressure range in which it operates, and each suited for certain kinds of investigation. Quenching furnaces have been used since the turn of the century; they are generally operated at one atmosphere total pressure. Various types of hydrothermal pressure apparatus, either externally or internally heated, have come into routine service with the advent of high P–T alloys, largely since the end of the Second World War; the range of pressures investigated is from a few tens of bars to 10 or 11 kilobars. High-pressure equipment employed for mineralogic-petrologic geochemistry is of a variety of designs, but the piston-cylinder device is by far the commonest; the capabilities of this apparatus range from slightly below 10 to more than 60 kilobars. A very brief introduction to some of the various types of equipment currently being employed is presented here; for a more comprehensive treatment, see Ulmer (1971).

In all the types of equipment to be described, the sample container is placed next to the hottest part of the furnace assembly. This placement provides the most stable and well-controlled temperature, as well as the highest temperature that can be attained for a given expenditure of power; this efficiency of operation thus maximizes the useful life of the equipment.

Quenching Furnaces

Figure 2.1 presents a schematic diagram of this type of one-atmosphere equipment. The anhydrous charge sample is placed in a nonreactive crucible or envelope, next to a thermocouple and suspended within the central cavity of the furnace. The anular core either is open to the atmosphere or is connected to a homogeneous gas reservoir or a gas-mixing apparatus, which provides a flow of a gas of constant composition, monocomponent or multicomponent, past the sample. The furnace core consists of a ceramic tube, around which resistance wire (e.g., platinum or nichrome) is wound; an insulating material, such as magnesia, alumina, or asbestos, is packed about this heating element, to shield the outermost furnace casing from high terperatures and to prevent excessive heat losses to the laboratory surroundings. Such equipment is safe, relatively trouble-free, and inexpensive to build except where platinum wind-

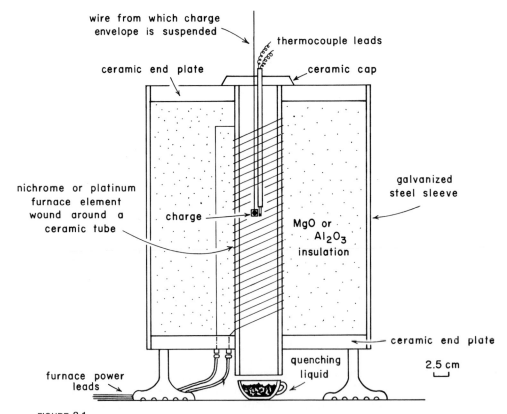

FIGURE 2.1.

One-atmosphere quenching furnace in simplified, schematic form. Control of atmosphere in ceramic tube is not shown.

ings must be used to achieve high temperatures (1,600°C±). At the termination of the experiment, the sample is removed from the furnace and allowed to cool to room temperature, sometimes by being dropped into a cold crucible containing a liquid of high heat capacity, such as water or mercury.

Quenching furnaces are used principally to elucidate one-atmosphere melting relationships. Because of the absence of fluxes, such as H_2O, which increase rates of recrystallization, the investigation of phase relations is generally confined to elevated temperatures; most dry silicate systems of interest melt at temperatures in excess of 1,000°C anyway. Lower-temperature, solid-solid reactions typically are much more sluggish for silicate systems. The principal petrogenetic applications resulting from use of quenching furnaces therefore involve crustal (and especially near-surface) magmatic processes.

Hydrothermal Pressure Apparatus

Figures 2.2 and 2.3 show two varieties of this type of equipment. In this apparatus, the sample is placed within the pressure-containing cell (the pressure vessel or "bomb") in a capsule of noble metal. The pressure medium employed is an inert gas, gas mixture, or fluid, such as argon, H_2O, or $CO_2 + H_2O$. Pressure, of course, is hydrostatic in such designs. Introduction of the pressure medium into the pressure cavity by pumping places the "bomb" in tensile stress; for this reason, alloys such as Haynes 25 and Réne 41, which possess great tensile strength even at temperatures of 700–900°C, are used to fabricate the pressure vessels. Two different designs are chiefly employed; (1) cold-seal or test-tube bombs; and (2) internally heated pressure vessels.

A common cold-seal bomb configuration is illustrated in Figure 2.2. The pressure vessel is placed partially within a furnace and is externally heated; the specimen capsule is situated in the hot end of the bomb, as close as possible to an external thermocouple. The pressure is maintained within the bomb by means of a cone-in-cone closure piece; to remain effective, the metals constituting the cone-against-cone bearing surface must remain cool enough to prevent excessive deformation, which would result in pressure loss. This type of equipment is quite stable and is virtually trouble-free. However, because the pressure vessel itself is heated to as high a temperature as the sample, the working range is limited by the tensile strength of the alloys used. Typical maximum values are 950°C at 500 bars, 900°C at 1,000 bars, 850°C at 2,000 bars and 700°C at 6,000 bars. A variety of crustal magmatic and metamorphic reactions thereby may be attacked by employing this type of equipment.

The design of an internally heated pressure vessel is shown in Figure 2.3. In this arrangement, a small resistance furnace is located within and coaxial to the pressure cavity. The charge container is centered in the anulus along with one or more thermocouples. The external pressure vessel is cooled by a circulating heat-exchange substance, such as water. Thus the pressure-confining jacket is maintained at low temperatures, where it possesses considerable ten-

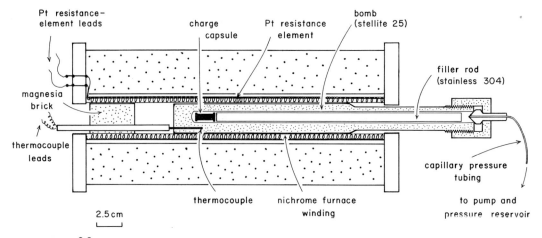

FIGURE 2.2.

Typical hydrothermal synthesis assembly, showing furnace, cold-seal pressure vessel, and sample container (through the courtesy of F. R. Boyd, Jr., as illustrated in Ernst, 1968).

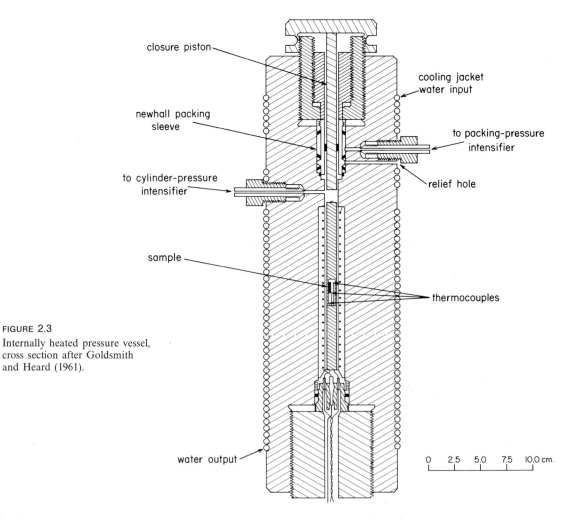

FIGURE 2.3

Internally heated pressure vessel, cross section after Goldsmith and Heard (1961).

sile strength and can withstand elevated pressures. High temperatures are localized in the vicinity of the run capsule, so both high temperatures and high pressures can be achieved simultaneously—on the order of 10 to 11 kilobars at 1,000 to 1,200°C. Drawbacks to this type of equipment result from the difficulty of avoiding pressure leaks in a system where not only power leads but thermocouples must be inserted through the closure piece and from the extensive pumping needed for very high compression of a large volume of the working gas. Because the success of the design depends on a major radial thermal gradient, and because of the small furnace size, moderate T fluctuations and strong axial and radial temperature gradients are characteristic. Therefore, although a wide range of crustal P–T conditions can be investigated with this type of apparatus, its general lack of long-term stability limits study to the more rapidly running reactions.

To prevent corrosion of the furnace assembly, an inert gas such as argon must be used in an internally heated pressure vessel. Therefore, the specimen container must be sealed, generally by welding. The charge capsule introduced into a test-tube bomb may also be sealed—and *must* be sealed where leaching of the sample, by an H_2O pressure medium, for instance, is a possibility. However, if one wants equilibrium between the charge and a multicomponent gas phase employed as the pressure medium (e.g., a CO_2–H_2O mixture of known composition), the charge container must be left crimped but unwelded, to prevent loss or contamination of the condensed assemblage but access of the fluid.

High-Pressure Equipment

A solid but weak pressure medium is used in apparatus capable of high pressures. What is probably the most typical design, the piston-cylinder device, is illustrated in Figure 2.4. As in the hydrostatic internally heated pressure vessel, the sample is enclosed in a precious metal capsule, and placed within the central core of a cylindrical resistance-furnace assembly. The latter, typically made of graphite, is placed in an outer cylinder of a solid pressure medium, such as talc, fired pyrophyllite, or silica glass, which is surrounded, in successively larger shells, by the boron carbide or tungsten carbide pressure vessel and a soft steel retaining ring. The entire assembly is mounted between two presses (end plates) for supporting pressure, and a piston, advanced by means of a hydraulic ram, provides the confining pressure. Thermocouple leads, which terminate next to the sample within the pressure cell, pass out axially through the support system. Sometimes the thermocouple circuitry is broken by deformation of the high-pressure cell; such deformation is passed along to the furnace, too, resulting in marked and somewhat irregular thermal and pressure gradients. Another drawback of this setup is that ram pressure is measured on the low-pressure side of the system, with calculation of run pressure based on the known ratio between piston surface areas on both sides, and on a reasonable correction for frictional resistance of the piston as it advances into (or is retracted out of) the pressure cell. The outstanding merits of this apparatus include its relatively long-term stability under high-pressure and high-temperature conditions, in the range from 9 or 10 to more than 60 kilobars, and up to 1,600 or 1,700°C range. Thus melting and solid-state reactions inferred to be taking place in the upper mantle can be investigated with this type of equipment. Experiments are usually run dry, with the capsule left unsealed unless there is danger of H_2O addition due to the dehydration of the solid pressure medium (e.g., talc); however, where a condition of fluid pressure equal to total pressure is desired, the capsule containing the charge plus the volatile constituent must be sealed.

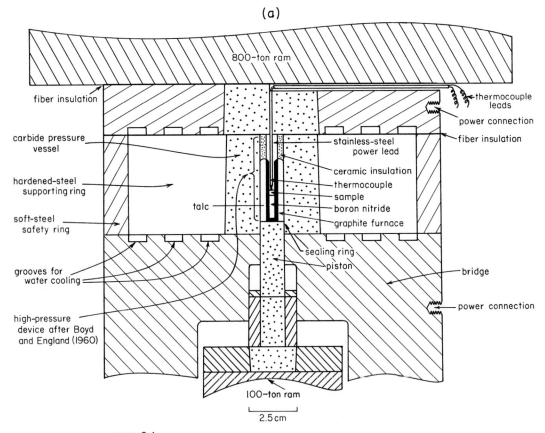

(a)

FIGURE 2.4.

Piston-cylinder device: (a) is over-all view of apparatus with carbide parts stippled, steel parts ruled, after Boyd and England (1960a); (b) shows details of various furnace, base plug, and sample geometries, taken from Johannes *et al.* (1971, Fig. 1).

(b)

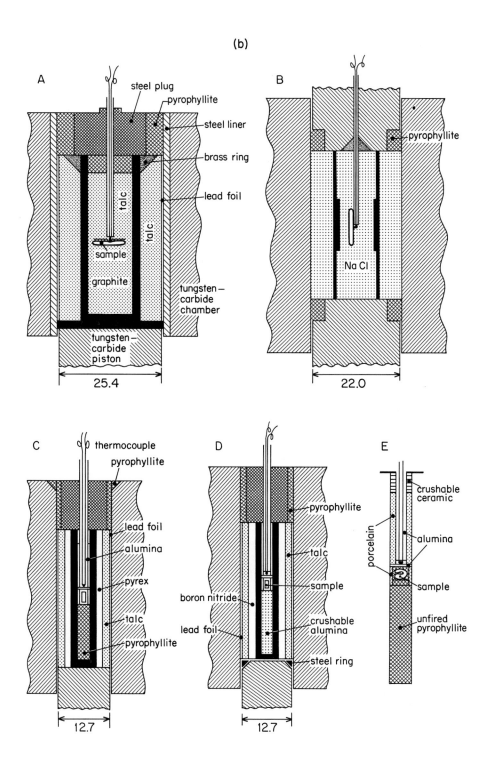

Control of Volatile Components

Volatile constituents are involved in many petrologic reactions of interest; hence the values of the chemical potential, activity, fugacity, or partial pressure of the volatile components influence the position of equilibrium. For this reason, methods have been devised for controlling such variables in the laboratory. In many devolatilization-type reactions, such as brucite = periclase + H_2O, albite + H_2O = aqueous melt, or calcite + quartz = wollastonite + CO_2, the fluid pressure may be kept equal to the total pressure by sealing an excess of the gaseous component in the hydrothermal or high-pressure charge capsule. For one-atmosphere quenching furnaces, the composition of the volatile phase may be controlled by introducing a proper mix of gases as a moving stream over the open charge container during the course of the experiment (e.g., see Darken and Gurry, 1953). A similar mixture of gaseous components may be continuously equilibrated with the open charge capsule of a hydrothermal apparatus, but geometric and strength restrictions preclude such control for currently available high-pressure devices. It is also possible to fix the proportions of CO_2 and H_2O in experiments employing sealed capsules by using a charge which contains oxalic acid or a metal oxalate. In general, the extent and nature of the reaction will influence the composition of the fluid phase; therefore the latter must be analyzed at the end of the experiment.

In certain oxidation-reduction reactions, oxygen fugacities are maintained at a low but undefined value by equilibration of the charge with a capsule of metallic iron; this arrangement is especially used in quenching furnaces and in the piston-cylinder apparatus. Of course, gas-component mixing provides any range of f_{O_2} desired for open capsule systems in all but high-pressure equipment. Two other methods are used for controlling (chiefly) hydrogen and oxygen fugacities in hydrothermal equipment. In the first, a semipermeable membrane of platinum separates the pressure chamber and the charge from a hydrogen reservoir, the pressure of which can be varied as desired (Shaw, 1967); this geometry allows the independent control of fluid (= total) pressure and the hydrogen partial pressure (P_{H_2} will be the same value within the platinum charge container, in the surrounding pressure medium, and in the controlling hydrogen reservoir). The other method, called the buffer technique, makes use of the fact that a univariant relationship (i.e., one independent variable) exists between certain assemblages that bear condensed volatiles, the temperature, and the fugacity of the volatile component, as shown in Figure 2.5 (see Eugster, 1957). For instance, the condensed pair hematite + magnetite defines f_{O_2} as a function of temperature. The charge capsule is enclosed in an outer capsule which contains the volatile buffer; chemical communication of volatile constituents (through the charge-capsule wall, which acts as a semipermeable membrane) allows equilibration of gas-component fugacities at the values appropriate for the condensed-buffer assemblage at the temperature of the experiment.

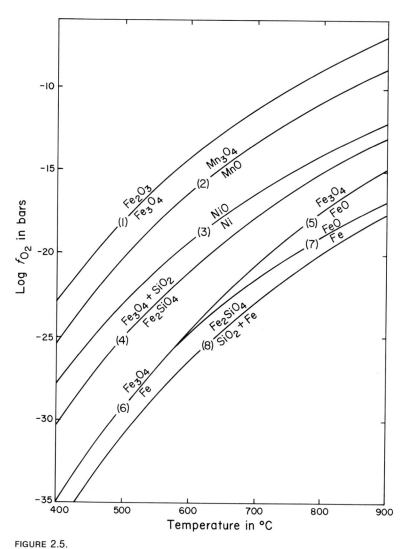

FIGURE 2.5.

A diagram of f_{O_2} against T for selected buffer assemblages (Eugster and Wones, 1962; Wones and Gilbert, 1969). The numbered oxidation reactions are as follows: (1) $4Fe_3O_4 + O_2 = 6Fe_2O_3$; (2) $6MnO + O_2 = 2Mn_3O_4$; (3) $2Ni + O_2 = 2NiO$; (4) $3Fe_2SiO_4 + O_2 = 2Fe_3O_4 + 3SiO_2$; (5) $6FeO + O_2 = 2Fe_3O_4$; (6) $3Fe + 2O_2 = Fe_3O_4$; (7) $2Fe + O_2 = 2FeO$; (8) $2Fe + SiO_2 + O_2 = Fe_2SiO_4$.

Many reactions of interest include the production or consumption of a volatile component, such as H_2O or CO_2. The brucite-periclase + H_2O equilibrium is a good example. Although this type of reaction will be discussed in more detail in Chapter 6, it is worth pointing out here that, because a volatile species is involved, its activity (or partial pressure, fugacity, or chemical potential) will influence the P–T locus of equilibrium between reactants and products. For instance, Figure 2.6 shows the hydrothermal stability reactions for $Mg(OH)_2$ and MgO as a function of a_{H_2O}: curve (a) shows the field boundary where a pure aqueous fluid is present at the operating total pressure (i.e., $a_{H_2O} \approx 1.0$); whereas curves (b) and (c) illustrate the thermal stability range of brucite at progressively greater degrees of contamination (by NaOH in this example) of the aqueous fluid (i.e., $a_{H_2O} < 1.0$).

DEMONSTRATION OF SYNTHETIC EQUILIBRIUM

Numerous macroscopic and microscopic investigations of rocks, performed during the course of more than a century, have amply demonstrated that the constituent minerals and phase assemblages often had departed markedly from a state of chemical equilibrium at the time of their formation—or even if they were formed under a close approach to equilibrium, conditions have subsequently changed significantly. Of course, we must thank the sluggishness of reactions for the near-surface persistence of mineral associations originally produced at great depths within the Earth, for most of these assemblages are clearly metastable under the presently observable conditions. Examples of disequilibrium in rocks include the following. Gradational or oscillatory compositional zoning is characteristic of many rock-forming minerals, such as the plagioclase and alkali feldspars, amphiboles, pyroxenes, garnets, epidotes, and olivines. Evidence of replacement of an earlier phase by a later phase—indicating chemical reaction—is widespread, especially among the alkali feldspars, chain silicates, sulfides, and carbonates. In addition, the presence of chemically incompatible phases in the same rock, such as corundum or magnesium-rich olivine with quartz, is not rare. Moreover, because electron microscopes (both scanning and transmission types) and electron microprobes are being more and more commonly employed in petrologic research, chemical heterogeneity is being recognized on ever-smaller scales; accordingly, petrologists, mineralogists, and geochemists will have to pay more attention to clearly defining what is the size of each domain of chemical equilibrium.

Since chemical phase inhomogeneity—hence departure from chemical equilibrium—is evident in most geologic materials, what is the point of worrying about the equilibrium state? The point is that, although the minimum Gibbs-free-energy configuration is seldom attained on more than a local scale in rocks, all reactions drive *toward* such a state. In other words, the observed reactions reflect a decrease in the total G for the system. It must be cautioned

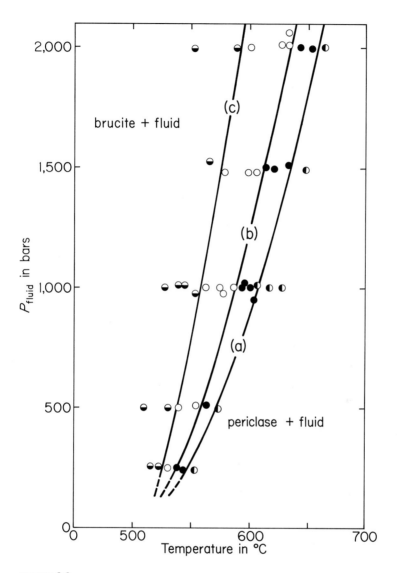

FIGURE 2.6.

A diagram of P_{fluid} against T for equilibrium among brucite, periclase, and fluid in the system MgO–NaOH–H$_2$O, after Barnes and Ernst (1963, Fig. 2). Curve (a) represents the field boundary where the aqueous fluid is pure H$_2$O; curves (b) and (c) represent 5.0 and 12.5 molal NaOH solutions, respectively. Although not investigated in these experiments or illustrated in the diagram, as P_{fluid} approaches zero, the thermal stability limit of the hydrous condensed phase—in this case, brucite—falls off rapidly. At $P_{fluid} = 0 = \alpha_{H_2O}$, dehydration curves of this sort intersect the temperature axis at 0°Kelvin.

that classical thermodynamics says nothing about the speeds of various reactions (to discover them, we must study the atomistics of such processes, i.e., by employing the concepts of statistical mechanics, the kinetic theory of reactions, and irreversible thermodynamics). Given an initial state and several alternative configurations, classical thermodynamics only allows us to predict which of the known transitions are possible and which are impossible. Because of this, we can never demonstrate that a certain Gibbs-free-energy configuration is truly minimal for a particular bulk composition under specified laboratory conditions. What we can prove is that, of several possible natural or synthetic phase configurations, one is the equilibrium assemblage (i.e., has a lower G) with respect to all the others. After we compare these assemblages with a range of natural occurrences, we are usually able to decide whether or not the particular state in question is typical for the inferred physical conditions. If it is, we then assume it to be the equilibrium assemblage that has minimum Gibbs free energy; an even lower G configuration might be possible in theory, but if it does not occur naturally or artificially, it is of little interest to anyone.

Granted that thus establishing the state of chemical equilibrium (actually, relative equilibrium) will have predictive and explanatory value for the study of observable petrologic reactions, how can equilibrium be demonstrated in the laboratory? First, the exact nature of the reaction must be elucidated by identifying the compositions of all participating phases. Once this is known, the approximate P–T–x* coordinates of the assemblage field boundaries must be discovered by means of reconnaissance experiments. Finally, reactant and product phase associations must be converted to product and reactant assemblages, respectively, to prove relative stabilities. The direction and extent of reaction typically are elucidated by examining the run products with a petrographic or scanning electron microscope or an electron microprobe, by employing x-ray diffraction techniques, or, in some favorable cases, by measuring weight loss or weight gain of a single crystal. Where there is a relatively large ΔH or change in electrical resistance during the reaction, differential thermal analysis (DTA) or electrical conductivity methods may be employed alternatively to signal the onset and nature of the reaction; however, these latter techniques can be applied only to rapid reactions. Mössbauer analysis is a valuable method for studying certain oxidation-reduction and order-disorder reactions involving iron-bearing phases. Where feasible, a variety of analytical methods should be used in conjunction because, in general, each provides a contrasting measure of the approach to equilibrium. For instance, textural relations and the presence of trace or minor phases normally can be established most conveniently by microscopic methods, whereas x-ray diffraction or electron microprobe techniques usually provide the best evaluation of phase chemistry.

To elucidate the nature of phase equilibrium studies, let us consider as an example the phase relationships among the aluminosilicate polymorphs, anda-

---

*Here x represents one or more chemical variables.

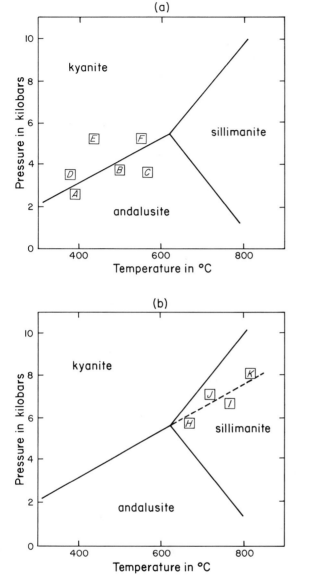

FIGURE 2.7.
Al_2SiO_5 polymorphism, after Richardson, Bell, and Gilbert (1968) and Richardson, Gilbert, and Bell (1969). Lettered boxes indicate the temperatures and pressures of individual experiments discussed in the text.

lusite, kyanite, and sillimanite, assuming strict stoichiometry (i.e., no systematic contrasts in element chemistry among these polymorphs). A topologically correct P-T diagram for Al_2SiO_5 is presented in Figure 2.7. We can locate the equilibrium boundary between andalusite and kyanite by demonstrating the complete or partial conversion of kyanite to andalusite under experimental conditions, such as A, B, and C, and the complete or partial conversion of andalusite to kyanite at, say, D, E, and F (see Figure 2.7a). These transformations demonstrate the reversibility of the reaction in question. Or, we could have proven the same thing by starting with a known mixture of kyanite and andalusite, and measuring the growth of the latter at the expense of the former at A, B, and C, and vice versa at D, E, and F. Note that such experiments tell us nothing about another possible energy configuration, sillimanite. Whether this third phase assemblage has a higher or lower G than andalusite and kyanite under physical conditions A–F would have to be verified by further studies (sillimanite is clearly metastable in this P–T region, as illustrated in Figure 2.7a). What the described experiments prove is that $G_{andalusite} < G_{kyanite}$ for the P–T range A–C, whereas inequality is reversed for the P–T range D–F; of course, the Gibbs free energies of andalusite and kyanite must be equal along the univariant curve where both phases are stable. Note also that if the experiments described above were carried out starting with sillimanite or a mixture of andalusite, kyanite, and sillimanite, decomposition of the metastable sillimanite might produce a phase possessing a lower Gibbs free energy outside of its own equilibrium field, such as kyanite at A or B; this situation could arise because, under these conditions, sillimanite has a higher G than either of the possible product assemblages. Inasmuch as run C lies on the sillimanite side of the metastable kyanite-sillimanite curve projected into the andalusite P–T field (not shown in Figure 2.7), we would expect metastable sillimanite either to persist or to transform to andalusite, but not to form kyanite. Therefore, to avoid ambiguity in locating an equilibrium field boundary, laboratory experiments must be performed on starting materials which constitute the actual equilibrium assemblages.

Metastable equilibria can also be investigated experimentally. Sometimes the reaction kinetics for the metastable portion of an equilibrium are fast enough to allow laboratory measurement, whereas the transformation proceeds too slowly along the stable portion for it to be detected in the laboratory. For instance, we can locate the metastable extension of the andalusite-kyanite P–T curve at physical conditions appropriate to the sillimanite stability field, provided that crystallization of the latter phase is suppressed. Newton (1966a) studied this particular metastable equilibrium (see dashed line in Figure 2.7b) employing known mixtures of andalusite and kyanite at physical conditions such as H, I, J, and K. The experiments were short enough so that sillimanite failed to nucleate and crystallize. Had sillimanite been produced, the resultant ratio of residual andalusite to kyanite could not have been used to locate the metastable equilibrium curve, because the rates at which sillimanite grows from these two phases need not reflect the Gibbs free energy differences.

One danger inherent in using experimentally investigated equilibria to solve petrologic problems is that laboratory conditions only *approximate* the natural environment. For instance, because of the sluggishness of most mineralogically interesting reactions, highly reactive starting materials, such as gels, glasses, mixtures of oxides, and unstable crystalline phases, are sometimes employed as reactants to speed up melting, crystallization, recrystallization, etc.; inasmuch as such substances are far from the equilibrium configuration, they may be replaced during the experiment by any number of more stable (but nevertheless metastable) assemblages. To demonstrate chemical equilibrium for the desired reaction, $r = p$, it must be shown that, under some physical conditions, p reacts to form r, and the field of stability of r (with respect to p) is thus indicated; whereas under contrasting conditions, r is replaced by p, so that the stability field of p (with respect to r) is indicated. Yet another configuration could exist which possesses an even lower Gibbs free energy, but unless it has been produced in the experiments, we have obtained no data about its stability field.

Furthermore, natural assemblages of interest usually contain components that are absent from the chemically simpler equilibria investigated in the laboratory. Where these components become participating species, the physical conditions of the natural reaction will differ from those of the synthetic equilibrium. Fortunately, the principles of classical thermodynamics can often be applied to evaluate what effect these additional constituents will have. Examples will be presented farther on as the need arises.

CHAPTER 3

Computational Approach
to Phase Equilibrium

THERMOCHEMICAL DATA

As consideration of the critical relationships in Chapter 1 shows, the way in which a particular reaction at equilibrium depends on the state variables, P and T, is itself a function of certain thermodynamic parameters, such as the extensive quantities ΔG, ΔV, ΔH, and ΔS.

The isobaric, isothermal Gibbs-free-energy change of a reaction is generally evaluated by means of equation (1.29), $\Delta G = \Delta H - T \Delta S$, provided that the enthalpy and entropy changes are already known. The Gibbs-free-energy change of a reaction taking place in solution can sometimes be derived directly, e.g., in the e.m.f. measurements of electrolytic cells, where it can be shown that ΔG is directly proportional to electrical potential.

Volume changes, of course, are calculated by means of simple physical measurements of reactant and product volumes. For condensed phases, the isobaric change of volume as a function of temperature (thermal expansivity) and the isothermal change of volume with pressure (compressibility) are relatively small compared to the over-all ΔV of condensed phases for most reactions. Moreover, given a change in the state variables P and T, the volume of condensed reactants will change with the same sign as, and by an amount similar to, the change in the volume of the condensed products. For this reason, the change in $\Delta V_{condensed}$ for a reaction within a moderate P–T range is generally ignored. Obviously, the thermal expansivities and compressibilities of volatile phases are relatively large numbers, and hence must be considered in any calculation involving gases.

Values for ΔH and ΔS of formation are established experimentally by various types of combustion or solution calorimetry. The enthalpy change of a reaction is simply the sum of the heats of formation of the products minus that sum for the reactants. In arriving at the heat of formation for a phase, we must choose a zero-value standard reference state arbitrarily. We will select the pure elements, such as Al, Si, and O, as our standards at one atmosphere; hence oxides, for instance, SiO_2 and Al_2O_3, in general have negative heats of formation, and phases such as Al_2SiO_5 typically possess even more negative values of ΔH^0 (sometimes written simply as H^0).

We can find the entropy change for a reaction under equilibrium conditions either by measuring the heat of the reaction, remembering from equation (1.29) that $\Delta S = \Delta H/T$, or by measuring the slope of the appropriate P–T curve and the volume change, remembering from equation (1.49) that $\Delta S = \Delta V\, dP/dT$. However, the value of total entropy for a participating species at some random P and T is more difficult to discover. As shown in equation (1.16), evaluation of the entropy requires: (1) knowledge of the $T_{\text{Kelvin}} = 0$ entropy, or assumption of perfect crystal behavior (i.e., the entropy at absolute zero is zero); (2) integration of the heat capacity from $0°K$ to the temperature of concern; and (3) evaluation of the heats of transition, if any, within the same temperature interval.

In general, we have thermochemical information for many mineralogically important phases at one atmosphere pressure and $25°C$; less abundant data are available for higher temperatures. We will use the values listed in Appendixes 2 through 6 to compute P–T–x conditions for certain reactions not yet investigated in the laboratory and, for others, to compare the results with direct phase-equilibrium experiments. The tables in the Appendixes are moderately extensive, to enable the student to address many more problems than can be set forth in this book. Because we generally deal with small absolute differences between very large numbers, it must be cautioned that relatively minor errors in the basic thermochemical data will create disproportionately large numerical errors in the calculated equilibrium values.

It should be emphasized that the values for S^0, ΔH^0, and ΔG^0 listed in Appendixes 2 and 3 refer to formation from the elements taken as standard state. However, the intrinsic entropies of the elements themselves are not zero, as is clear from these tabulations. Hence for the proper application of equation (1.29), $\Delta G^0 = \Delta H^0 - T\Delta S^0$, the entropies of the constituting elements, such as Si and O_2, must be subtracted from that given for the corresponding compound, such as α quartz. Of course, in most chemical reactions of geologic interest, we are concerned about the entropy change among compound species rather than the total value of S^0 for any participating phase; so this problem seldom need be addressed by the student.

APPLICATIONS OF THERMOCHEMICAL DATA

Isobaric Compositional Variation

Let us consider what happens during the isobaric decomposition of a sample assemblage if we add a second constituent. An appropriate example is the melting behavior of a crystalline phase. In most situations of petrological interest, silicate components are miscible in all proportions in molten solutions (melts, magmas). Likewise, we know that volatile constituents of a reaction are miscible in all proportions in the gaseous state. However, not all minerals of grossly contrasting chemistries exhibit complete or even extensive solid solution. We need to find a general thermodynamic relationship between the compositional variation of the phases and a state variable, say, temperature, at constant pressure. Remember from equation (1.77) that for a complete reaction

$$\left[\frac{\partial \ln K}{\partial \left(\frac{1}{T} \right)} \right]_P = - \frac{\Delta H^0}{R}.$$

At equilibrium the activities of a component or participating species i, which occurs in the crystalline phase as i' and the liquid phase as i'', can be substituted for K in the partial reaction $i' \rightarrow i''$ (see equation 1.73); hence

$$\left[\frac{\partial \ln (a_i''/a_i')}{\left(\partial \frac{1}{T} \right)} \right]_P = - \frac{\Delta H_i^0}{R}, \tag{3.1}$$

where the standard-state heat of reaction (actually, the heat of fusion in our example) is simply $(H_i'')^0 - (H_i')^0$, namely, the difference in molar heat content of i in the melt and in the solid, respectively. For simplicity we have assumed the same stoichiometry of participating species in both liquid and crystalline states. Equation (3.1) provides the sought-for relationship, between the change of activity (composition) of the phases in question with melting temperature, as a function of the standard-state heat of fusion, assuming ΔH of fusion to be constant and equal to the standard-state value. To the extent that ideal solution behavior is obeyed, activities may be replaced by the corresponding mole fractions of component i in the liquid and solid phases. Then in a $\log (X_i''/X_i')$ versus $1/T$ plot, the slope of the equilibrium line between the coexisting phases is equal to $-\Delta H_i/2.303R$ (here we have substituted base 10 for natural logarithms); this will be a straight line only if the heat of fusion is constant within the temperature interval examined.

   Now let us consider, for an example, the one-atmosphere melting behavior of diopside when we add a second constituent, such as $Na_2O \cdot Al_2O_3 \cdot 6SiO_2$. From Appendix 3 we find that pure diopside melts at $1,391°C$ ($= 1,664°K$)

and that the standard-state heat of fusion is 18.5 kcal/mole. Experiments have demonstrated that a small amount of Al_2O_3 is accommodated in the clinopyroxene, and hence the system actually departs from strictly binary behavior; however, the effect is rather small and will not concern us here. Therefore, because to a first approximation diopside exhibits no solid solution toward albite, equation (3.1) may be simplified to

$$\left[\frac{\partial \log a_{Di}''}{\partial\left(\frac{1}{T}\right)}\right]_P = -\frac{\Delta H_{Di}^0}{2.303R}, \tag{3.2}$$

because the activity, a_{Di}', of pure, crystalline diopside is unity. Assuming that the heat of fusion is constant for $CaMgSi_2O_6$, we have

$$\left[\frac{\partial \log a_{Di}''}{\partial\left(\frac{1}{T}\right)}\right]_P = -\frac{18,500}{2.303 \cdot 1.987} = -4043. \tag{3.3}$$

Provided that the activity of the component **Di** in the melt is equal to its mole fraction, this expression may be evaluated for any value of X_{Di}'' between 0 and 1; for instance, for melts with **Di** mole fractions of 0.9, 0.5, and 0.3, the temperatures of equilibrium are computed as 1,360, 1,211, and 1,096°C, respectively. These computed points are shown in Figure 3.1, along with the experimentally established phase relations. The discrepancy between observed and calculated curves probably indicates that ΔH is actually a function of composition and temperature, and that $\gamma_{Di} \neq 1.0$ throughout the compositional range investigated; the deviation of the fusion reaction from ideality is perhaps due to nonbinary behavior. The data of Appendix 3 may be employed similarly to calculate the lowering of the high (i.e., disordered) albite melting point. Of course, the intersection of the two melting curves, which is the eutectic temperature (i.e., that value for which a specific composition of the melt is in equilibrium with both diopside and high albite), may also be computed by using the data of Appendix 3 in a log X versus $1/T$ diagram; here the eutectic is simply the common temperature at which the $(X_{Di}'' + X_{Ab}'')$ sum is equal to unity.

Although our discussion has focused on equilibrium between a crystalline phase of fixed composition and a melt of variable composition, equation (3.1) is of general importance in describing the decomposition of any phase assemblage' to assemblage", regardless of the state of aggregation, in the presence of an additional component. This added constituent may be miscible in either ' or " or both; we arbitrarily selected an example in which a_i' was constant and a_i'' variable. For fusion, however, the assumption of constancy of heat of reaction (implying no interaction of participating species in the liquid phase) severely limits the usefulness of the expression. Computation of crystal-melt equilibria at mantle temperatures and pressures will be discussed in Chapter 5.

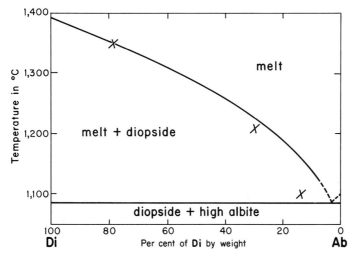

FIGURE 3.1.

Experimentally established one-atmosphere isobaric diopside liquidus in the system CaO·MgO·2SiO$_2$–Na$_2$O·Al$_2$O$_3$·6SiO$_2$, given in terms of per cents by weight of **Di** and **Ab** (Bowen, 1915). Note that, for the calculations performed in the text (values shown by X), the **Di** mole fractions of 0.9, 0.5, and 0.3 are 79, 29, and 15 per cent **Di** by weight, respectively.

P–T Variation at Constant Composition

Here we will consider a class of heterogeneous reactions in which we seek to evaluate the effect of pressure on the temperature of equilibrium at constant composition. Because they are simpler, we will first discuss reactions involving only condensed phases, such as the transition of aragonite to calcite, or the anhydrous melting of albite. For such equilibria, the isothermal compressibilities and thermal expansions of the participating species are small, and differences are even smaller; hence they can be ignored safely, and the ΔV of the reaction may be taken as constant and equal to the standard-state value at room temperature (i.e., $\Delta V = \Delta V^0$).

The pressure P attending equilibrium between reactants and products may be evaluated at any temperature, provided the one-atmosphere or standard Gibbs free energies (hence ΔG^0 for the reaction) are known, by the following method. Equation (1.33) gives the effect of pressure increment on reactant r and product p assemblages:

$$dG_r = V_r\,dP, \text{ and } dG_p = V_p\,dP.$$

The difference between dG_p and dG_r (i.e., $\Delta V^0\,dP$) at the pressure of equilibrium, P, must be equal in magnitude, but of opposite sign, to ΔG^0 in order for the over-all ΔG of the reaction to be zero, inasmuch as we know that

$$\Delta G = \Delta G^0 + \Delta V^0 \, dP. \tag{3.4}$$

At equilibrium, then,

$$-\Delta G^0 = \Delta V^0 \, dP, \tag{3.5}$$

or

$$dP = -\Delta G^0 / \Delta V^0. \tag{3.6}$$

Because dP is merely P minus one atmosphere,

$$P - 1 = -\Delta G^0 / \Delta V^0. \tag{3.7}$$

Let us use this relationship and the thermochemical data from Appendix 2 to find the pressure at which calcite and aragonite can stably coexist at 298°K. For the reaction aragonite = calcite, $\Delta G^0 = -230$ cal/mole and $\Delta V^0 = 2.784$ cc/mole. Substituting these values in equation (3.7), we have

$$P - 1 = \frac{230 \cdot 41.86}{2.784} = 3{,}458 \text{ bars,}$$

where the factor 41.86 converts calories to cm^3 bars (see Appendix 1). The indicated value of P, 3,459 bars, is in excellent agreement with phase-equilibrium experiments, which yield extrapolated equilibrium pressures on the order of 3 or 4 kilobars at this temperature (see Figure 3.2). The agreement seems even more remarkable when it is realized that the ΔG^0 for the reaction lies within the estimated uncertainties for the Gibbs free energies of formation of both calcite (± 330 cal) and aragonite (± 350 cal). Of course, the earlier high-pressure experiments were taken into account in evaluating the thermochemical data for the $CaCO_3$ polymorphs (Robie and Waldbaum, 1968, p. 243), so internal consistency is to be expected.

Having obtained one point on a univariant P–T curve, we can now define the curve at other temperatures for which the standard Gibbs free energies of reactants and products are available, for instance, from Appendix 3. Another method now to be described allows us to calculate the P–T slope for the reaction at any temperature by employing the Clapeyron equation (1.49),

$$\frac{dP}{dT} = \frac{\Delta S}{\Delta V},$$

provided the entropies of all participating species are known. The value of dP/dT is nearly constant within a moderate range of physical conditions, because the difference in entropy between products and reactants does not change rapidly with P and T. (This does not mean that the entropy is nearly constant for individual participating phases, as may be seen from perusal of Appendixes 2 and 3. For instance, the entropy of calcite at 298°K is 22.15 cal/deg/mole,

whereas that of aragonite is 21.18 cal/deg/mole; at 1,000°K, the corresponding values for calcite and aragonite are 52.85 and 51.01 cal/deg/mole. In spite
of the great increase in the entropy of each phase, the ΔS changes only from a
value of 0.97 cal/deg/mole at 298°K to 1.84 cal/deg/mole at 1,000°K.) Accordingly, heterogeneous equilibria that involve only condensed phases will
approximate straight equilibrium P–T lines within moderate ranges of pressure and temperature.

As an example, let us use the Clapeyron equation to calculate the P–T slope
for the aragonite + calcite equilibrium near 298°K:

$$\frac{dP}{dT} = \frac{0.97 \cdot 41.86}{2.784} = 14.6 \text{ bars/°C.} \tag{3.8}$$

Or, for each kilobar of pressure increment, the temperature of the aragonite
+ calcite equilibrium is elevated about 69C°. The calculated and experimentally established curves, which are very similar, are compared in Figure 3.2.

Let us now turn our attention to the class of heterogeneous equilibria in
which a volatile component takes part in the reaction. A simple example is
6 hematite = 4 magnetite + O_2; another is brucite = periclase + H_2O. In the
first example, the partial pressure of oxygen is so low that, to a first approximation, ideal gas behavior may be assumed—but, to be more rigorous, oxygen
fugacity is employed. At very low pressures, pure hematite and magnetite can

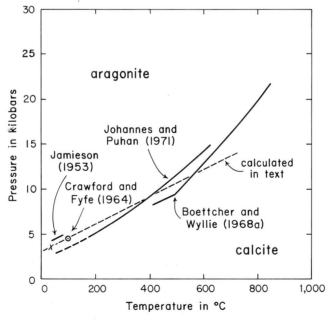

FIGURE 3.2.
The aragonite-calcite equilibrium experimentally established by various authors,
and calculated by using thermochemical data from Appendix 2.

be in equilibrium with oxygen gas, the fugacity of which is a univariant function of temperature, as was illustrated in Figure 2.5 (in a two-component system, $F = 2 - 3 + 2 = 1$). From the van't Hoff reaction isotherm (equations 1.71 and 1.73), we may write

$$\Delta G^0 = -RT \ln f_{O_2}, \qquad (3.9)$$

because the activities of the condensed phases hematite and magnetite are unity. Rearranging this expression, and transforming to base 10 logarithms, we have

$$\log f_{O_2} = \frac{-\Delta G^0}{2.303 \; RT} \qquad (3.10)$$

This expression is readily evaluated at, say, 298°K by using the data of Appendix 2. For example,

$$\log f_{O_2} = \frac{(4 \cdot -243{,}094 + 0 - 6 \cdot -177{,}728)}{2.303 \cdot 1.987 \cdot 298} = -68.91. \qquad (3.11)$$

Obviously, the fugacity of oxygen in equilibrium with hematite and magnetite at room temperature is a very small number! Even at temperatures on the order of 1,000°C, where the position of equilibrium has been established by direct vapor-pressure or gas-composition measurements, values of $\log f_{O_2}$ approach -5.

At total pressures exceeding one atmosphere, the equilibrium oxygen fugacity is displaced slightly because the activities of the pure solid phases depart slightly from unity. Or, if we redefine the standard state to be at the new, high pressure being considered, the increment in pressure influences the ΔG^0 of the reaction, as shown in equation (1.33),

$$\left(\frac{\partial \, \Delta G^0}{\partial P} \right)_T = \Delta V^0,$$

because ΔV^0 is not zero. The effect is slight, however; so for our purposes it will not be considered further.

Treatment of the reaction brucite = periclase + H_2O is slightly more complicated than the previous example, but only because here we wish to establish the locus of equilibrium throughout the large P–T interval in which the volatile component is at the operating total pressure. Unlike that of condensed phases, the Gibbs free energy of supercritical H_2O changes rapidly with P because of its great compressibility, especially at low pressures and high temperatures. So, in contrast to the situation for the aragonite-calcite equilibrium, here we cannot regard ΔV as independent of pressure. And, unlike the situation for the low-pressure devolatilization reaction of hematite = magnetite + oxygen, here the effect of increased pressure on the activities of the participating condensed phases cannot be ignored.

Let us begin our analysis of the reaction brucite = periclase + H_2O by dividing it into two portions: condensed reactant and product; and noncondensed product (in this case, H_2O). At the equilibrium pressure P and at some conveniently selected temperature T, the over-all Gibbs-free-energy difference for the reaction must be zero:

$$\Delta G = 0 = \Sigma \, G_{\text{condensed products}} + G_{H_2O} - \Sigma \, G_{\text{condensed reactants}}. \quad (3.12)$$

But the Gibbs-free-energy change of the condensed portion of the reaction is simply the sum of the one-atmosphere value, $\Delta G^0_{\text{condensed}}$, and the increment due to elevated pressure, $\Delta V^0_{\text{condensed}} \, dP$ (see equation 3.4). Hence

$$0 = \Delta G^0_{\text{condensed}} + \Delta V^0_{\text{condensed}} \, dP + G_{H_2O}. \quad (3.13)$$

Values of $\Delta G^0_{\text{condensed}}$ and G_{H_2O} are presented for specific temperatures in Appendixes 2–4. However, since both these terms are functions of the state variables, the easiest way to locate the equilibrium values is graphically, as illustrated in Figure 3.3. The temperature at which P_{fluid} is one atmosphere (i.e.,

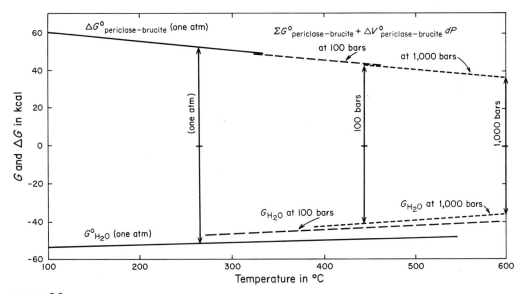

FIGURE 3.3.

Isobaric diagrams of Gibbs free energy against temperature for MgO–$Mg(OH)_2$, that is, $\Delta G_{\text{condensed}}$, and for H_2O at one atmosphere, 100 bars, and 1,000 bars. The free energies of the condensed assemblage and the fluid become equal at 265°C (one atmosphere), 445°C (100 bars), and 601°C (1,000 bars).

standard-state conditions) for equilibrium between brucite and periclase is seen to be about 265°C; here $\Delta G^0_{\text{periclase-brucite}} = -G^0_{\text{H}_2\text{O}}$, and the $\Delta V^0\, dP$ term drops out ($dP = 0$). The Gibbs free energy of formation for H_2O increases isothermally with rising pressure as can be seen from Appendix 4 and Figure 3.3; this is a consequence of equation (1.33), which indicates that the change of G with respect to P is proportional to the volume. In contrast, the decrement in Gibbs-free-energy change of the condensed assemblage with pressure is very slight. For instance, at 1,000 bars,

$$\Delta V^0_{\text{condensed}}\, dP = 999\,(11.248 - 24.63) \cdot 2.389 \cdot 10^{-2}$$
$$= -\,320 \text{ cal,} \tag{3.14}$$

where the factor $2.389 \cdot 10^{-2}$ converts cm^3 bars to calories. Accordingly, the zero over-all Gibbs-free-energy change for the reaction is computed to lie at 445°C at 100 bars and at 601°C at 1,000 bars. The 1,000-bar value is compared with the experimentally established curve in Figure 3.4. The agreement is remarkable, and is probably quite fortuitous, especially since the computed equilibrium for 100 bars appears to lie at a relatively low temperature compared to the extrapolated low-pressure portion of the laboratory curve.

Lest the unwary reader too enthusiastically embrace the computation of such curves from the calorimetric data listed in Appendixes 2–4, he should attempt to determine the P–T location for the equilibrium muscovite + quartz = sanidine + andalusite + H_2O. He will discover sadly that at elevated pressures, the computed curve lies within a few hundred degrees C of room temperature, far below values thought to be appropriate for many natural occurrences, and equally far from the experimentally established equilibrium

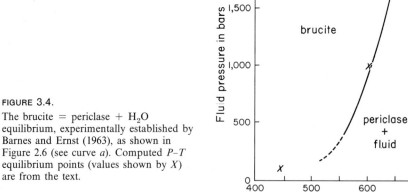

FIGURE 3.4.

The brucite = periclase + H_2O equilibrium, experimentally established by Barnes and Ernst (1963), as shown in Figure 2.6 (see curve a). Computed P–T equilibrium points (values shown by X) are from the text.

(see Figure 6.24b). This situation probably reflects the difficulty of obtaining high-accuracy calorimetric data for the aluminosilicates.

Another troublesome point for some readers may be the obvious increase in Gibbs free energy for H_2O as a function of temperature shown in Figure 3.3; this relationship, although it reflects the systematic increase in molecular dissociation of H_2O with elevated temperature, seems to contradict equation (1.32). The phenomenon, however, is a consequence of the standard state chosen. The absolute Gibbs-free-energy values of the elements (arbitrarily selected as standard state) H_2 and $\frac{1}{2}O_2$ decrease with T faster than does that of H_2O; hence the latter assemblage exhibits a Gibbs-free-energy increase relative to the elements. Of course, had we chosen oxides as our standard state, G_{H_2O} could not vary as a function of temperature, since, by definition, it would have a zero value for G^0 of formation.

Now let us turn to consideration of a situation in which the volatile phase either does not consist exclusively of the species participating in the reaction or, if it is a pure phase, is nevertheless at a lower total pressure than that which is acting on the rest of the system. It is geologically reasonable, for instance, that in some circumstances the dehydration of brucite could take place in the presence of aqueous fluid diluted by CO_2, or adjacent to fissures where P is maintained at, say, hydrostatic pressure rather than at the lithostatic value to which the condensed phases are subjected. In such a situation, the activity, chemical potential, fugacity, or partial pressure of H_2O would be less than in the example just considered, where H_2O pressure equalled P_{total}. In the absence of a volatile contaminant or in a situation in which rock strength is completely lacking (hence fissures cannot be supported), a separate fluid phase may not even be present; nevertheless, the chemical potential, activity, or fugacity of the volatile component may be specified by equilibrium between condensed volatile-bearing reactants and devolatilized condensed products. How do departures of P_{total} from $P_{volatile}$ affect equilibrium? (For extensive discussions of the problem, see Thompson, 1955, and Greenwood, 1961.)

We may calculate the influence of raising total pressure at a constant equilibrium pressure of the volatile constituent, or $P_{E_{volatile}}$ (the value of fluid pressure at which condensed reactants and condensed products can coexist stably), by recognizing that the pressure increment operates on the condensed assemblage but not on the volatile constituent. Recall that, for equilibrium between reactants and products at any given T and P (i.e., total pressure), from equation (1.31) we have $dG = V\,dP - S\,dT$. Rewriting the complete reaction and employing the stipulation that the equilibrium pressure depart from total pressure (normally it is less than $P_{total}*$) yields:

---

*It is possible for fluid pressure to exceed the lithostatic value, but only by an amount equal to the tensile strength of the rocks being subjected to the attendant confining (lithostatic) pressure; at greater values of P_{fluid}, the enclosing rocks would fracture, permitting a decrease in the fluid pressure.

$$d\Delta G = 0$$

$$= (\Delta V_{condensed}) \, dP_{total} + (\Delta V_{volatile}) \, dP_{E_{volatile}} \qquad (3.15)$$

$$- (\Delta S_{total}) \, dT.$$

Here, at constant $P_{E_{volatile}}$, we have

$$(\Delta V_{condensed}) \, dP_{total} = (\Delta S_{total}) \, dT, \qquad (3.16)$$

and finally

$$\left(\frac{\partial P_{total}}{\partial T}\right)_{P_{E_{volatile}}} = \frac{\Delta S_{total}}{\Delta V_{condensed}}. \qquad (3.17)$$

Again we will assume negligible isothermal compressibilities and thermal expansivities for the condensed assemblage; so $\Delta V^0_{condensed}$ may be substituted for $\Delta V_{condensed}$. If calorimetric data are not available, but we do have access to experimentally established curves for $P_{volatile} (= P_{total})$ against T, the entropy change of the reaction may be evaluated by measuring the curve slope (i.e., dP/dT) at the locus where $P_{E_{volatile}} = P_{total}$, because, from the Clapeyron equation (1.49), we have

$$\left(\frac{dP}{dT}\right) \Delta V_{total} = \Delta S_{total}. \qquad (3.18)$$

Substituting equation (3.18) into (3.17) yields:

$$\left(\frac{\partial P_{total}}{\partial T}\right)_{P_{E_{volatile}}} = \left(\frac{dP_{total}}{dT}\right) \frac{\Delta V_{total}}{\Delta V^0_{condensed}}. \qquad (3.19)$$

From this relationship, the effect of raising lithostatic pressures above those characteristic of the volatile component may be evaluated. We have, of course, assumed that the volatile component has a fixed volume at P–T conditions removed from the locus where the dP_{total}/dT slope was measured; because this assumption is not strictly valid, equation (3.19) is only approximately correct. However, for our purposes it is sufficiently accurate to demonstrate the effect and its magnitude. For quantitative treatment of this problem, see Greenwood, 1961.

For an example, let us consider the reaction calcite + quartz = wollastonite + CO_2. At 50 bars P_{CO_2}, the experimentally established equilibrium temperature is at 504°C. Here the curve slope, dP_{total}/dT, is 1.26 bars/°C. Employing the volume data of Appendixes 2 and 6, we find that $\Delta V^0_{condensed}$ is -19.69 cc, whereas V_{CO_2} is 1303 cc. Thus, from equation (3.19),

$$\left(\frac{\partial P_{total}}{\partial T}\right)_{P_{E_{CO_2}} = 50} = 1.26 \left(\frac{1283}{-19.69}\right) = -82 \text{ bars/}°\text{C}. \qquad (3.20)$$

Relationships between the P_{total}–T curve and several calculated $P_{E_{CO_2}}$ isobars are illustrated in Figure 3.5. Of course the $P_{E_{CO_2}}$–T curves possess negative slopes because, although ΔV_{total} is positive, the sign of $\Delta V^0_{condensed}$ is negative.

Schreinemakers' Treatment of Invariant Points and Radiating P–T Curves

According to the phase rule (equation 1.82), in a system consisting of n components, $n + 2$ phases (e.g., u, v, w, x, y, and z for a four-component system) can be stable together only at an invariant point; here the pressure and temperature, as well as the compositions of all the phases, are uniquely specified by the assemblage. For the general case, $n + 2$ univariant P–T curves, along each of which $n + 1$ phases are stable, radiate out from such a point. Pairs of these univariant curves define $n + 2$ divariant sectors, or P–T fields. (There are as many divariant fields as univariant curves and, in general, as many univariant curves as phases.) For simplicity, we will designate a univariant equilibrium curve of the sort $u + v = w + x + y$ by placing square brackets around the phase which takes no part in the reaction, thus $[z]$.

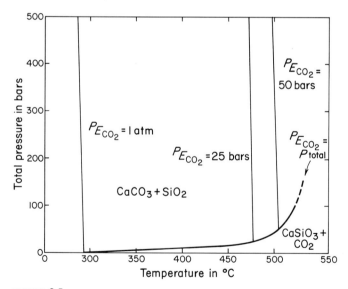

FIGURE 3.5.
The calcite + quartz = wollastonite + CO_2 reaction at $P_{CO_2} = P_{total}$, and at $P_{E_{CO_2}}$ isobars of I atm, 25 bars, and 50 bars. The equilibrium curve where carbon-dioxide pressure equals total pressure is taken from Harker and Tuttle (1956) and Greenwood (1967a).

Invariant points in geologically significant systems are important because they restrict the range of physical conditions at which specific phase associations are possible. The P–T locations of some of these mineralogic invariant points have been estimated from field relationships or computed from thermodynamic data, whereas others have been established experimentally. Many yet remain to be studied systematically, however. Here we will describe only briefly the graphical treatment of invariant points, which was elucidated more than half a century ago by Schreinemakers. For references, as well as a lucid and extended exposition of the theory, method of calculation, and examples, see Zen (1966) and Zen and Roseboom (1972).

Any univariant P–T curve describes the locus of conditions for which the Gibbs free energy of a reaction is zero; within the higher temperature (and higher entropy) divariant field, products possess a lower aggregate free energy than reactants, whereas the converse is true at temperatures below the curve. The P–T curve marks the contrasting stability limits for each *assemblage,* but gives no information about the conditions in which individual phases of an association may be stable. Naturally, the stability field of the phase compatibility $u + v \ldots + y$ must lie entirely within the overlapping portions of the P–T stability fields for all the individual phases.

The intersection of any two univariant curves must be at an angle of 180° or less; depending on the number of components in the system, at least one additional curve is generated within the larger sector by this intersection, and no matter how few or many curves are generated, no P–T sector can exceed 180°. The reason for this relationship, known as the Morey-Schreinemakers rule, is as follows. As we have seen, any univariant curve divides P–T space into complementary portions where products have a higher Gibbs free energy than reactants, and vice versa. As shown in Figure 3.6, the metastable extension of a curve, such as $u + v = w + x + y$, into a P–T sector larger than 180°, within which the stability of $u + v$ is tentatively assumed, would indicate both metastability and stability for the assemblage $u + v$; obviously this contradiction may be avoided by recognition that the stable assemblage $u + v$ is confined to the P–T sector in which the angle between curves [z] and [w] is less than 180°.

Now let us consider the arrangement of the univariant curves in P–T space disposed about a reaction [z], of the sort $u + v = w + x + y$. This equilibrium defines a particular phase-compatibility shift in chemographic space; to analyze it qualitatively, we need not know the exact stoichiometry of the reaction, but we must have positive amounts of participating reactants and products, so that they appear on the proper sides of the equation (e.g., negative amounts of a reactant would indicate that this species is actually a product). Any such equilibrium divides P–T space into two portions, each of which is characterized by specific phase compatibilities, as well as forbidden associations. For instance, on one side of curve [z], assemblages (u,z) and (v,z) are stable, but (w,z), (x,z), and (y,z) are metastable, and conversely. The convention employed

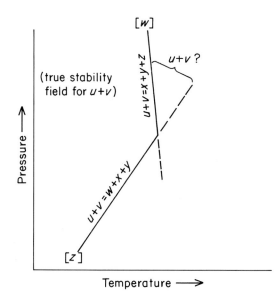

FIGURE 3.6.

Intersection of two univariant curves, [w] and [z], which pair limits the P–T stability range of the relatively low-temperature assemblage $u + v$. If $u + v$ are assumed to be stable in the P–T sector which exceeds 180°, the metastable extension of [z] to higher pressures (on the low-temperature side of which $u + v$ are supposedly incompatible) lies within the presumed $u + v$ stability field defined by curve [w]. Metastable extensions of the curves are shown by dashes.

here is that (u,z) signifies the assemblage from which u and z are absent, namely $v + w + x + y$. The P–T sector (w,z) lies between curves [w] and [z], within the angle of intersection less than 180°. Reaction [w], $u + v = x + y + z$, also divides P–T space into two portions, one side characterized by the phase associations (u,w) and (v,w), the other by (w,x), (w,y), and (w,z). The sector (w,z), bounded by the curves [w] and [z], includes phase assemblage (w,z) and others as well, such as (x,z), (y,z), (w,x), and (w,y). Assemblages (w,z) must occur throughout the area bounded by the curves [w] and [z]. The others listed above will be stable in at least a portion of this P–T sector; where other reaction curves intervene, some of these associations may become metastable. If enough of the reactions at the invariant point are known, the sequence of all possible curves can be specified. To determine the sense of rotation (i.e., which is the order, of the two possible orders, in which the curves are met), the P–T slopes of the curves, and the P–T location of the invariant point itself, we need to have experimental or thermochemical data for two or more of the reactions.

As an example, let us examine phase relations in a portion of the four-component system $K_2O \cdot Al_2O_3 \cdot 6SiO_2 - Al_2O_3 - SiO_2 - H_2O$. The phases of interest as shown in Figure 3.7 are: sanidine, abbreviated **san**; andalusite, **and**; quartz, **q**; muscovite, **ms**; aqueous melt, **l**; and aqueous fluid, **f**. We will ignore equilibria involving corundum, other polymorphs of alumina, aluminosilicate, and silica, pyrophyllite, and clay minerals. Three of the reactions have been studied experimentally, and their P–T locations are known (these curves could have been calculated from thermochemical data, although with less accuracy).

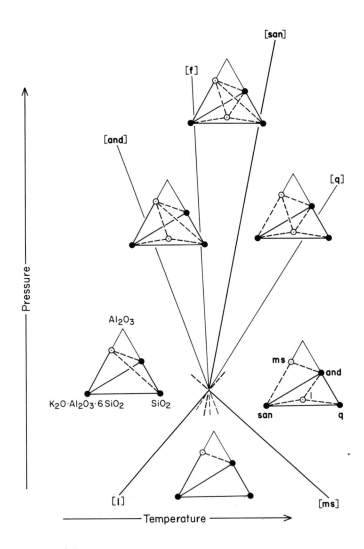

FIGURE 3.7.

Schreinemakers' array of univariant curves for the invariant point where **ms, and, san, q, l,** and **f** coexist, projected from the H_2O apex of the tetrahedron (i.e., the quaternary system) $K_2O \cdot Al_2O_3 \cdot 6SiO_2$–$Al_2O_3$–$SiO_2$–$H_2O$ onto the anhydrous base. Experimentally located curves are shown in heavy lines. Hydrous phases, muscovite, and liquid, are indicated by open circles, anhydrous phases by solid circles. Metastable extensions of the curves are shown by dashes.

They are:

$$[\mathbf{l}] \qquad \mathbf{ms} + \mathbf{q} = \mathbf{san} + \mathbf{and} + \mathbf{f};$$

$$[\mathbf{san}] \qquad \mathbf{ms} + \mathbf{q} = \mathbf{and} + \mathbf{l} + \mathbf{f};$$

and

$$[\mathbf{ms}] \qquad \mathbf{san} + \mathbf{and} + \mathbf{q} + \mathbf{f} = \mathbf{l}.$$

Using the preceding discussion as a guide, we see that [**l**] divides P–T space into two portions; at lower temperatures and higher pressures, the assemblages $\mathbf{ms} + \mathbf{q} + \mathbf{and} + \mathbf{f}$ (san,l), $\mathbf{ms} + \mathbf{q} + \mathbf{san} + \mathbf{f}$ (and,l), and $\mathbf{ms} + \mathbf{q} + \mathbf{san} + \mathbf{and}$ (f,l) are stable; on the other side of the curve, $\mathbf{q} + \mathbf{san} + \mathbf{and} + \mathbf{f}$ (ms,l) and $\mathbf{ms} + \mathbf{san} + \mathbf{and} + \mathbf{f}$ (q,l) are stable phase compatibilities. Similarly, on the low-temperature side of [**san**], the associations $\mathbf{ms} + \mathbf{q} + \mathbf{l} + \mathbf{f}$ (san,and), $\mathbf{ms} + \mathbf{q} + \mathbf{and} + \mathbf{f}$ (san,l), and $\mathbf{ms} + \mathbf{q} + \mathbf{and} + \mathbf{l}$ (san,f) are stable, whereas the equilibrium associations $\mathbf{q} + \mathbf{and} + \mathbf{l} + \mathbf{f}$ (ms,san) and $\mathbf{ms} + \mathbf{and} + \mathbf{l} + \mathbf{f}$ (q,san) are confined to temperatures exceeding this curve. Comparable P–T restrictions of the various phase assemblages may also be demonstrated for curve [**ms**]. Using this technique, three other curves can be located in the proper topological sequence. As illustrated in Figure 3.7, they are:

$$[\mathbf{f}] \qquad \mathbf{ms} + \mathbf{q} = \mathbf{san} + \mathbf{and} + \mathbf{l};$$

$$[\mathbf{q}] \qquad \mathbf{ms} + \mathbf{l} = \mathbf{san} + \mathbf{and} + \mathbf{f};$$

and

$$[\mathbf{and}] \qquad \mathbf{ms} + \mathbf{q} + \mathbf{san} + \mathbf{f} = \mathbf{l}.$$

To specify the arrangement shown, we have assumed that the melt phase contains only a little H_2O, so that the line connecting **and** with **l** passes underneath (i.e., possesses less H_2O than the $\mathbf{ms} + \mathbf{q}$ line at their projected intersection). On the other hand, melt has been assumed to contain enough H_2O so that its composition lies above the $\mathbf{ms} + \mathbf{q} + \mathbf{san}$ plane. In order to evaluate the P–T curve slopes for these reactions, however, we need quantitative information about the stoichiometry and volumes of all the phases, including the aqueous melt.

Thus far the univariant reactions which have been considered involve $n + 1$ phases in a system of n components. Univariant reactions which take place among fewer than $n + 1$ phases do so because of the nongenerality of phase compositions within the system. For instance, in a system of at least two components, two or more phases may possess the same composition (hence show unary behavior), whereas in a system of three or more components, three phases may lie along a line in compositional space (hence are binary), or four phases may be coplanar in compositional space (hence are ternary), etc. Another way of stating this relationship is that some of the phases may be described in terms of fewer than n components. Systems exhibiting such compositionally nongeneral phases are referred to as degenerate. A result of this phenomenon is the stable-to-stable, or stable-to-metastable, coincidence of certain P–T curves in a Schreinemakers' array, as will now be discussed.

Phases which are not involved in the univariant degenerate reactions are termed indifferent phases. In some systems, an indifferent phase may not take part in any of the equilibria—for instance, consider the indifferent role of quartz in the binary system SiO_2–Al_2O_3·SiO_2 adjacent to the aluminosilicate triple point (see Figure 2.6). In contrast, indifferent phases such as aqueous fluid and analcime, abbreviated **f** and **am** (and also albite, **ab**, and nepheline, **ne**), do take part in some—but not all—of the reactions of the ternary system Na_2O·Al_2O_3·$2SiO_2$–Na_2O·Al_2O_3·$6SiO_2$–H_2O. As illustrated in Figure 3.8, two compositional coincidences are present here. Jadeite, **jd**, lies along the binary join **Ab**–**Ne**; hence the reaction

[am,f]   2 jadeite = albite + nepheline

passes through the invariant point without inflection. Analcime lies along the binary line connecting jadeite and fluid, but only within the P–T field where jadeite is stable; hence the reaction

[ab,ne]   jadeite + fluid = analcime

terminates at the invariant point. The only truly ternary reaction in the vicinity of the invariant point is the decomposition of analcime,

[jd]   2 analcime = albite + nepheline + 2 fluid.

In this discussion, all phases are assumed to have fixed stoichiometric compositions, but in fact, analcime exhibits extensive solid solution toward the component **Ab**.

An alternative display of a Schreinemakers' array, in compositional rather than in P–T space, is also instructive. In an isobaric, isothermal section through P–T chemographic space, we may select two variable chemical parameters as coordinates. For a reaction of the sort $aA + bB = lL + mM$ (here uppercase letters designate the participating species, lowercase letters the number of moles of each), the van't Hoff reaction isotherm (see equation 1.72) reduces at equilibrium to

$$\log \frac{a_L{}^l a_M{}^m}{a_A{}^a a_B{}^b} = - \frac{\Delta G^0}{2.303 RT}. \tag{3.21}$$

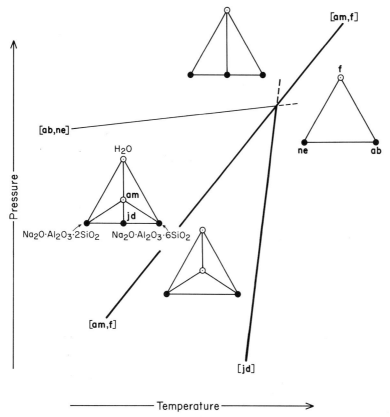

FIGURE 3.8.

Schreinemakers' array of univariant curves for the degenerate invariant point where **ab**, **ne**, **am**, **jd**, and **f** coexist, in the ternary system $Na_2O \cdot Al_2O_3 \cdot 2SiO_2$–$Na_2O \cdot Al_2O_3 \cdot 6SiO_2$–$H_2O$. Experimentally established curves are shown in heavy lines. Metastable extensions of the curves are shown by dashes.

Here, two variable activities or fugacities may be selected as ordinate and abscissa; activities of the other participating species then become dependent variables, and their values either remain fixed or change as a consequence of variation in the determining activities. Alternatively, since there is a direct relationship between activity or fugacity and the chemical potential (see equations 1.66 and 1.62), the chemical potentials may be employed instead as the chemical parameters.

To illustrate the method, let us consider equilibria in the system Fe–O–S, at one atmosphere total pressure and 25°C, among the phases hematite, magnetite, and pyrite, abbreviated **hem**, **mt**, and **py**, respectively. The ordinate is selected as log sulfur fugacity, the abscissa as log oxygen fugacity. The following reactions must be considered:

$$[\textbf{py}] \qquad 6Fe_2O_3 = 4Fe_3O_4 + O_2;$$

$$[\textbf{hem}] \qquad 3FeS_2 + 2O_2 = Fe_3O_4 + 3S_2;$$

and

$$[\textbf{mt}] \qquad 4FeS_2 + 3O_2 = 2Fe_2O_3 + 4S_2.$$

Because the condensed assemblage consists exclusively of pure phases exhibiting no solid solution, the activities of these phases are unity. Equilibrium between hematite and magnetite does not involve S_2; so the value of f_{O_2} may be obtained directly from the data of Appendix 2. As we have already seen from equation (3.11), $\log f_{O_2} = -68.91$. Now let us examine the pyrite–magnetite equilibrium. Here we have, from equation (3.21),

$$\log \frac{(f_{S_2})^3}{(f_{O_2})^2} = -\frac{-243{,}094 + 3 \cdot 38{,}296}{2.303 \cdot 1.987 \cdot 298} = 93.99, \qquad (3.22)$$

or

$$3 \log f_{S_2} - 2 \log f_{O_2} = 93.99. \qquad (3.23)$$

The $\log f_{S_2}/f_{O_2}$ slope for this reaction is positive as is evident from equation (3.22) and has a numerical value of 2/3. Locations of the curve may be established by arbitrarily assigning numbers to one of the two fugacities and solving for the other (e.g., where $\log f_{O_2} = -68.91$, $\log f_{S_2} = -14.61$, and where $\log f_{O_2} = -80.00$, $\log f_{S_2} = -22.00$). In the same way, the $f_{S_2} - f_{O_2}$ location of the pyrite–hematite equilibrium may be computed. Calculated curves and their metastable extensions are shown in Figure 3.9. Similar curve locations for higher T and P may be obtained by using the Gibbs-free-energy data for elevated temperatures from Appendix 3 and by applying pressure corrections as given in equations (1.75) and (3.6).

As is evident from this illustration, only three curves radiate out from the invariant point, because both P and T have been specified arbitrarily; hence $F = 3 - p + 0$. Or, looked at another way, because the fugacities of the two volatile components are employed as the independent variables, equilibria defined by absence of one of the gaseous constitutents cannot be represented (unless degenerate, as is the case for the binary hematite–magnetite equilibrium). Strictly speaking, if we confine our consideration to the ternary system Fe–O–S as originally stated, the existence of a total pressure of one atmosphere precludes the occurrence of a separate gas phase in the low $f_{O_2} - f_{S_2}$ region for which the calculations were made (nonetheless, values for the fugacities, or chemical potentials, of oxygen and sulfur are specified by the univariant curves and the invariant point shown in Figure 3.9). A separate gas phase

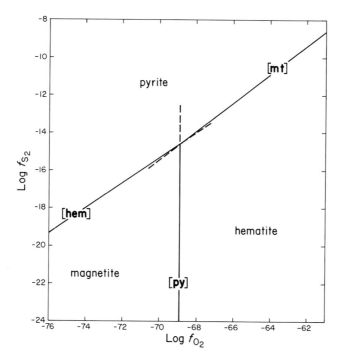

FIGURE 3.9.

Diagram of $\log f_{S_2}$ against $\log f_{O_2}$ computed for a part of the system Fe–O–S at one atmosphere total pressure and 25°C. Metastable extensions of the curves are shown by dashes.

which contains these volatile components at the computed values will be present if we introduce an inert gas component such as argon or nitrogen to the system; this nonreactive contaminant may be regarded as part of the container if we choose to ignore the minor amounts of S_2 and O_2 in our counting of phases, but must be recognized as a component if we count the $S_2 + O_2$ as a separate phase.

One of the desirable properties of isothermal, isobaric fugacity-fugacity, activity-activity, or μ-μ diagrams is that the slopes of the univariant equilibria are governed by the reaction stoichiometries, and hence may be obtained even in the absence of experimental and thermochemical data (e.g., see Korzhinskii, 1959).

CHAPTER 4

Predominantly Liquidus Diagrams and Crustal Igneous Petrology

CHEMICAL DIVERSITY AND THE CLASSIFICATION OF IGNEOUS ROCKS

The bulk compositions of igneous rocks naturally reflect the chemistry of their constituent minerals. Although oxides, sulfides, carbonates, and sulfates occur, they are definitely minor phases in most igneous rocks (we are ignoring certain rare varieties—e.g., carbonatites). All the major igneous minerals are silicates; hence the chief component present in such rocks is SiO_2. Other principal oxides include Al_2O_3, TiO_2, Fe_2O_3, FeO, MgO, MnO, CaO, Na_2O, K_2O, H_2O, CO_2, and P_2O_5.

    Although numerous minerals are present in varying proportions in the different igneous rocks, the spectrum of bulk compositions is relatively narrow. Ultramafic rocks, for instance, carry on the order of 35 to 40 per cent silica by weight; whereas the most felsic types of igneous rock contain 75 to 80 per cent SiO_2 by weight, barely twice as much. Other oxides show even more restricted ranges, from a maximum on the order of 5 to 20 per cent down to nearly zero.

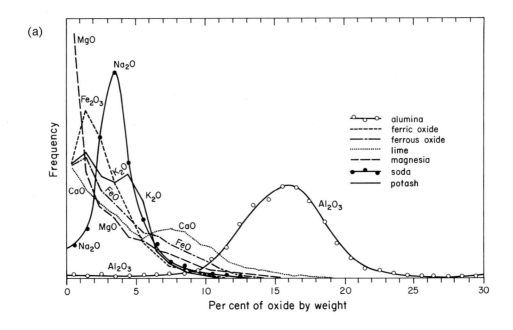

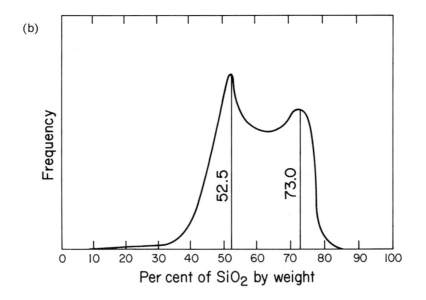

FIGURE 4.1.

The frequency distribution of percentages by weight for various oxides of igneous rocks, taken from Richardson and Sneesby, 1922 (after Washington, 1917). Diagram (a) presents major oxides exclusive of silica; diagram (b) shows silica abundance only.

The gradational nature of the compositional range of igneous rock is illustrated in Figure 4.1. It is apparent that there is a continuous spectrum of values between rather well-defined limits. The bimodal distribution evident in Figure 4.1(b) has perplexed petrologists for a half a century. It may be merely a sample bias reflecting the concentration of low-density, silicic igneous rocks in the uppermost crust, but this is by no means certain. The explanation for the observed range and the absence of more extreme chemical variation has to do with the origin and diversity of igneous rocks. Whatever their ultimate source, present-day magmas (and undoubtedly those of the geologically recognizable past, too) must have been derived from the partial fusion of pre-existing solid materials, since the Earth is in a largely molten state only in its outer core. If a small amount of liquid were derived from the partial fusion of different crustal or mantle source rocks, or from the same protolith under vastly different P–T conditions, such as would be occasioned by partial melting at different depths, then magmas of contrasting composition would be generated (see Chapter 5). Crystal-melt equilibrium (i.e., the production of liquids enriched in certain elements, depleted in others, that were present in the original protolith) evidently is the dominant process in the generation of magma from a solid precursor during a heating stage.

Differentiation of a homogeneous primary magma on cooling has been abundantly documented, and is also partly responsible for the range in compositions of igneous rocks. Although formerly some petrologists called upon processes such as assimilation of country rocks, liquid immiscibility, or hydrothermal bubble-linked diffusion (gaseous transfer), these mechanisms now are generally regarded as minor processes, incapable of producing the magnitude of the observed differentiation (e.g., Bowen, 1928, Chap. 2). Crystal fractionation—the limited interaction, or nonreaction, between crystals and melt of contrasting compositions—is thought to be responsible for much, but not all, of the chemical diversity observed in igneous-rock series. This latter process results from a failure of early-formed solid phases to maintain chemical equilibrium with the residual, lower-temperature melt. Reasons for this phenomenon include crystal armoring, gravitative settling of crystals, crystal sorting due to laminar flow or convection of the magma, and filter pressing (i.e., the squeezing out of interstitial liquid from a largely crystalline accumulation). In all these cases, the early-formed solid phases are physically separated from the melt by distances exceeding those of the diffusion paths for the time available, and hence crystals and liquid cannot remain in chemical communication.

The range of observed compositions of igneous rocks is thus a function of crystal-melt equilibrium (and disequilibrium). First, it is due to the generation of magma from various protoliths by partial fusion, sometimes under contrasting physical conditions; and second, it is the result of crystal fractionation accompanying solidification of the melt.

From what has been said, it is apparent that any classification of igneous rocks will necessarily be arbitrary, for the constituent mineral compositions—

and hence bulk compositions—exhibit complex and often continuous gradations. Numerous schemes for igneous-rock classification have been proposed.
We will present a very simple one, based on modal mineralogy, and specifically
based on the amount of quartz present, the **An** content of the plagioclase, and
the proportion of total feldspar which is alkali feldspar (solid solution between
$KAlSi_3O_8$ and $NaAlSi_3O_8$ in which the $CaAl_2Si_2O_8$ component makes up less
than 10 mole per cent). The classification and mineralogic range of natural
compositions are illustrated in Figure 4.2. In practice, the nature of mafic
minerals, and their proportions also, are a function of the feldspar compositions
and quartz content, as shown in Table 4.1. For reference, the average compositions of some common plutonic types of igneous rock, both magma types
and crystal accumulates, are presented in Tables 4.2 and 4.3. A much more

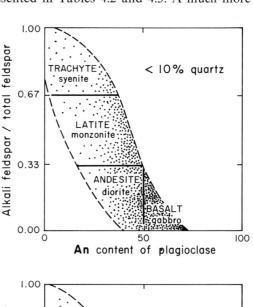

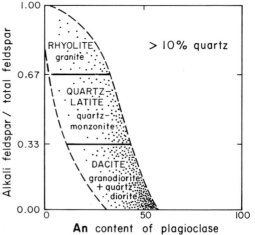

FIGURE 4.2.

A simplified mineralogic
classification of igneous rocks
(from Ernst, 1969, Figure 5.7).
Uppercase letters indicate
extrusive rocks, lowercase letters
intrusive rocks. Stippled pattern
reflects the approximate color
index (proportion of mafic
minerals); see also Table 4.1.

TABLE 4.1.

Typical mafic minerals and color indexes of common igneous rocks (after Ernst, 1969, Table 5.1); sphene and an iron oxide phase commonly present in minor quantities.

Rock type[a]	Mafic minerals	Color index (i.e., per cent by volume of mafic minerals)
Less than 10 per cent quartz		
BASALT, gabbro	olivine, augite ± hypersthene (or pigeonite)	35–65
ANDESITE, diorite	hypersthene ± augite ± hornblende	20–45
LATITE, monzonite	hornblende ± biotite	10–30
TRACHYTE, syenite	biotite ± hornblende ± sodic amphibole (or sodic pyroxene)	0–20
More than 10 per cent quartz		
DACITE, granodiorite, quartz diorite	hornblende ± augite ± biotite	20–50
QUARTZ LATITE, quartz monzonite	hornblende ± biotite	10–25
RHYOLITE, granite	biotite ± hornblende ± sodic amphibole (or sodic pyroxene)	0–15

[a]Capital letters designate volcanic rocks; lowercase letters stand for plutonic equivalents.

TABLE 4.2.

Average chemical compositions in per cent by weight of some igneous intrusive rocks (after Nockolds, 1954).

Oxide	Gabbro	Diorite	Monzonite	Syenite	Granio-diorite	Quartz monzonite	Granite
SiO_2	48.36	51.86	55.36	59.41	66.88	69.15	72.08
TiO_2	1.32	1.50	1.12	0.83	0.57	0.56	0.37
Al_2O_3	16.84	16.40	16.58	17.12	15.66	14.63	13.86
Fe_2O_3	2.55	2.73	2.57	2.19	1.33	1.22	0.86
FeO	7.92	6.97	4.58	2.83	2.59	2.27	1.67
MnO	0.18	0.18	0.13	0.08	0.07	0.06	0.06
MgO	8.06	6.12	3.67	2.02	1.57	0.99	0.52
CaO	11.07	8.40	6.76	4.06	3.56	2.45	1.33
Na_2O	2.26	3.36	3.51	3.92	3.84	3.35	3.08
K_2O	0.56	1.33	4.68	6.53	3.07	4.58	5.46
H_2O+	0.64	0.80	0.60	0.63	0.65	0.54	0.53
P_2O_5	0.24	0.35	0.44	0.38	0.21	0.20	0.18

TABLE 4.3.

Average chemical compositions in per cent by weight of some igneous crystal accumulates (after Nockolds, 1954)

Oxide	Dunite	Peridotite	Pyroxenite	Anorthosite
SiO_2	40.16	43.54	50.50	54.54
TiO_2	0.20	0.81	0.53	0.52
Al_2O_3	0.84	3.99	4.10	25.72
Fe_2O_3	1.88	2.51	2.44	0.83
FeO	11.87	9.84	7.37	1.46
MnO	0.21	0.21	0.13	0.02
MgO	43.16	34.02	21.71	0.83
CaO	0.75	3.46	12.00	9.62
Na_2O	0.31	0.56	0.45	4.66
K_2O	0.14	0.25	0.21	1.06
H_2O+	0.44	0.76	0.47	0.63
P_2O_5	0.04	0.05	0.09	0.11

comprehensive and complex descriptive system for categorization of igneous rocks has been presented by Streckeisen (1967); Carmichael, Turner, and Verhoogen (1974, Chap. 2) have provided a detailed classification based on thermodynamic and phase-equilibrium principles.

Now that the classification and chemical diversity of igneous rocks have been briefly mentioned, we will proceed to a discussion of phase relationships in one-component systems, then in two-, three-, and more complex multi-component systems.

UNARY DIAGRAMS

Let us now turn briefly to consideration of a one-component phase diagram, that for the system H_2O. Pressure-temperature relationships are illustrated in Figure 4.3. Phase relationships can be readily understood by use of the Clapeyron equation (1.49),

$$\frac{dP}{dT} = \frac{\Delta S}{\Delta V}.$$

The atomic arrangement in ice is much more ordered than in water; so the melting of ice proceeds with an entropy increase. On the other hand, at low pressures, ice polymorph I is less dense than liquid water. Hence the reaction $ice_I \rightarrow water$ has a negative P–T slope; this phenomenon is unlike the melting behavior of most other solids, which typically are more dense than their fused equivalents. Another interesting aspect of the diagram concerns the critical

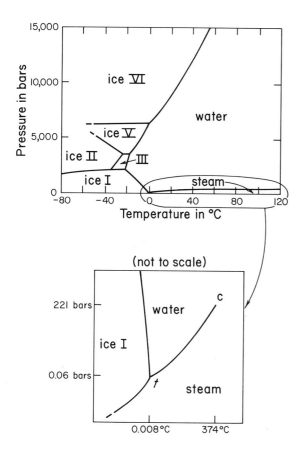

FIGURE 4.3.

Phase diagram for the unary system H_2O, after Kennedy and Holser (1966, Figure 16.1). The water-steam curve has been offset slightly toward higher pressures for clarity.

point c. Liquid water and water vapor, or steam, have well-defined, contrasting properties (e.g., density) at low pressures and temperatures. However, at higher pressures and temperatures approaching 200 bars and 300°C, these phases become more nearly alike. At the critical point of 221 bars and 374°C, the properties become identical, and it is then impossible to distinguish between liquid and gaseous H_2O. For this reason, at temperatures and pressures in excess of this value, we speak of a supercritical aqueous *fluid* for phase compositions close to pure H_2O, rather than of water or steam. According to the phase rule, the variance (degrees of freedom) of the triple point, t, is zero, because there are three phases present in a one-component system—but only at a single value of temperature and pressure.

A second example of a one-component phase diagram, the system SiO_2, is presented as Figure 4.4. As is readily seen, at atmospheric pressure α quartz is stable up to 573°C, β quartz between 573 and 867°C, tridymite between 867 and 1,470°C, and cristobalite from 1,470 up to the melting point at about 1,713°C. Within the diagram, each P–T field is characterized by the stability of a single phase. Recalling from the phase rule for a unary system that

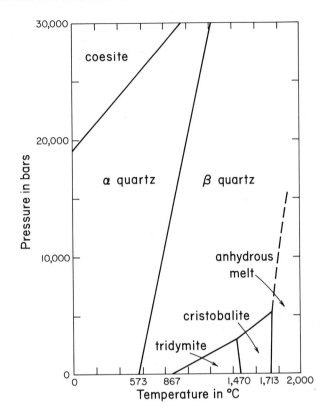

FIGURE 4.4.

Phase diagram for the unary system SiO_2, after Boyd and England (1960b).

$F = 1 - p + 2$, we can see that the variance within individual fields is two; both temperature and pressure can be specified arbitrarily without contradicting the given state of the system. Curvilinear lines define equilibrium between two phases such as β quartz + tridymite. Obviously, the phase rule here dictates only a single degree of freedom; either temperature or pressure may be randomly selected, but then the other is defined by the observed equilibrium mineral association. The triple point where β quartz, tridymite, and cristobalite coexist stably is invariant; here both P and T are fixed by the phase assemblage (i.e., $F = c - p + 2 = 1 - 3 + 2 = 0$).

At temperatures less than 573°C, α quartz has a lower Gibbs free energy than β quartz, but the reverse is true above 573°C; furthermore, at 573°C the molar Gibbs free energies of these two phases must be equal. At some higher temperature—say, 700°C—β quartz is more stable than α quartz at one atmosphere pressure. However, because the molar volume of β quartz slightly exceeds that of α quartz, increased pressure will result in a more rapid increase in $G_{\beta\,\text{quartz}}$ than in $G_{\alpha\,\text{quartz}}$, because from equation (1.33) we know that

$$\left(\frac{\partial G}{\partial P}\right)_T = V.$$

To a first approximation we assume that the compressibilities and thermal expansivities of condensed phases are negligible, or at least tend to cancel one another out, and so can be ignored in our considerations. At some higher pressure, the one-atmosphere difference in G per mole of SiO_2 as α quartz versus β quartz has been overcome. The conditions of equilibrium for the over-all reaction ($\Delta G = 0$) is given by equation (3.7):

$$P - 1 = -\frac{\Delta G^0}{\Delta V^0}.$$

These curves may also be checked by means of the Clapeyron equation. Since the molar volume of tridymite exceeds that of cristobalite, whereas tridymite has a smaller entropy than the high-temperature polymorph, the univariant curve along which both are stable has a negative slope. On the other hand, the anhydrous-silica melting curve possess a positive P–T slope because ΔV is positive, anhydrous melt having a larger molar volume than cristobalite.

Now let us consider the effect on the system SiO_2 of adding a second component, H_2O. Because the various polymorphic transitions indicated in Figure 4.4 do not involve an aqueous fluid phase, the presence or absence of H_2O will not influence P–T curve locations; hence the subsolidus transitions retain unary behavior. However, inasmuch as H_2O is soluble in siliceous melts, the chemical potential, activity, partial pressure, or fugacity of H_2O is an important variable to take into account wherever fusion takes place. As a matter of fact, the presence of H_2O drastically lowers the melting temperature of the SiO_2 polymorphs, as shown in Figure 4.5. Why should this be so?

The answer may be seen by considering the reaction involved, silica polymorph + fluid = aqueous melt, in light of the Clapeyron equation. As written, the reaction entails an entropy increase proceeding to the right, but a substantial volume decrease because, whereas H_2O occupies a very large volume in the fluid phase, in the aqueous melt it possesses a partial molar volume approaching that of liquid water. Thus dP/dT is negative. Because H_2O fluid is highly compressible, the absolute value of ΔV is very large at low pressures, but becomes progressively smaller at higher pressure. For this reason, the melting curve slope is very gentle (and negative) at low pressures, but steepens considerably (sign immaterial) at elevated pressures.

The melting curve illustrated in Figure 4.5 for the binary system SiO_2–H_2O is that appropriate for an H_2O-saturated siliceous liquid. If, however, the aqueous fluid contains appreciable amounts of another dissolved species which is nearly insoluble in the melt, such as CO_2, the activity of H_2O will be diminished in both the tenuous fluid and the liquid; accordingly, the onset of melting at any particular value of P_{fluid} will occur at a higher temperature for a contaminated aqueous fluid phase than for the pure SiO_2–H_2O system shown in Figure 4.5. This situation is analogous to that portrayed in Figure 2.6 for the brucite–periclase equilibrium. For the melting phenomenon, lowered a_{H_2O}

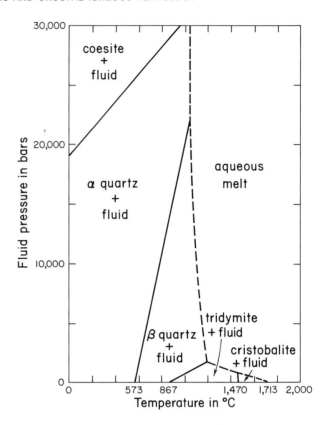

FIGURE 4.5.

Effect of P_{fluid} on the system $SiO_2(-H_2O)$, modified from Tuttle and England (1955). Where sufficient H_2O is present to saturate the silicate liquid, an aqueous fluid phase is associated with the H_2O-bearing melt.

causes the stability field of the anhydrous condensed assemblage (compared to that of the H_2O-bearing melt) to expand to higher temperatures.

Thus far we have considered reactions in which an excess of fluid—either pure or multicomponent—was present. Suppose, however, that in the simple binary system SiO_2–H_2O, there is insufficient volatile component to saturate the mass of siliceous liquid at—say—5,000 bars confining pressure. Below the melting temperature, small amounts of an aqueous fluid phase will be present at the operating total pressure. With the commencement of fusion, a little H_2O-saturated melt will be produced, thereby exhausting the system in free H_2O (fluid). At this stage, β quartz and liquid coexist stably. Any further increment of heat absorbed will result in an elevation of temperature and further melting of the silica polymorph; this will give rise to an H_2O-undersaturated liquid. The activity of H_2O in the melt progressively declines, therefore, as the amount of silicate liquid increases with rising temperature, in the absence of a separate aqueous fluid phase. Throughout this discussion we have focused our attention on the unary component SiO_2, but where H_2O is soluble in one of the phases (melt), behavior is of course binary in nature.

Another phase diagram for a unary system, polymorphism of the alumino-silicates, has already been presented as Figure 2.7. In this system, three $P-T$ fields, each characterized by the stability of a single aluminosilicate polymorph, are bounded by pressure-temperature curves along which two phases are mutually stable. All three polymorphs are in equilibrium only at the triple point. Although not especially germane to igneous petrology, this system does illustrate unary relations analogous to those described above for the polymorphism of silica. A check on the experimental results may be performed by calculating the $P-T$ locations and slopes of the univariant curves for the indicated equilibria, employing the data of Appendix 2.

BINARY DIAGRAMS

We now turn our attention to a consideration of various sorts of binary crystal-melt equilibria. Useful terminology which needs to be introduced at this point and is applicable to multicomponent systems in general includes the following:

Liquidus

That $P-T-x$ (x is one or more chemical variables) line or surface along which the compositions of melt in equilibrium with a crystalline phase are defined; all such melt compositions are saturated with respect to the crystalline phase. At temperatures exceeding those of the liquidus, no solid phase can be in equilibrium with melt.

Solidus

That $P-T-x$ line or surface along which the compositions of one or more crystalline phases in equilibrium with melt are defined. Melt is unstable at temperatures below the solidus.

Transition Loop

A transition loop consists of a pair of $P-T-x$ lines or surfaces, along each limb of which the composition of one or more phases is specified. A binary liquidus + solidus provides a good example of a transition loop.

Solvus

That $P-T-x$ line or surface defined by the compositions of coexisting crystalline (or immiscible liquid) phases of similar structures which exhibit at least limited mutual solubility as a function of the state variables. Where the participating species possess contrasting atomic structures, the feature is not a true solvus but is rather a transition loop.

Tie Line

A line located in chemographic (compositional) space at fixed P–T conditions by joining the compositions of two coexisting phases in chemical equilibrium with one another.

N-Phase Polyhedron

A polyhedron circumscribed in chemographic space at fixed P–T conditions by the combination of all limiting tie lines which link coexisting equilibrium phases.

Eutectic

The lowest temperature point on the liquidus at which a unique melt of fixed composition is in equilibrium with two or more crystalline phases, the chemistries of which are also invariant. Solidus and liquidus share a common P–T value here, but contrast in phase chemistry. In compositional space, the liquid lies within the n-phase polyhedron defined by the associated solids. A eutectic-type melting reaction is simply: crystalline phases (in the eutectic proportions) = eutectic melt. For an example, see Figure 4.6.

Peritectic

A relatively low-temperature inflection point on the liquidus at which a unique melt of specified composition is in equilibrium with two or more crystalline phases, the chemistries of which are also invariant. In compositional space, the liquid lies outside the n-phase polyhedron defined by the associated solids. It is therefore an inflection on the liquidus at supersolidus temperature marked by the nonstoichiometric melting of one or more phases. A peritectic-type melting reaction is of the sort: one or more crystalline phases (not in the peritectic proportions) = peritectic melt + a different crystalline phase assemblage. See Figure 4.10 for an example.

Minimum

This relationship is similar to that of the eutectic type (see above) except that, in addition, the relative proportions or partitionings of at least two components in a crystalline phase are identical to that in the liquid. Solidus and liquidus share a common tangent. See Figure 4.16 for an example.

Join

Any line, plane, or n-apex polyhedron (n corners in $n - 1$ dimensions) connecting two, three, or n components or phases in compositional space. An

Alkemade line, surface, or *n*-apex polyhedron connects the compositions of two, three, or *n* solid phases which coexist with the same liquid.

Four principal types of binary crystal-melt equilibria will be distinguished: (1) a simple eutectic showing no solid solution; (2) a peritectic exhibiting no solid solution; (3) complete solid solution; and (4) a eutectic with limited solid solution. In all examples considered, melt occurs as a single, homogeneous phase; liquid immiscibility occurs in some systems, but because this phenomenon is largely confined to petrologically uninteresting compositions, it will not be treated. As a convenience we will study chiefly one-atmosphere isobaric sections; more data are available at such conditions, and many phase diagrams are topologically similar at somewhat higher pressures anyway. When we discuss type (3) equilibrium, complete solid (as well as liquid) solution, the solvus relationship will be introduced, because the principle elucidated will be applied to type (4) equilibrium, eutectic with limited solid solution.

(1) Simple, Binary Eutectic Showing No Solid Solution

The system $CaO \cdot MgO \cdot 2SiO_2$–$CaO \cdot Al_2O_3 \cdot 2SiO_2$ presented in Figure 4.6 exemplifies this type of equilibrium. This diagram is characterized at high temperatures by a one-phase melt field and three two-phase fields, namely, melt + diopside, melt + anorthite, and, at temperatures below the eutectic *e*, diopside + anorthite. In the melt field there are two degrees of freedom for

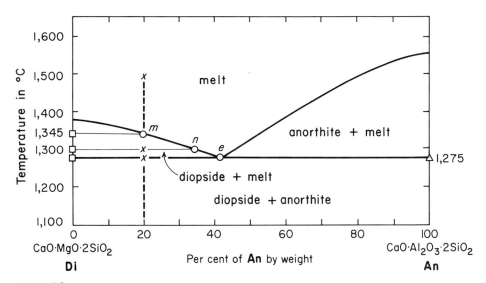

FIGURE 4.6.

One-atmosphere isobaric temperature-composition diagram for the binary system $CaO \cdot MgO \cdot 2SiO_2$–$CaO \cdot Al_2O_3 \cdot 2SiO_2$, after Osborn (1942). Dashed line indicates a portion of the cooling path followed by bulk composition *x* (see text for discussion).

this isobaric system: both temperature and a chemical parameter (e.g., **Di** content of the liquid) may be independently chosen. In contrast, the two-phase fields are characterized by only one independent variable: if the temperature is selected at random, the compositions of the pair of phases are determined and vice versa (or, for the two separate crystal + melt fields, choosing the composition of the melt that is in equilibrium with—hence saturated with respect to—a crystalline phase automatically specifies the temperature). The eutectic, *e*, is an isobaric invariant point: here three phases of fixed compositions coexist only at a unique temperature.

Let us follow the crystallization history of some arbitrary but general composition lying in this binary system, say, *x* in Figure 4.6. We will postulate that initially the material is at a very high temperature, such as 1,500°C, and thus completely molten. As heat is withdrawn, the temperature of the liquid declines well below the value of 1,391°C, at which pure diopside would crystallize, before the first clinopyroxene crystals form, at a temperature of about 1,345°C. At this temperature the bulk composition of the sample is the sum of the melt composition (still *x* = *m*), plus an infinitesimal amount of crystalline diopside. Further subtraction of heat results in the crystallization of an appreciable quantity of diopside; hence the melt composition is impoverished in the **Di** component, and with falling temperature moves towards the **An** side of the diagram. At any particular temperature—say, 1,300°C—the tie line connecting the compositions of diopside crystals and melt, *n*, must pass through the bulk composition *x*. The relative proportions of the phases (in mole or weight proportions, depending on the units employed in the phase diagram) is a function of how closely they approach the chemistry of the bulk composition. As a mechanical analogue, consider the bulk composition as the center of balance, with the chemically distinct phases acting as point masses and disposed about it in chemographic space. In the binary example chosen, the bulk composition would be the fulcrum, with lever arms to the compositions of the two phases; the closer a phase is to *x*, the greater is its proportion.

At the eutectic temperature, the melt *e* has become saturated with respect to crystalline anorthite as well as diopside; so the withdrawal of heat results in the precipitation of both solid phases together. At equilibrium, the heat subtracted is matched by the combined heats of crystallization of these two phases; hence the temperature remains constant as long as melt *e* is present. This behavior is exactly analogous to that of an ice + water mixture at fixed pressure. For the isobaric case, $F = c - p + 1 = 2 - 3 + 1 = 0$; invariancy means that the coexistence of diopside + anorthite + melt *e* can occur in this system only at the eutectic temperature, 1,275°C. After the melt *e* is totally consumed, continued subtraction of heat simply causes the temperature of the crystalline assemblage to decline.

The heating path for this bulk composition, diopside + anorthite, in the appropriate proportions to equal *x*, would be precisely the same, but in reverse sequence, provided crystals and melt remain in equilibrium. Addition of heat would result in temperature increment up to the eutectic value. At 1,275°C, diopside and anorthite react in the eutectic ratio to produce melt *e* as heat is supplied. Only after the last trace of anorthite disappears, however, does temperature again commence to rise. The melt gradually becomes more **Di**-rich and increases in amount. It achieves the bulk composition $x(= m)$ at about 1,345°C, as the last trace of diopside disappears. Further addition of heat will now merely cause the temperature of the melt to rise. If, however, the early-formed eutectic melt is removed from the system as it is generated by the supply of heat, a different situation, *fractional fusion*, obtains: for the bulk composition *x*, liquid *e* is formed on heating the diopside + anorthite assemblage to 1,275°C. After much of the pyroxene and all the plagioclase has melted, subtraction of this eutectic liquid will leave a residual, monomineralic diopside assemblage, which will itself be fused only at the much higher temperature of 1,391°C. For a general discussion of fractional fusion, see Presnall (1969).

But why is the melting point of a pure crystalline phase such as diopside depressed by the addition of a second component, in this case, **An**? The added species is much more soluble in the melt than in the solid phase; therefore the molar Gibbs free energy of the melt is reduced relative to that of the crystalline assemblage; if **An** component were actually added to the system, spontaneous mixing, with a decrease in *G*, would occur in the liquid relative to the solid assemblage. This relationship is illustrated schematically in Figure 4.7, an

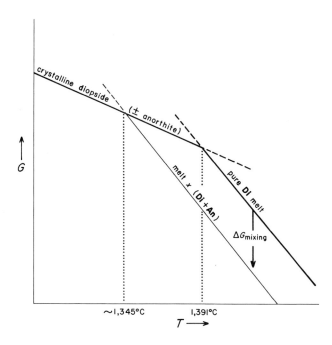

FIGURE 4.7.

One-atmosphere isobaric diagram of molar Gibbs free energy versus temperature for the compositions $CaO·MgO·2SiO_2$ and *x* in the system **Di–An** (see Figure 4.6). Heavy lines are for the unary system **Di**, light lines for the binary composition *x*. The slight concavity to the temperature axis is not shown.

isobaric G–T section. Recalling from equation (1.32) that

$$\left(\frac{\partial G}{\partial T}\right)_P = -S,$$

and realizing that the entropy of any phase is greater than zero, we can see that the Gibbs free energy must decrease with increasing temperature. Moreover, because the entropy of all phases increases with rising temperature, G–T curves are concave relative to the temperature axis; this slight curvature is not shown in the illustration.

First, examine the isobaric G–T curves for the component $CaO \cdot MgO \cdot 2SiO_2$, shown as heavy lines in the figure. At low temperatures the Gibbs free energy of pure crystalline diopside is less than that of pure **Di** melt; hence crystalline diopside is stable relative to the corresponding liquid (the reverse is true at elevated temperatures). The isobaric rate of change of G as a function of T for the crystalline phase is less than that for the melt, because melt is more disordered—hence has a higher entropy—than the solid. The Gibbs free energies of diopside and liquid are identical at the intersection of the two G–T curves, thus locating the equilibrium melting temperature for pure diopside at 1,391°C.

Now consider the addition of a second component, such as $CaO \cdot Al_2O_3 \cdot 2SiO_2$. If we imagine a mechanical mixture of pure diopside + pure anorthite at low temperatures, and a mechanical combination of pure **Di** melt + pure **An** melt at high temperatures, the relative disposition of the G–T curves and of the temperature of intersection will be unchanged, regardless of the value for the molar G_{An}, because both curves would be displaced by the same amount. However, although diopside and anorthite do not exhibit marked solid solution (in fact, diopside is slightly aluminous for bulk compositions lying along the **Di**–**An** join), liquid **Di** and liquid **An** are miscible in all proportions. Therefore a mechanical mixture of **Di** melt + **An** melt will spontaneously homogenize, reflecting a decrease in G (equation 1.29):

$$\Delta G_{\text{mixing}} = \Delta H_{\text{mixing}} - T\Delta S_{\text{mixing}}.$$

For ideal solutions, the heat of mixing is zero by definition, but because the degree of configurational disorder increases, the entropy of mixing is positive. (Liquid immiscibility only occurs in systems within compositional ranges in which the enthalpy of mixing is positive and exceeds the value of the quantity $T\Delta S_{\text{mixing}}$). Thus, for intermediate compositions, such as the illustrated composition x, the G–T curve for **Di** + **An** melt intersects that for pure diopside (+ pure anorthite) at a lower temperature—say, about 1,345°C—than the melting point of diopside in the unary system **Di**.

An analogous explanation for this melting-point depression may be given in terms of an isobaric μ–T diagram, as shown in Figure 4.8. From equation

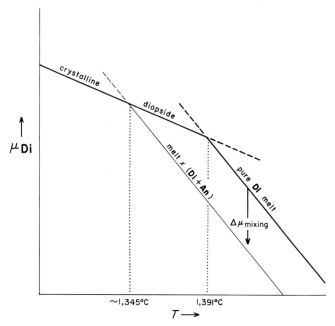

FIGURE 4.8.
One-atmosphere isobaric diagram of chemical potential versus temperature for the compositions CaO·MgO·2SiO$_2$ and x in the system **Di–An** (see Figure 4.6). Heavy lines are for the unary system **Di**, light lines for the binary composition x.

(1.66), we have

$$\mu_{\textbf{Di}} = \mu_{\textbf{Di}}{}^0 + RT \ln a_{\textbf{Di}}.$$

At any temperature, activity of **Di** in the pure crystalline phase is fixed, of course, but its activity in the liquid is proportional to the concentration of **Di** in the liquid. As **An** component is added to the melt, $a_{\textbf{Di}}$ (hence $\mu_{\textbf{Di}}$) decreases because of dilution, and the $\mu_{\textbf{Di}}$–T curves intersect at a lower temperature than for the unary system CaO·MgO·2SiO$_2$. This decrease in $\mu_{\textbf{Di}}$ for the melt at any temperature is simply

$$(\mu_{\textbf{Di}})_{\text{soln}} - (\mu_{\textbf{Di}})_{\text{pure}} = (\mu_{\textbf{Di}}{}^0 + RT \ln a_{\textbf{Di}})_{\text{soln}} - (\mu_{\textbf{Di}}{}^0 + RT \ln a_{\textbf{Di}})_{\text{pure}},$$

which reduces to

$$(\Delta \mu_{\textbf{Di}})_{\text{mixing}} = RT \ln \frac{(a_{\textbf{Di}})_{\text{soln}}}{(a_{\textbf{Di}})_{\text{pure}}}.$$

The value of $(\Delta\mu_{Di})_{mixing}$ is negative, as schematically illustrated in Figure 4.8, because the activity quotient is less than unity.

Perhaps the clearest way to explain the phase relationships in the system **Di–An**, however, is to use isothermal isobaric G–x diagrams. Three different one-atmosphere isotherms at 1,400, 1,275, and 1,000°C are illustrated in Figures 4.9a, 4.9b and 4.9c, respectively. Let us examine some properties of these diagrams. Any mechanical mixture of crystalline diopside + crystalline anorthite will possess a molar G lying along the chord connecting the Gibbs free energies of these two pure solid phases (e.g., see Figure 4.9c). Likewise, any mechanical mixture of (immiscible) droplets of **An** melt + droplets of **Di** melt

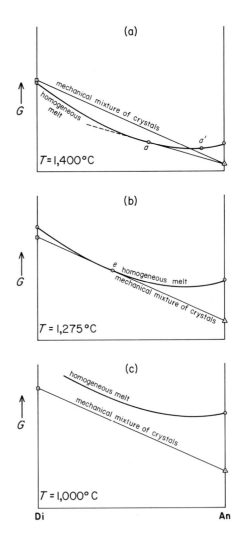

FIGURE 4.9.
One-atmosphere isobaric diagram of Gibbs free energy versus composition for the binary system CaO·MgO·2SiO$_2$– CaO·Al$_2$O$_3$·2SiO$_2$ at (a) 1,400°C; (b) 1,275°C; and (c) 1,000°C. The heavy line indicates the molar Gibbs free energy for various melt compositions.

would, if not mixed, have a molar G lying along a chord (not illustrated in Figure 4.9) connecting the values of G for these pure melts. However, from the results of the experimental investigation depicted in Figure 4.6, it is known that such a two-phase melt will spontaneously homogenize, with concomitant decrease in Gibbs free energy as indicated by equation (1.29). For this reason, binary **Di–An** melts have Gibbs free energies less than that defined by the chord connecting $(G_{\mathbf{Di}})_{\text{melt}}$ with $(G_{\mathbf{An}})_{\text{melt}}$. The magnitude of ΔG_{mixing} achieves a maximum where the configurational disorder is greatest, in general, where the miscible components are at equal concentrations—see also equation (1.27). Thus, where spontaneous mixing takes place, G–x curves for the solution are convex relative to the compositional coordinate (e.g., see Figure 4.9c).

The 1,400°C isotherm shown in Figure 4.9a illustrates the situations at temperatures above the melting point of pure diopside; here $(G_{\mathbf{Di}})_{\text{melt}}$ has a slightly lower value than G_{diopside}. In contrast, $(G_{\mathbf{An}})_{\text{melt}}$ exceeds $G_{\text{anorthite}}$, since this temperature is less than that at which pure anorthite melts. For compositions in the $CaO \cdot MgO \cdot 2SiO_2$-rich portion of the diagram, the lowest possible Gibbs free energy is achieved by a single melt phase, the composition of which ranges from pure **Di** to a melt a. On the other hand, a two-phase mixture of melt a + anorthite gives the minimum G for $CaO \cdot Al_2O_3 \cdot 2SiO_2$-rich compositions. The tie line connecting melt a and anorthite is tangent to the melt G–x curve at a. It is clear that in this portion of the binary system (a–**An**), the two-phase assemblage has a lower Gibbs free energy than any other single-phase or two-phase mixture. As an exercise, the interested student may construct other (metastable) tie lines connecting the G–x curve for melt with that of anorthite, for instance, that connecting melt a' with anorthite; one soon realizes that the lowest possible Gibbs free energy for **An**-rich compositions is obtained by constructing the tangent to the melt curve.

Before proceeding farther, let us consider the physical meaning of any tangent to a G–x curve, such as that of melt in Figure 4.9a. The tangent gives the rate of change of Gibbs free energy as a function of compositional change, or $(\partial G/\partial x)_{T,P}$. As comparison of this expression with equation (1.41) shows, this expression closely approximates the definition of the chemical potential of component i, the partial molar Gibbs free energy. The two expressions would be exactly equivalent only if $\partial x = \partial n_i$; these latter variables are directly proportional but are not identical. Extensions of any such tangent (for instance, at point a) to the side lines **Di** and **An** in Figure 4.9a give the values of $\mu_{\mathbf{Di}}$ and $\mu_{\mathbf{An}}$, respectively, these being the partial molar Gibbs free energies of the components in the phase a. Of course, at equilibrium for the 1,400°C tie line considered, $(\mu_{\mathbf{An}})_{\text{anorthite}} = (\mu_{\mathbf{An}})_{\text{melt}}$. The chemical potential of **Di** is defined in the liquid phase, but not in the pure phase anorthite. In the mechanical mixture of crystalline phases, the tangent to the chord connecting the molar Gibbs free energies of diopside and anorthite is the chord itself; hence

$$(\mu_{\mathbf{Di}})_{\text{diopside}} = G_{\text{diopside}}; \text{ and } (\mu_{\mathbf{An}})_{\text{anorthite}} = G_{\text{anorthite}}.$$

The 1,275°C isotherm shown in Figure 4.9b presents $G–x$ relationships at the eutectic temperature. Here the melt curve is tangent to the chord representing the molar Gibbs free energy of a mechanical mixture of crystals—both diopside and anorthite—only at the eutectic composition. At this unique composition of melt, e, the following relationships obtain:

$$(\mu_{Di})_{diopside} = (\mu_{Di})_{melt}; \text{ and } (\mu_{An})_{anorthite} = (\mu_{An})_{melt}.$$

This is simply another way of saying that melt e is saturated with respect to both **Di** and **An** components.

At even lower temperatures, such as the 1,000°C isotherm illustrated in Figure 4.9c, the $G–x$ curve for melt lies at higher values (and thus is metastable) compared to any combination of crystalline phases in the system **Di–An**. Clearly, at 1,000°C only crystalline phases are in chemical equilibrium. Comparing Figures 4.9a, 4.9b, and 4.9c, we note that the isobaric $G–x$ curves rise with decreasing temperature as a consequence of equation (1.32), with the Gibbs free energy of melt elevated at a greater rate than that for the solid assemblage because of the larger entropy of the former.

(2) Binary Peritectic Exhibiting No Solid Solution

The one-atmosphere isobaric system $2MgO\cdot SiO_2–SiO_2$ presented in Figure 4.10 illustrates this type of equilibrium. In such systems, a compositionally intermediate crystalline phase is present which decomposes to melt + another crystalline phase. Such behavior is called incongruent melting, because the liquid produced has a different composition from the preexisting solid. In the system **Fo–Si**, the intermediate compound is a polymorph of $MgSiO_3$. The diagram is characterized by a divariant melt field (as well as a region for two immiscible liquids at high silica concentrations), several fields for melt + one solid phase, and several fields for two solid phases. All the two-phase areas have a single degree of freedom: arbitrary selection of either the temperature or the composition of one of the phases of variable composition (i.e., melt) automatically specifies the other. Two isobaric invariant points, p and e, are present in Figure 4.10, each characterized by the coexistence of two solid phases and a liquid, all of fixed compositions.

As an aid in understanding the diagram, let us examine the equilibrium crystallization path followed by some bulk composition, such as x, in Figure 4.10. Since x coincides with the composition $MgO\cdot SiO_2$, it is not a general case, but it will still be instructive. Imagine first that the temperature is very high—say, 1,800°C. Although below the melting point of forsterite at this temperature, bulk composition x falls well within the homogeneous melt field. As heat is subtracted from the system, the temperature declines until, at a temperature of about 1,600°C, the first crystals of forsterite appear in the liquid m ($= x$).

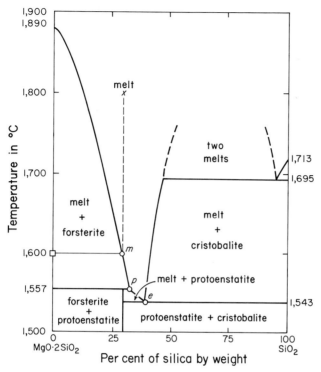

FIGURE 4.10.

One-atmosphere isobaric temperature-composition phase relations for the binary system $2MgO \cdot SiO_2$–SiO_2, slightly revised, after Bowen and Anderson (1914) and Greig (1927, Figure 3). The structural modification of $MgSiO_3$ is protoenstatite under these physical conditions (see Boyd and Schairer, 1964; Boyd, England and Davis, 1964).

Further withdrawal of heat will result in the precipitation of more forsterite; hence as temperature falls, the melt follows the liquidus path toward the binary eutectic. The melt reaches the peritectic composition at 1,557°C, at which point the system consists of two phases, melt p and forsterite. Continued subtraction of heat will result in the reaction of these two phases at constant temperature to form protoenstatite. If equilibrium is maintained, this isothermal process will continue as long as heat is removed, and the last drop of melt p will react with the last crystal of forsterite—at which point bulk composition x will consist of but a single phase, protoenstatite. This phenomenon is known as the *discontinuous-reaction relation,* because the process takes place only at specific *P–T* conditions, giving rise to a discontinuity in the phase assemblage.

Further subtraction of heat will merely cause the temperature of the proto-enstatite crystals to drop.

Now suppose that the bulk composition had been richer in Fo than x was. In this case, the liquidus would have been reached at a temperature greater than about 1,600°C, but then the same equilibrium crystal-melt history would have ensued as described above. However, at the peritectic, reaction of all the melt p with forsterite to produce protoenstatite would not have used up all the forsterite. Accordingly, after the last drop of melt had been consumed, the bulk composition would be represented by a mixture of forsterite + protoenstatite, which, on further withdrawal of heat, would cool down as expected.

Or suppose that the bulk composition lay between the values of x and p. Initially at elevated temperatures, such a homogeneous melt on cooling would reach the liquidus at a temperature less than 1,600°C and not far above 1,557°C; then, as forsterite precipitated, the melt would migrate to composition p. During subtraction of heat at the peritectic, all the forsterite would react with most of the melt p to produce protoenstatite, but a small amount of p would remain after the forsterite had disappeared. Because of this, further withdrawal of heat would allow the crystallization of more protoenstatite, and accordingly the liquid would continue its compositional migration toward $MgO \cdot SiO_2$-depleted, (hence SiO_2-enriched) values. The melt would very shortly reach the eutectic composition, e, at 1,543°C, where cristobalite would join protoenstatite in crystallization at the expense of the liquid phase.

Heating paths would be exactly the reverse of the cooling histories described above, provided crystals and melt remain in equilibrium with one another. As discussed for the binary eutectic **Di–An** (Figure 4.6), however, separation of liquid and solid phases will allow for the discontinuous generation of melt on heating. The reader should explore several heating paths in order to become acquainted with this phenomenon.

Yet another, rather important point remains to be considered. We have assumed that at the reaction point p, equilibrium is achieved between crystals and the melt. In natural situations, however, this is not always the case; so we need to consider the phase behavior where equilibrium is not maintained. When liquid has reached the peritectic composition, for all bulk compositions lying to the left of p, forsterite has precipitated in quantities exceeding its stoichiometric proportions; in other words, modal forsterite is greater than normative forsterite. If this phase fails to react with the liquid for some reason (e.g., rapid cooling, gravitative settling, phase armoring), the melt behaves as if it represents a new bulk composition for the system; on further cooling, protoenstatite crystallizes from it, and the melt migrates with decreasing temperature down to the eutectic—thus chemical differentiation, or crystal fractionation, has occurred. (For a general discussion of crystal fractionation, see Bowen, 1928, Chaps. 3 and 4.) Here a silica polymorph appears, ending the possibility for further change in liquid composition. Such a cooling history for the bulk composition x, for instance, might result in a solid product consisting of forsterite phenocrysts armored by protoenstatite set in a matrix of protoenstatite + cristobalite in the eutectic proportions. Of course, melting would ensue on heating such a disequilibrium assemblage to the eutectic temperature appropriate for the pressure (e.g., one atmosphere).

Now let us consider this one-atmosphere isobaric system in terms of a G–x diagram. Three different isotherms, at 1,600, 1,557, and 1,543°C are presented in Figures 4.11a, 4.11b, and 4.11c, respectively. The compositional parameter along the abscissa has been highly distorted for clarity. Tangents to the G–x melt curve emanating from cristobalite and forsterite define a pair of two-phase regions separated by an intervening homogeneous liquid area in Figure 4.11a. At 1,600°C the Gibbs free energy of protoenstatite exceeds that of liquid

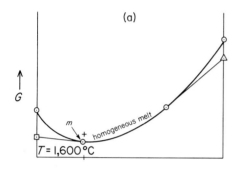

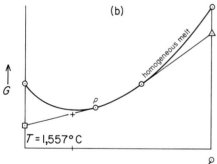

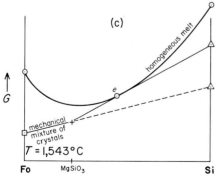

FIGURE 4.11.

One-atmosphere isobaric diagram of Gibbs free energy versus composition for the binary system $2MgO \cdot SiO_2$–SiO_2 at (a) 1,600°C; (b) the peritectic temperature, 1,557°C; and (c) the eutectic temperature, 1,543°C. The heavy line indicates the molar G for various melt compositions. The composition and G of the $MgSiO_3$ polymorph, protoenstatite, is indicated by +.

of the same composition ($= m$), as well as that of any forsterite + silica-poor melt assemblage. However, at some slightly lower temperature (not illustrated), the molar Gibbs free energy for protoenstatite will fall on the melting curve, signifying its metastable melting temperature; here protoenstatite melts congruently but metastably, because the two-phase assemblage consisting of forsterite + liquid possesses an even lower molar G. At the still lower temperature of 1,557°C depicted in Figure 4.11b, forsterite, protoenstatite, and the peritectic melt p are colinear in G–x dimensions, and

$$(\mu_{Fo})_{forsterite} = (\mu_{Fo})_{protoenstatite} = (\mu_{Fo})_{melt}.$$

Clearly, forsterite, protoenstatite, and liquid p represent an isobaric invariant assemblage; thus the T is fixed. As is evident from Figure 4.11c, at temperatures below the peritectic, assemblages with minimum Gibbs free energy include forsterite + protoenstatite, protoenstatite + melt, and cristobalite + melt (and, at 1,543°C, also protoenstatite + cristobalite + melt e), but the G of forsterite is too high for it to be stable with melt.

As an aside we will note, too, that the intersection of two G–x chords at the value for protoenstatite in Figure 4.11c and in the entire subsolidus region indicates that this phase must be capable of different chemical potentials for the same component at fixed pressure and temperature. Consider, for instance, the value of $(\mu_{SiO_2})_{protoenstatite}$. In equilibrium with cristobalite, the value is obviously high and equivalent to that possessed by the polymorph of silica, the triangle on the righthand side line; on the other hand, where protoenstatite coexists with forsterite—a phase incompatible with cristobalite, so that $(\mu_{SiO_2})_{forsterite} \neq (\mu_{SiO_2})_{cristobalite}$—its chemical potential is equal to that in the forsterite, as shown by the dashed extension of the chord to low values on the righthand side of the diagram. This puzzling situation can be resolved if we suppose that protoenstatite shows compositional variability, hence μ variation, as a function of its equilibrium phase association. This range in chemistry would have to be infinitesimal, since no obvious variations in SiO_2 versus MgO have ever been detected in protoenstatite. Compositional change, no matter how slight, must, of course, be possible for the chemical potential of any arbitrarily chosen component to have physical meaning, as is clear from the definition of μ presented in equation (1.41).

Another example of the peritectic type of phase equilibrium is the topologically similar binary system $K_2O \cdot Al_2O_3 \cdot 4SiO_2$–$SiO_2$. Crystal-melt relations at atmospheric pressure are illustrated in Figure 4.12. As is evident from this diagram, the intermediate binary compound, K-feldspar, bears a reaction relation to the melt just as protoenstatite does in the analogous system **Fo–Si**.

(3) Complete Binary Solid Solution

In this type of phase-equilibrium diagram, complete miscibility exists not only within the melt, but within the crystalline assemblage as well. Two slightly

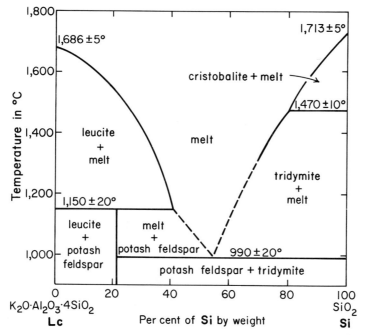

FIGURE 4.12.

One-atmosphere isobaric temperature-composition phase relations for the binary system $K_2O \cdot Al_2O_3 \cdot 4SiO_2$–$SiO_2$, from Schairer and Bowen (1947a, Figure 3). The structural modification of $KAlSi_3O_8$ is sanidine under these physical conditions. Dashed lines indicate extrapolation from the dry, one-atmosphere ternary system K_2O–Al_2O_3–SiO_2 (e.g., see Schairer and Bowen, 1947b, 1955).

different types are important in petrology: (a) one in which the lowest melting temperature is situated along a unary side line, for example, the intermediate olivines; and (b) one in which the lowest melting temperature is at an intermediate composition along the binary join, for example, the alkali feldspars.

The one-atmosphere isobaric phase diagram for the system $2MgO \cdot SiO_2$–$2FeO \cdot SiO_2$, representing type (a), is presented as Figure 4.13. Here the fields for homogeneous one-phase melt and homogeneous one-phase olivine solid solution are separated from one another by a transition loop. Within this T–x region, two phases of specified composition coexist: a ferrous iron-rich liquid, and a magnesium-rich crystalline phase. A single degree of freedom exists within the transition loop, since the arbitrary selection of temperature or composition for one of the phases automatically specifies the other variable. Of course, within each single-phase field, two degrees of freedom exist: both temperature and composition of the phase may be selected, practically at random.

We will now trace the equilibrium crystallization path of a general bulk composition, such as x, in Figure 4.13, to gain an appreciation for this type of diagram. Suppose that the system is at some very high temperature, say,

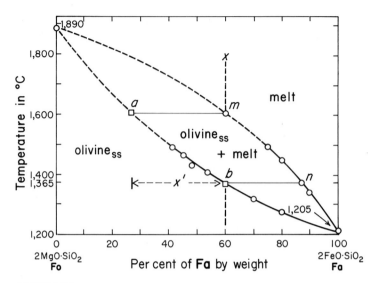

FIGURE 4.13.

One-atmosphere isobaric temperature-composition phase relations for the binary system 2MgO·SiO$_2$–2FeO·SiO$_2$, after Bowen and Schairer (1935, Figure 7). Dashed portions of liquidus and solidus curves are extrapolated to the known melting point of pure forsterite. Abbreviation: ss = solid solution.

1,800°C; although this temperature is below the melting point of pure forsterite, our system is still completely molten. It will remain so as heat is withdrawn until the temperature has fallen to about 1,600°C; at this value, the liquidus curve has just been reached and the first tiny crystals of magnesium-rich olivine of composition a appear. The melt m still possesses the bulk composition x, of course, because only trace amounts of the second phase are present. However, as heat is continually subtracted from the system, more of the rather forsteritic olivine crystallizes, and the liquid, being selectively depleted in **Fo**, is thereby enriched in **Fa** component. However, if equilibrium is maintained as the melt becomes more **Fa**-rich at lower temperatures, it reacts continuously with the early-formed olivine crystals to make them over into a more **Fa**-rich solid phase. (Note that, in general, an increased concentration of a constituent in one phase must be matched at equilibrium by a proportional increase of that component in every other coexisting phase of variable composition; this follows from the law of mass action, as presented in equation 1.73.) Thus, as temperature declines, the liquid gradually diminishes in amount and becomes more ferrous, while concomitantly the solid gradually increases in amount and also becomes more iron-rich.

This behavior is called the *continuous-reaction relation* because it takes place during the entire course of crystallization. The melt always maintains a greater

Fe/Mg ratio than does the coexisting crystalline phase. Under equilibrium conditions, the last drop of melt of composition n is used up at a temperature of about 1,365°C, bringing the composition of the homogeneous olivine solid solution (now representing 100 per cent of the phases in the system) to the bulk composition b ($= x$). Further subtraction of heat will merely result in a drop in temperature for the crystalline assemblage.

If crystal fractionation occurs, the general T–x course followed by the liquid in Figure 4.13 will be the same down to the temperature of about 1,365°C. However, if early-formed Mg-rich olivines have been prevented from reequilibrating with the melt during cooling, a continuous range of solid compositions from a to b will have precipitated in the course of crystallization. Perhaps armoring has occurred, in which case each grain will have a magnesian core and successively more ferrous shells. Although melt n is in equilibrium with the crystal rims of composition b, the aggregate composition of the solids at this temperature is at some value such as x' (see Figure 4.13); hence an appreciable amount of liquid is still present in the system. Only when the average composition of all the crystals reaches x will the melt be exhausted; somewhat as happened with nonequilibrium at a peritectic point, chemical differentiation or crystal fractionation has occurred. For this system, with perfect fractionation, a vanishingly small amount of liquid could reach the **Fa** composition on the unary side line.

As in all other systems, equilibrum heating paths are the exact opposite of equilibrium cooling paths. For any bulk composition, the temperature at which melt first appears, of course, depends on whether the solid assemblage is the product of equilibrium crystallization or crystal fractionation (i.e., is a disequilibrium assemblage). In the system **Fo**–**Fa** at one atmosphere (Figure 4.13), and assuming the bulk composition x to be initially at subsolidus temperatures, if liquid is removed as rapidly as it is produced by the heating (i.e., perfect fractional fusion), the last crystal will be end-member forsterite in composition, and will disappear only at a temperature of 1,890°C.

The 1,600°C isothermal, one-atmosphere isobaric liquidus-solidus phase relations in the system **Fo**–**Fa** are presented in G–x terms in Figure 4.14. The curves for homogeneous melt and for the crystalline solution are both concave upward because of the negative term for the Gibbs free energy of mixing. They must intersect at this intermediate temperature of 1,600°C because, although the $(G_{Fo})_{melt}$ is greater than $G_{forsterite}$, the situation is reversed for the **Fa** end-member liquid and solid assemblages. The two G–x curves have a common tangent, chord am (tie line am in Figure 4.13). It is apparent from the G–x diagram that, for bulk compositions lying within the am compositional interval, a two-phase assemblage will have a lower molar Gibbs free energy than either liquid or solid alone. The extensions of the am chord to the **Fo** and **Fa** unary side lines as illustrated in Figure 4.14 allow evaluation of the values of μ_{Fo} and μ_{Fa}, respectively, in the two coexisting phases.

Another petrologically important example of this type of equilibrium is

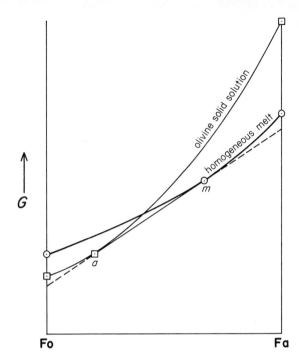

FIGURE 4.14.

One-atmosphere isobaric Gibbs free energy versus composition for the binary system $2MgO \cdot SiO_2 - 2FeO \cdot SiO_2$ at 1,600°C. The heavy line indicates the molar $\bar{G}$ for various melt compositions.

the system $Na_2O \cdot Al_2O_3 \cdot 6SiO_2 - CaO \cdot Al_2O_3 \cdot 2SiO_2$. Phase relationships at one atmosphere total pressure are shown in Figure 4.15.

Let us now consider a second type of continuous-reaction relation, that of type (b); phase equilibria are characterized by a binary minimum melting temperature as well as by both complete solid solution and complete liquid miscibility. An example significant for geology, the system $Na_2O \cdot Al_2O_3 \cdot 6SiO_2 - K_2O \cdot Al_2O_3 \cdot 6SiO_2$ at 3 kb P_{fluid}, is shown in Figure 4.16. It is in fact only pseudobinary, because H_2O is soluble in the liquid phase; this high aqueous-fluid pressure lowers the melting temperatures (see Figure 4.5 and the discussion of this effect in the text) and suppresses the incongruent melting of K-feldspar shown in Figure 4.12. Ignoring subsolidus relations for the moment, we can see that the system **Ab–Or**(–H_2O) is topologically similar to those illustrated in Figures 4.13 and 4.15, except that both liquidus and solidus curves have been depressed in it to form an internal pseudobinary minimum melting point, m. At this invariant point, the **Ab–Or**(–H_2O) liquid and the homogeneous alkali feldspar solid solution have identical compositions except for the H_2O content of the melt. Ignoring this aqueous component, both liquid and solid solutions can be described at and near this temperature by a single constituent, the minimum melting composition m; hence for the purposes of the phase rule, it may be regarded as a one-component system. Crystal–melt equilibria along this pseudobinary join enrich the melt in the low-melting constituent m, whereas the crystals concetrate either component **Ab** or **Or**, depending on the initial bulk composition. For instance, at a temperature of about 780°C, the tie lines presented in Figure 4.16 illustrate the fact that melts are enriched in m relative to **Ab**- and **Or**-rich crystalline phases.

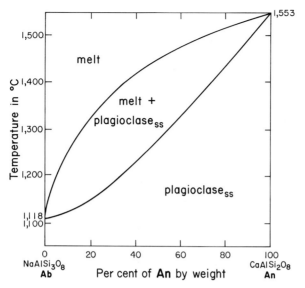

FIGURE 4.15.
One-atmosphere isobaric temperature-composition phase relations for the binary system $Na_2O \cdot Al_2O_3 \cdot 6SiO_2 - CaO \cdot Al_2O_3 \cdot 2SiO_2$, after Bowen (1913). Abbreviation: ss = solid solution. Simple structural formulae are used along the abscissa for the plagioclase$_{ss}$ in order to preserve the molar proportions on an atomic basis. In contrast, if we had employed the oxide end-members $Na_2O \cdot Al_2O_3 \cdot 6SiO_2$ and $CaO \cdot Al_2O_3 \cdot 2SiO_2$, you can see that plagioclase having a Na/Ca ratio of unity (= labradorite) would have to be represented as $Ab_{33}An_{67}$.

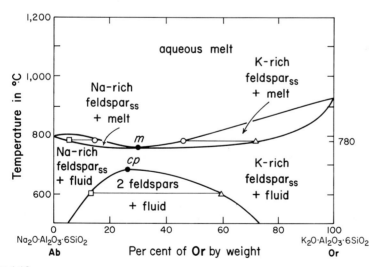

FIGURE 4.16.
Three-kilobar P_{fluid} isobaric temperature-composition phase relations for the pseudobinary system $Na_2O \cdot Al_2O_3 \cdot 6SiO_2 - K_2O \cdot Al_2O_3 \cdot 6SiO_2(-H_2O)$ projected onto the binary side line **Ab–Or**.
The liquid phase contains a small proportion of H_2O, whereas the fluid phase is nearly pure H_2O. Where sufficient H_2O is present to saturate the silicate liquid, an aqueous fluid phase is associated with the H_2O-bearing melt. Abbreviations: ss = solid solution; m = minimum on the solidus; cp = solvus critical point. Liquidus-solidus relations are taken from Tuttle and Bowen (1958, Figure 17), solvus relations from Orville (1963, Figure 9) and Luth and Tuttle (1966, Figure 6).

A schematic isothermal isobaric $G–x$ diagram for the system $Na_2O \cdot Al_2O_3 \cdot 6SiO_2–K_2O \cdot Al_2O_3 \cdot 6SiO_2$ at 3 kb P_{fluid} ($\approx P_{H_2O}$) and 780°C is presented in Figure 4.17; curvatures have been exaggerated for clarity. The isotherm chosen lies above the minimum melting temperature m (the lowest temperature on the solidus) but below the melting points of both pure albite and pure K-feldspar. The pair of intersections of the $G–x$ curves for homogeneous liquid and alkali feldspar solid solutions generates two tie lines connecting melt and the crystalline phases (see also Figure 4.16). Evidently the ΔG_{mixing} for the liquid is a larger negative number than that for the solid solutions. Analogously, a system in which the absolute value of the negative Gibbs free energy of mixing of the crystalline solid solution exceeded that of the completely miscible liquid would be characterized by a phase diagram possessing an internal maximum.

You have undoubtedly noticed the subsolidus two-phase region at intermediate compositions in the **Ab–Or**(–H_2O) system shown in Figure 4.16. This curve, separating homogeneous alkali feldspar$_{ss}$ from the two-phase crystalline assemblage, is termed the solvus, as defined earlier in this chapter. Although reactions depicted here are totally subsolidus, and so could equally well be presented in Chapter 6, the phenomenon will be discussed now because solvus phase relations can, under some conditions, affect crystal–melt equilibria. At any appropriate temperature, say, 600°C, bulk compositions lying within the solvus under equilibrium conditions will consist of an **Ab**-rich feldspar solid solution and an **Or**-rich feldspar solid solution ($\pm$ aqueous fluid). With increasing temperature, the compositions of these coexisting solids approach one another, and at the crest of the solvus, the so-called critical point, cp (the maximum solvus temperature), the phases become physically (e.g., crystallographically) and chemically indistinguishable; hence we have a single homogenous solid solution. It should be pointed out here that alkali–feldspar-solvus relationships are strictly binary, since H_2O does not enter into the reaction. Increase in pressure raises the temperature of the solvus on the order of 10–15°C per kilobar because intermediate alkali feldspar solid solutions exhibit a slight positive ΔV_{mixing} (e.g., see Orville, 1967, Figure 6); this elevation is a function of P_{total} no matter whether the pressure medium is H_2O, CO_2, methane, or simply the confining pressure provided by anhydrous phases surrounding the alkali feldspar(s).

Exsolution (i.e., separation of two or more phases from a single, homogeneous precurser at subsolvus temperatures) results when the assemblage **Ab**-rich feldspar$_{ss}$ + **Or**-rich feldspar$_{ss}$ has an aggregate molar Gibbs free energy less than that of a homogeneous alkali feldspar solid solution alone. Isobaric $G–x$ relations at several different isotherms, namely, the critical point (≈ 680°C), 600°C, and about 500°C, are depicted in Figure 4.18. Note that at intermediate feldspar compositions for the 600°C isotherm, for instance, a mechanical mixture of pure crystalline $NaAlSi_3O_8$ + pure crystalline $KAlSi_3O_8$

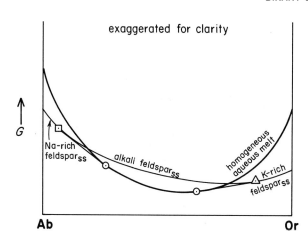

FIGURE 4.17.
Three-kilobar P_{fluid} isobaric diagram of Gibbs free energy versus composition for the pseudobinary system $Na_2O \cdot Al_2O_3 \cdot 6SiO_2$– $K_2O \cdot Al_2O_3 \cdot 6SiO_2(–H_2O)$ at about 780°C. Concavities of the G–x curves have been greatly accentuated to render them distinguishable from the two-phase tangents. Symbols correspond to those of Figure 4.16.

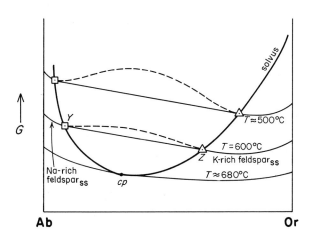

FIGURE 4.18.
Isobaric (e.g., three kilobars) diagram of Gibbs free energy versus composition for the binary system $Na_2O \cdot Al_2O_3 \cdot 6SiO_2$– $K_2O \cdot Al_2O_3 \cdot 6SiO_2$ at the solvus critical temperature $\approx 680°C$, and at 600 and 500°C. The compositions of alkali feldspar solid solutions y and z correspond to those of Figure 4.16. Heavy curve indicates G–x values for solvus; dashed lines signify metastable alkali feldspar solid solutions.

would have a higher molar G than the complete solid solution—which itself would be metastable relative to the two-phase assemblage of **Ab**-rich solid solution y and **Or**-rich solid solution z.

An interesting type of metastable exsolution is that termed *spinodal decomposition*, which arises because of a composition-related energy barrier against the onset of exsolution. The spinodal solvus lies within the equilibrium solvus in T–x coordinates, coincides with a portion of it in G–x coordinates, and is defined by the pair of inflection points (e.g., g' and h') within the latter, as shown in Figure 4.19. It can be seen from Figure 4.19b that any local chemical reorganization for a composition external to the spinodal, as at *i*, would result in a local two-phase assemblage (e.g., *g* + *g'*) of higher molar *G*; within the spinodal, as at *j*, any exsolution would produce a net decrease in the local Gibbs free energy (e.g., *g'* + *h'*).

The preceding discussion of spinodal decomposition assumes a dimensional misfit, or discontinuity in crystal structure, at the boundary between the two exsolving phases; although both exsolving phases possess the same atomic arrangement (crystal structure), their compositional contrasts result in systematic differences in structural scale, hence in a misfit. This type of spinodal—a metastable solvus—may be termed a *chemical spinodal*, calling attention to the compositional differences across the bounding interface. Another variety of exsolution may be conceived of in which the three-dimensional atomic periodicity is maintained across the phase boundary by gradients (strain) in the lattice spacings. Such a phase boundary is termed coherent, and the metastable solvus a *coherent spinodal*. The strain energy in the vicinity of the boundary raises the local molar Gibbs free energy; hence the crest of the coherent spinodal (not shown) as well as the limbs fall well within the equilibrium solvus and chemical spinodal illustrated in Figure 4.19.

(4) Binary Eutectic with Limited Solid Solution

One may think of this type of binary equilibrium as the necessary modification of phase relationships in the situation where the solvus curve intersects the solidus curve of an internal-minimum type of transition loop. Referring back to Figure 4.16, consider what happens in the pseudobinary system **Ab**–**Or**(–H_2O) as P_{fluid} increases: the temperature of the solidus falls to lower values (of course, if H_2O were not soluble in the melt, pressure increment would raise the fusion temperatures), whereas the solvus temperature rises. At elevated pressures approaching 5 kb P_{fluid}, the two curves intersect. For fluid pressures in excess of this value, the binary minimum relationship depicted in Figure 4.16 is replaced (after some possible minor complications which we need not consider here; e.g., see the discussion of Figure 4.40 on p. 129) by a eutectic-type equilibrium, characterized by limited solid solution, as presented in Figure 4.20. As is apparent from the depicted isobaric phase relationships, at high, but declining temperatures the miscibility of the **Or** component in **Ab**-rich feldspar increases, because the melt is becoming enriched in **Or**; for potassic bulk compositions, the Na-content of K-rich feldspar increases similarly as the liquid becomes more **Ab**-rich. The maximum mutual solid solution occurs at the isobaric invariant point, eutectic *e*. At temperatures below the eutectic in the subsolidus region, the extent of solid solution again diminishes with falling temperature.

As one might anticipate, isobaric, isothermal *G*–*x* diagrams for this type of equilibrium combine many of the features of Figures 4.17 and 4.18. Schematic

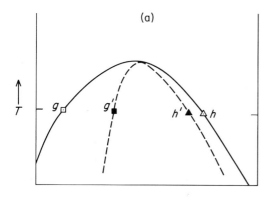

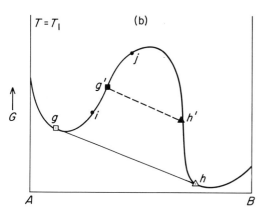

FIGURE 4.19.

Isobaric diagrams for the binary system AB showing the equilibrium solvus and the chemical spinodal solvus: (a) is a $T–x$ diagram showing the relationship between the two solvi: (b) is a $G–x$ diagram at temperature T_1, showing the solvus curve $gig'jh'h$ and the metastable spinodal $g'jh'$. Not shown in the coherent spinodal, which lies entirely within the equilibrium two-phase region of (a), and in fact within the chemical spinodal. For a discussion of chemical and coherent spinodal solvi, and the thermodynamics of a coherent interface, see: Yund and McCallister (1970, especially Figures 5 and 6); Fletcher and McCallister (1974); and Robin (1974).

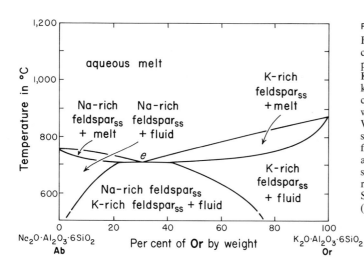

FIGURE 4.20.

Five-kilobar P_{fluid} isobaric temperature-composition phase relations for the pseudobinary system $Na_2O\cdot Al_2O_3\cdot 6SiO_2$–$K_2O\cdot Al_2O_3\cdot 6SiO_2(-H_2O)$. As at three kilobars (Figure 4.16), the liquid contains moderate amounts of H_2O, whereas the fluid is nearly pure H_2O. Where sufficient H_2O is present to saturate the silicate liquid, an aqueous fluid phase is associated with the aqueous melt. Abbreviations: ss = solid solution: e = eutectic point. Phase relations from Yoder, Stewart, and Smith (1957, Figure 38) and Morse (1970, Figure 2).

isobaric G–x sections at $T > T_e$, T_e, and $T < T_e$ are illustrated in Figures 4.21a, b, and c, respectively. Note that superheated solid solutions could recrystallize at high temperatures to metastable Na-rich and K-rich crystalline phases under conditions in which homogeneous melt is the stable, lowest-Gibbs-free-energy assemblage (Figure 4.21a), and that supercooled solid solutions could melt metastably at slightly subsolidus temperatures under conditions where **Ab**-rich + **Or**-rich crystalline phases constitute the stable assemblage (Figure 4.21c).

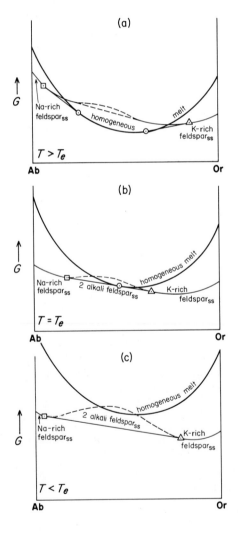

FIGURE 4.21.

Five-kilobar P_{fluid} isobaric diagrams of Gibbs free energy versus composition for the pseudobinary system $Na_2O \cdot Al_2O_3 \cdot 6SiO_2$–$K_2O \cdot Al_2O_3 \cdot 6SiO_2(-H_2O)$ at temperatures: (a) just above the eutectic; (b) at the eutectic; and (c) just below the eutectic. Concavities of the G–x curves have been greatly accentuated for clarity. Tie lines connecting two alkali feldspar solid solutions are tangent to the homogeneous crystalline$_{ss}$ curve.

TERNARY DIAGRAMS

We are now ready to consider phase equilibria involving slightly more complex chemical systems. Fortunately, the types of phase relationships just described for unary and binary systems carry over into these more complicated systems, because the same principles are applicable no matter how many components are involved.

For graphic treatment in the two-dimensional space of a page of this book, only ternary phase relations (= two degrees of chemical freedom) can be simply and unambiguously represented; such treatment necessitates fixing P and T as constant, or holding one of the physical conditions constant and projecting the other variable onto the plane of the paper by means of numerical values or contours. For systems with more than three components, $n - 3$ components must also be projected onto the plane of the paper for two-dimensional graphic representation. Because most igneous (and metamorphic) rocks contain on the order of six to ten components, graphic treatment becomes rather cumbersome, and some workers now advocate vectorial or strictly mathematical treatments (e.g., see Korzhinskii, 1959, Chap. 2; Greenwood, 1967b, 1968). Since the more abstract vectorial and mathematical methods assume familiarity with chemographic relationships, for simplicity we will use the more intuitively obvious graphic techniques here.

Let us consider several different types of ternary system: (1) simple eutectic, the crystalline phases of which display no solid solution; (2) complete binary solid solution; (3) binary peritectic relations; and (4) combinations of limited or complete solid solution with the discontinuous-reaction relation. All diagrams to be presented are at one atmosphere total pressure; temperatures on the liquidus surface are shown by means of isothermal contours.

(1) Simple Ternary Eutectic System Showing No Solid Solution

The most readily understood isobaric ternary phase diagram is doubtless the ternary eutectic. Such a system involves a single homogeneous melt and three crystalline phases of invariable composition; here the composition of the liquid lies within the compositional triangle defined by the solid phases. An example, the system **Fo–Sp–Lc**, is illustrated in Figure 4.22. It can be seen that the binary eutectic relations exhibited along the side lines of Figure 4.22 carry over in the ternary system; each binary eutectic point (e.g., a) is one termination of a ternary *cotectic* curve (e.g., ae), the other end of which is located at the ternary eutectic, e.

Let us trace the equilibrium crystallization path of some random bulk composition, x, initially at a very high temperature, in order to understand this diagram. For convenience it is redrawn as Figure 4.23. As heat is subtracted

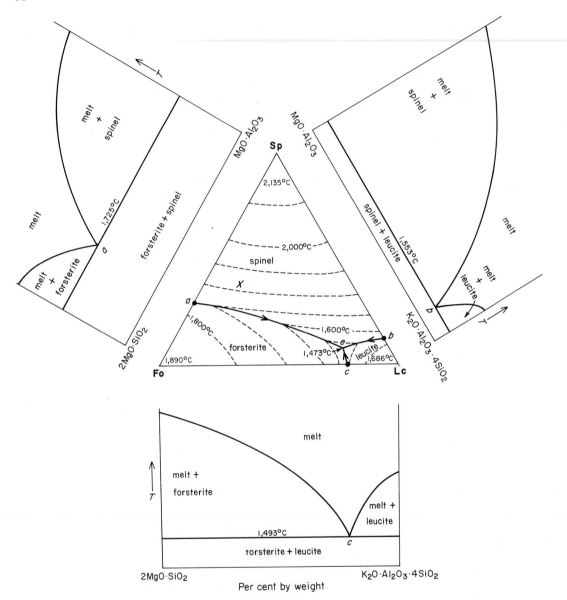

FIGURE 4.22.

One-atmosphere isobaric phase diagram for the ternary system $2MgO \cdot SiO_2$–$MgO \cdot Al_2O_3$–$K_2O \cdot Al_2O_3 \cdot 4SiO_2$, after Schairer (1955). Proportions are in percentages by weight. Isotherms on the liquidus are shown for 100°C intervals. The bounding one-atmosphere binary phase relations are also presented. The crystalline phase in equilibrium with melt at liquidus temperatures is also indicated.

from the homogeneous liquid, temperature declines until the liquidus surface is reached at about 1,840°C. At this value the melt has become saturated with spinel, and any further subtraction of heat will result in crystallization of this phase. As $MgAl_2O_4$ precipitates, the composition of the residual liquid migrates directly away from the **Sp** apex (the complete compositional path followed by melt is shown by the heavy line in Figure 4.23a). At some lower temperature, say, 1,750°C, the two-phase assemblages will consist of a small proportion of spinel and a large proportion of melt *h*. Eventually, with continued cooling, the composition of the liquid, which continues to decrease in amount, reaches the cotectic curve *ae* at point *i*. Here the melt also has reached saturation with respect to Mg_2SiO_4; so further subtraction of heat results in the crystallization of both spinel and forsterite. The compositions of spinel, melt *i*, and forsterite outline the highest-temperature three-phase triangle for the bulk composition *x*, and the highest three-phase triangles for all other bulk compositions lying along the chord spinel–*i*; since the bulk composition lies along the leading edge spinel–*i*, the amount of Mg_2SiO_4 present is infinitesimal. However, as

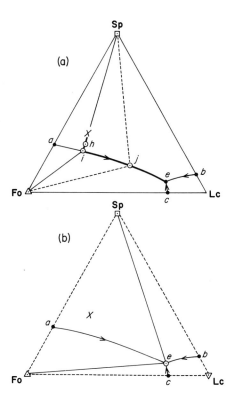

FIGURE 4.23.

Simplified reproduction of Figure 4.22, the one-atmosphere isobaric diagram for the system $2MgO·SiO_2$–$MgO·Al_2O_3$–$K_2O·Al_2O_3·4SiO_2$, illustrating the crystallization history of a general bulk composition, *x*. Diagram (a) shows the compositional migration of melt in heavy lines, the highest temperature three-phase triangle, spinel–forsterite–melt *i*, in solid lines, and a somewhat lower-temperature three-phase triangle, spinel–forsterite–melt *j*, in dashed lines. Diagram (b) gives the three-phase triangle at the eutectic temperature prior to the appearance of crystals of leucite, spinel–forsterite–melt *e*, shown in solid lines, and after the disappearance of the last drop of liquid, spinel–forsterite–leucite shown in dashed lines.

heat is withdrawn from the system, the proportion of forsterite (+ spinel) increases as the amount of liquid decreases and as the latter migrates at successively lower temperatures toward the ternary eutectic, *e*. As long as these phases remain in chemical equilibrium, the melt composition is constrained to a value along the cotectic curve. Thus at successively lower temperatures, the three-phase triangle enlarges, since the trailing edge, forsterite–spinel, remains fixed along the binary side line at the left, whereas the leading apex—such as melt *j*— migrates to the right toward *e*. Eventually the melt, which now represents only a small proportion of the system, reaches the eutectic composition at a temperature of 1,473°C, where it finally becomes saturated with respect to leucite as well as to forsterite and spinel (see Figure 4-23b). Continued withdrawal of heat at this stage results in the precipitation of all three crystalline phases in the eutectic proportions. Under isobaric equilibrium conditions, the temperature must remain constant as liquid is consumed, because the assemblage is invariant (for this isobaric three-component system consisting of four phases, equation 1.82 reads: $F = 3 - 4 + 1 = 0$). When the last drop of melt disappears, the bulk composition is contained within the three-phase triangle spinel–forsterite–leucite; further withdrawal of heat merely lowers the temperature of this assemblage. The line of liquid descent (the compositional path followed by melt during solidification) will be unaffected by the removal of early-formed crystals from the system, provided that equilibrium is maintained between the melt and the solid phase(s) with which it is saturated.

The course of equilibrium fusion for bulk composition *x* would be the exact reverse of the cooling history. But suppose that, either episodically or as it was produced, liquid left the system (i.e., fractional fusion took place). The following phase sequence would then be observed. On heating the crystalline assemblage spinel + forsterite + leucite at one atmosphere total pressure, partial melting would take place at 1,473°C until all the leucite and some of the forsterite and spinel had been consumed. If the liquid then separated from the crystals (perhaps because of its lower density, hence buoyancy), we would be left with a solid assemblage consisting of spinel + forsterite. With continued supply of heat, further melting of these crystals would ensue only at the temperature of the binary eutectic *a* (see Figure 4.22), namely, 1,725°C; at this stage, all the forsterite and some of the spinel would be fused. Subtraction of this binary eutectic melt from the system would leave a residual crystalline assemblage of monomineralic $MgAl_2O_4$. Continued heating would not produce any additional liquid until the melting point of spinel was achieved at 2,135°C. Clearly, although equilibrium crystallization and equilibrium fusion are identical processes in a reverse sense, crystal fractionation and fractional melting are strongly contrasting phenomena, as stressed by Presnall (1969).

In order to illustrate how phase assemblages depend on bulk composition, let us examine an isothermal section for this system at 1,500°C. As shown in Figure 4.24, there is a small divariant region of homogeneous liquid, *qrst*, at temperatures which exceed the liquidus values. Two three-phase triangles,

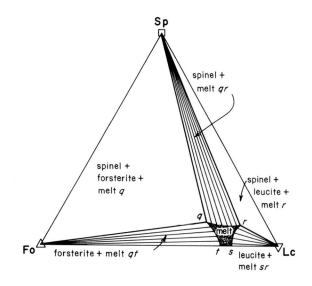

FIGURE 4.24.

One-atmosphere isobaric phase diagram for the system $2MgO \cdot SiO_2 - MgO \cdot Al_2O_3 - K_2O \cdot Al_2O_3 \cdot 4SiO_2$ at 1,500°C (i.e., a temperature exceeding that of the ternary eutectic). One-phase melt field (black) is bounded by the 1,500°C isotherm on the liquidus (see Figure 4.22). Three-phase fields are separated from one another by two-phase melt + crystal fields in which tie lines are shown connecting the solid with various liquid compositions.

spinel–forsterite–liquid q and spinel–leucite–liquid r, define the compositions of cotectic melts q and r, respectively; the variance is zero for these assemblages, and the compositions of all the phases are fixed. Three two-phase fields exist, each characterized by one solid phase and a liquid of variable composition (e.g., spinel + melt qr). Such assemblages are univariant because one compositional variable (of the liquid, in this case) must be specified in order to define exactly the chemical potentials of all the components in the coexisting phases. At temperatures between 1,473 and 1,493°C, a third three-phase triangle, involving the assemblage forsterite + leucite + cotectic melt, would appear. As is evident from a comparison of Figures 4.23 and 4.24 and the discussion here, these three three-phase triangles approach one another with declining temperature and achieve a common apex, e, at the eutectic. Here, with the subtraction of heat, all three are replaced by a fourth three-phase triangle, the subsolidus association (as shown in Figure 4.23b).

Before leaving this system, we will briefly consider quantitative evaluation of the composition of material that is being subtracted from an initially homogeneous phase—such as crystallization of a cooling melt. It is obvious that, starting with a one-phase liquid, as a second phase begins to precipitate out, the residual melt must move directly away from the second phase in compositional coordinates. Where the second phase is of invariant composition, the compositional path described by the melt is a straight line (e.g., line xhi in Figure 4.23a). However, if the composition of the second phase varies in sympathy with the changing chemistry of the liquid, or if an additional solid phase joins in crystallization and the proportions of the two solids vary with changing composition of the liquid, then the line of liquid descent will be curved (e.g., line *ije* in Figure 4.23a). The geometry is shown schematically in Figure 4.25. In this latter situation, the bulk composition of the material being subtracted from the melt lies somewhere along the tangent to the curve. This tangent provides the instantaneous direction of migration of melt *y* in chemographic space. For the system chosen, since the crystalline phases are of fixed composition, the proportions of these two phases being removed together from the liquid is given by point *z*, the intersection of this tangent with the spinel–forsterite side line.

Just as binary (*T*–*x*) phase equilibria carry over into more complex systems, *G*–*x* diagrams for *n* components display many of the same phenomena which have been illustrated in simple binaries (see Figure 4.9). In the example just discussed, of a simple ternary eutectic exhibiting no solid solution, the description of ternary, isobaric, isothermal *G*–*x* relations is as follows. The liquidus surface is always concave upward (i.e., concave towards increasing Gibbs free energy); whereas tangent planes to two (or three) solid phases of invariant composition in general are inclined in *G*–*x* space and are rigorously planar. Below the eutectic temperature, the *G*–*x* surface for melt lies everywhere at greater *G* values than that for the corresponding subsolidus crystalline assemblage, spinel–forsterite–leucite. At the isothermal *G*–*x* section for the eutectic temperature, this plane and the liquidus surface have a common tangent (at composition *e* of Figure 4.22). For an isothermal section at temperatures slightly above the eutectic, three planes of tangency to the liquid *G*–*x* surface, defined by the assemblages spinel–forsterite–cotectic melt, spinel–leucite–cotectic melt and forsterite–leucite–cotectic melt, are separated from one another by *G*–*x* regions of only one solid phase plus liquid of variable composition, and by a small field of homogeneous, one-phase melt (see Figure 4.24 for an analogue at a slightly higher temperature.)

FIGURE 4.25.

Schematic representation of the cotectic curve (heavy line with curvature greatly exaggerated) along which melt is stable with respect to both spinel and forsterite in the system $2MgO \cdot SiO_2$–$MgO \cdot Al_2O_3$–$K_2O \cdot Al_2O_3 \cdot 4SiO_2$. Tangent to this curve, *zy* (dashed line) gives the bulk composition of the crystalline material *z* being removed from melt *y*.

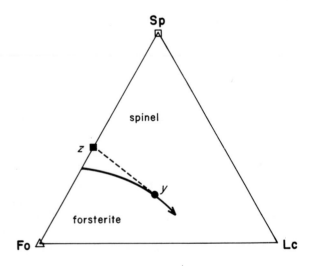

(2) Ternary System Exhibiting Complete Binary Solid Solution

We now turn our attention to a somewhat more complicated type of ternary phase equilibrium, one involving complete binary solid solution. Here we will use the system **Di–Ab–An** as an example. Because the clinopyroxene contains small amounts of aluminum and has a Mg/Ca ratio slightly greater than unit value, the system actually is quinary, involving the five components CaO, MgO, Al_2O_3, Na_2O and SiO_2, as shown by Kushiro (1973a). However, for simplicity we will ignore the compositional variation of diopside and melt, and will consider phase relations to be essentially ternary. The one-atmosphere phase diagram is presented in Figure 4.26. Here it can be seen that primary crystallization fields for clinopyroxene and plagioclase adjoin along a cotectic curve, the maximum temperature of which is 1,275°C at the bounding binary eutectic along the **Di–An** side line, and the minimum temperature of which is about 1,100°C at the bounding **Di–Ab** binary eutectic. Evidently the eutectic nature of these binary relationships as well as the plagioclase solid-solution series carry over into the more complicated ternary system.

Consider first an isothermal section in this system at some arbitrary temperature, say, 1,300°C. The chemographic relationships are presented in Figure 4.27. At this temperature the broad "thermal valley" of the cotectic curve is a single phase melt in the region *abcd*-**Ab**. More **Di**-rich bulk compositions consist of two phases, a clinopyroxene of fixed composition (actually a solid solution) and a melt ranging in composition from *c* to *d*, depending on the bulk composition. For bulk chemistries rich in normative plagioclase, the situation is more complicated, because both liquid and plagioclase have variable compositions in this two-phase region. As is obvious from the **Ab–An** binary diagram at the bottom of Figure 4.27, melt *a* is in equilibrium with plagioclase *w*. Equally clear from the **Di–An** binary eutectic (not shown), melt *b* is stable with end-member anorthite crystals. Liquids intermediate in composition between *a* and *b* are saturated with plagioclase intermediate in composition between *w* and **An**; for example, for bulk composition *x*, plagioclase *j* is in equilibrium with melt *s* at 1,300°C and one atmosphere total pressure. The **Ab** component is concentrated preferentially in the melt relative to the coexisting plagioclase, as seen from the disposition of the tie lines linking equilibrium phases in this two-phase region.

Just to make sure that we understand the nature of this system, let us construct another one-atmosphere isobaric section, this time at 1,200°C, as shown in Figure 4.28. Naturally enough, at lower temperatures the field of homogeneous melt has further contracted to the most fusible compositional portion of the diagram near the **Ab** corner. Again there are (isobaric, isothermal) univariant regions where diopside and intermediate plagioclase ranging in composition from *y* to *z* coexist with liquids of compositional ranges *gh* and *fg*, respectively. At such a low temperature, there is a unique melt, *g*, which lies on the cotectic curve, hence is in equilibrium with two crystalline phases of

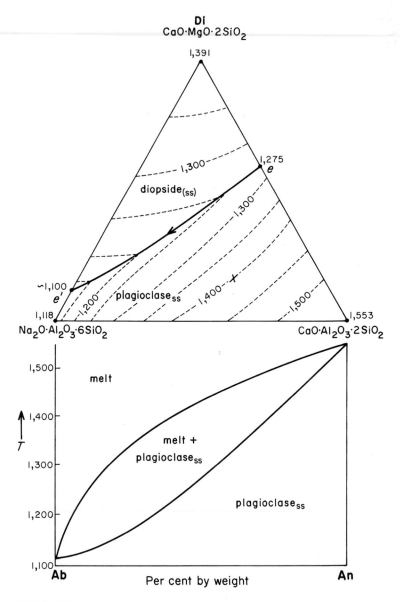

FIGURE 4.26.

One-atmosphere phase diagram for the ternary system CaO·MgO·2SiO₂–Na₂O·Al₂O₃·6SiO₂–CaO·Al₂O₃·2SiO₂, after Bowen (1915), Osborn (1942), Schairer and Yoder (1960), and Kushiro (1973a). Proportions are in percentages by weight. Isotherms on the liquidus are shown for 50°C intervals. The bounding one-atmosphere **Ab–An** binary is also illustrated. Clinopyroxene and therefore coexisting melt show small compositional departures from the ternary system. The crystallization sequence of bulk composition x is treated in Figure 4.29.

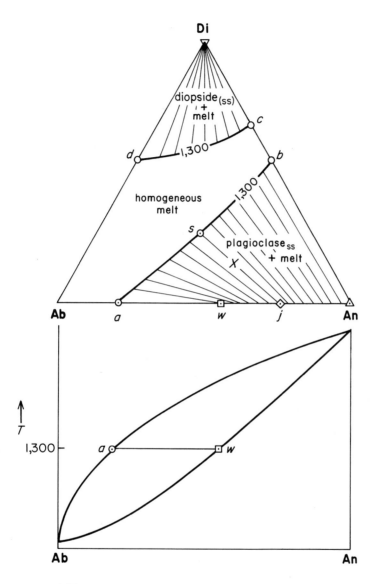

FIGURE 4.27.
Isothermal section at 1,300°C for the one-atmosphere isobaric **Di–Ab–An** system, after Figure 4.26. Melt *s* and plagioclase *j* are stable for bulk composition *x*.

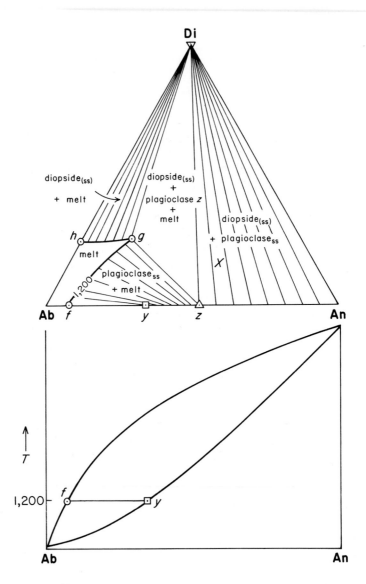

FIGURE 4.28.
Isothermal section at 1,200°C for the one-atmosphere isobaric **Di–Ab–An** system, after Figure 4.26.

invariant composition, diopside and plagioclase z; these three phases outline the only three-phase triangle at this temperature. For more **An**-rich bulk compositions such as x, the assemblage is subsolidus, and consists of clinopyroxene and plagioclase solid solution ranging from z to pure anorthite, depending on the bulk composition being considered.

Now imagine the spectrum of isothermal sections within the temperature interval characterized by a cotectic melt. At the highest temperature, 1,275°C, where melt e (Figure 4.26), diopside, and anorthite are colinear, the three-phase triangle is present in a degenerate, collapsed condition as a straight line. At lower temperatures the three-phase triangle opens and sweeps progressively to the left across the ternary diagram, with the cotectic melt as the leading apex. Finally, the three-phase triangle collapses to a line at the low temperature termination of $\sim 1,110$°C, where melt e', diopside, and albite are colinear.

We are now ready to follow the course of equilibrium crystallization of a simplified "magma," the composition of which lies in the system $CaO \cdot MgO \cdot 2SiO_2 - Na_2O \cdot Al_2O_3 \cdot 6SiO_2 - CaO \cdot Al_2O_3 \cdot 2SiO_2$. If we select a bulk composition which lies in the primary crystallization field of clinopyroxene but is initially entirely molten, a rather simple phase sequence is observed as it cools. First, diopside$_{(ss)}$ appears, and as it precipitates, the melt moves directly away from the **Di** until it reaches the cotectic curve. At this point a rather calcic plagioclase solid solution joins clinopyroxene in crystallizing. As heat is continually removed, both solid phases increase in amount at the expense of liquid. Both melt and the plagioclase$_{ss}$ become enriched in **Ab** until the melt has diminished to zero; at this precise stage, the original bulk composition lies on the trailing edge of the three-phase triangle, namely, the tie line diopside$_{(ss)}$–plagioclase$_{ss}$.

As might be expected, a more complicated history is presented by the crystallization of normative plagioclase-rich bulk compositions. Let us select at random such a homogeneous melt, x, as shown in Figure 4.29a, assuming the maintenance of perfect equilibrium. As heat is withdrawn from a totally molten liquid, temperature will decline until, at 1,400°C, the first crystal of **An**-rich plagioclase solid solution appears (see Figure 4.26). The exact composition of this crystal, i, is not clear from Figure 4.26, but may be ascertained from construction of a 1,400°C isothermal section. Melt x begins to move directly away in compositional space form plagioclase i as the latter precipitates out. In other words, tie line xi is a tangent to the curve representing the liquid line of descent. At any temperature, the bulk composition is made up of two phases, liquid plus crystals; so for the equilibrium situation x lies on the tie line so defined. For instance, as we have already observed, at 1,300°C, melt s and plagioclase j are connected by the tie line passing through x (compare Figures 4.27 and 4.29a). As heat is continually subtracted from this system, the liquid decreases in amount as it becomes richer in **Ab** + **Di**; concomitantly the solid plagioclase proportion increases as it becomes more sodic. Eventually the liquid reaches the cotectic curve at t, where the equilibrium plagioclase composi-

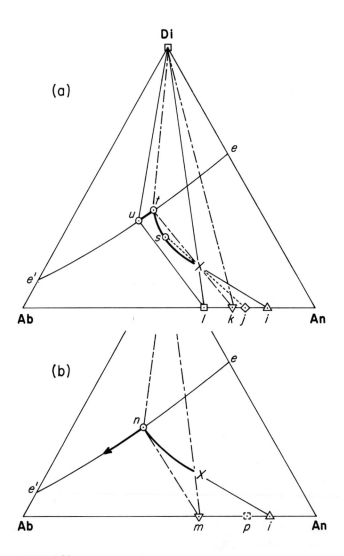

FIGURE 4.29.

Compositional migration of melt and crystals on cooling of the
bulk composition x in the ternary one-atmospheric isobaric system
Di–Ab–An, assuming: (a) equilibrium conditions; and (b) perfect
crystal fractionation. The liquid line of descent is shown in heavy
solid lines, the compositions of various crystalline assemblages
by symbols. With perfect fractionation (case b), a small amount
of melt will actually achieve the composition e' before solidification
is completed.

tion is *k*. Here the melt achieves saturation in **Di**, and clinopyroxene joins plagioclase solid solution in crystallization; because the bulk composition *x* lies along the leading edge, *tk*, of the three-phase triangle **Di**-*t*-*k*, only infinitesimal amounts of diopside are present. With further subtraction of heat, the melt moves along the cotectic curve, decreasing further in amount. Under equilibrium conditions it has been entirely consumed when it reaches the composition *u*; now the bulk composition *x* lies along the trailing edge, **Di**–*l*, of the three-phase triangle. Further subtraction of heat merely decreases temperature; at 1,200°C, for instance, bulk composition *x* is clearly subsolidus (see Figure 4.28).

To recapitulate the equilibrium crystallization sequence of the intially homogeneous high-temperature melt on cooling, the liquid composition follows the path *xst*, whereas plagioclase continuously reacts with the melt and changes composition from *i* through *j* to *k*. With the additional coprecipitation of clinopyroxene, the melt follows the cotectic curve from *t* to *u*, whereas plagioclase crystals become more sodic and finally achieve the composition *l*.

But what about the case of perfect fractional crystallization, in which the early-formed crystals fail to react with the melt? The line of liquid descent is illustrated in Figure 4.29b. As in the equilibrium cooling, the first plagioclase to precipitate out at the liquidus of melt *x* has the composition *i*. However, once crystallized this solid never reacts with the liquid. As the melt becomes depleted in composition *i*, it moves directly away from it, becoming relatively enriched in **Ab** (and **Di**). Either the early-formed crystals are physically separated from the melt or they are armored by later, less refractory rims; in any event, there is no chemical communication or exchange between solid and liquid. Therefore, as crystallization proceeds, a spectrum of plagioclase compositions precipitates out of the liquid, displacing it in the manner shown in Figure 4.29b toward an eventual intersection with the cotectic curve. However, since the melt composition is no longer constrained by continuous reaction with plagioclase (in effect, at every point homogeneous melt represents a new bulk composition for the "system"), the line of liquid descent is less strongly curved (concave to the **Di** apex) than in the equilibrium case. It is nevertheless curved because, as it becomes more sodic, the plagioclase$_{ss}$ concomitantly being formed varies sympathetically in composition. Eventually the homogeneous melt reaches the cotectic curve at *n*, where it is saturated with plagioclase *m*. Here a complete gradation in plagioclase crystals from *i* to *m* is present, the aggregate bulk composition of which is some intermediate value, such as *p*. (The bulk composition *x* must be situated along the line connecting *n* and *p*.) With further subtraction of heat, clinopyroxene$_{(ss)}$ and plagioclase$_{ss}$, becoming progressively more sodic than *m*, precipitate out until the melt is finally used up at *e'*. Of course, the subsolidus plagioclase composition must average out to the composition *l* given in Figure 4.29a, but with perfect fractionation it can range in composition from *i* all the way to end-member albite. If limited reaction occurs, however, the over-all compositional range will be more restricted, and this is the most common situation for igneous rocks.

From a comparison of Figures 4.29a and b, you can see that, within the primary crystallization field of plagioclase$_{ss}$, the line of liquid descent is more strongly curved for equilibrium cooling than for situations involving crystal fractionation. At this stage it would be good experience for the reader to construct heating paths for bulk composition x, assuming, first, equilibrium, and, second, fractional fusion.

(3) Ternary System Exhibiting Binary Peritectic Relations

We now turn our attention to the system $2MgO \cdot SiO_2 - CaO \cdot Al_2O_3 \cdot 2SiO_2 - SiO_2$; one-atmosphere phase equilibria are presented in Figure 4.30. Melting behavior is ternary here except for compositions near the **Fo–An** side line, where

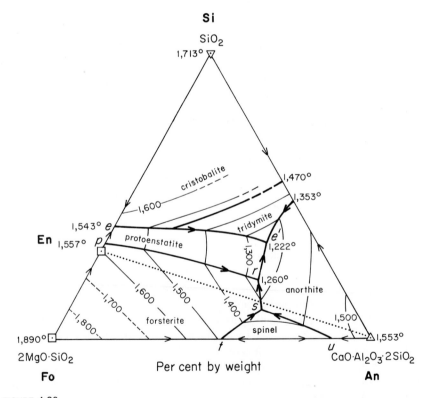

FIGURE 4.30.

One-atmosphere isobaric phase diagram for the pseudoternary system $2MgO \cdot SiO_2 - CaO \cdot Al_2O_3 \cdot 2SiO_2 - SiO_2$, after Anderson (1915, Figure 10). Temperatures are in °C; isotherms on the liquidus are shown for 100°C intervals. The crystalline phase in equilibrium with melt at liquidus temperatures is also indicated. The dotted line signifies a join, **En–An**, along which no solid solution exists.

spinel, $MgAl_2O_4$, appears on the liquidus. The join **En–An**, shown by dots, divides the system into the forsterite-normative and quartz-normative subsystems **Fo–En–An** and **Si–En–An**, respectively. The bounding binary $2MgO \cdot SiO_2 – SiO_2$ involves the reaction relation between forsterite and melt to produce protoenstatite (see Figure 4.10). This behavior carries over into the ternary system, as is clear from an examination of Figure 4.30: the primary crystallization field of forsterite completely overlaps **En**-rich portions of the **En–An** join, indicating that, for all such bulk compositions, and for the adjacent band of quartz-normative bulk compositions, forsterite rather than protoenstatite is stable on the liquidus.

The binary peritectic and eutectic points p and e of Fig. 4.10 have three-component analogues in r and e', the ternary *reaction point* and ternary eutectic, respectively of Figure 4.30. The line pr is a *reaction curve* along which forsterite reacts with melt to produce protoenstatite; the compositional path being followed by liquid at any particular temperature is given by the tangent to the appropriate point on the reaction curve, which in turn is the sum of the vectors representing the subtraction of protoenstatite from the melt plus the "addition" of forsterite (actually the withdrawal of silica from the liquid to convert Mg_2SiO_4 into $MgSiO_3$). The line ee' is a cotectic curve along which a silica polymorph and protoenstatite coprecipitate from the melt (this curve is different from pr by virtue of the fact that the line of liquid descent here is the vector sum defined by the removal of both SiO_2 polymorph and protoenstatite). Another reaction point, s, exists where spinel reacts with melt to produce forsterite and anorthite. In the chemographic region tsu, phase relations are actually quaternary, because spinel and the melt in equilibrium with it do not lie in the composition plane **Fo–An–Si**.

Let us examine the equilibrium crystallization history of an initially molten bulk composition x lying along the **En–An** join, as diagrammed in Figure 4.31a. On subtraction of heat, the melt will cool until, at a temperature of about $1,530°C$, the first crystallites of forsterite appear. With continued removal of heat, additional forsterite precipitates out, and the melt composition moves directly away from $2MgO \cdot SiO_2$ along the line xa. On reaching the reaction curve pr, liquid a is in equilibrium with both forsterite and protoenstatite, and with further cooling the melt follows path ar. A tangent to the reaction curve at, say, a (or at any other locus on this curve) gives the instantaneous direction of migration of melt toward the **An–Si** side line, and would intersect the **Fo–Si** binary (i.e., the composition of the substance being removed from the liquid) on the silica-rich side of $MgO \cdot SiO_2$. Because the crystalline phases which are stable with all reaction-curve melts are forsterite and protoenstatite, but not a silica polymorph, the material being subtracted from the liquid consists of protoenstatite and SiO_2 component. As discussed above, the latter species actually is silica in the melt reacting with early-formed Mg_2SiO_4 to convert the forsterite into protoenstatite. With maintenance of equilibrium, when the

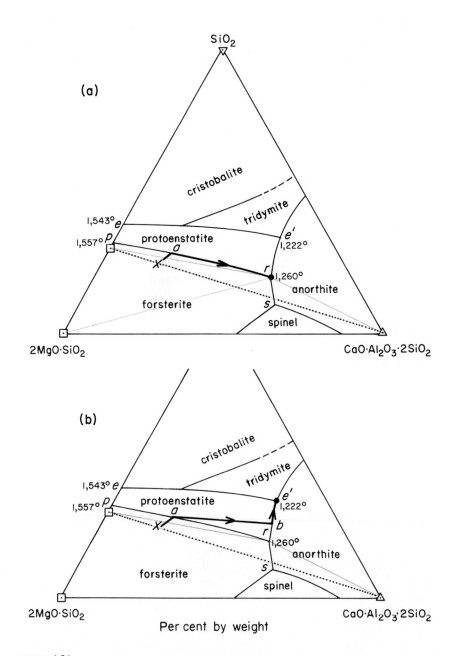

FIGURE 4.31.

The system **Fo–An–Si** at one atmosphere total pressure, reproduced from Figure 4.30. The line of liquid descent during crystallization of the bulk composition *x*, which contains neither normative forsterite nor normative quartz, is illustrated, assuming: (a) equilibrium; and (b) fractionation.

melt initially achieves composition r, the assemblage consists of the phases forsterite, protoenstatite, and liquid. As heat is lost, all the remaining forsterite reacts with all the residual melt to produce anorthite and additional crystalline $MgSiO_3$ at constant T. When the last traces of melt and forsterite are consumed, temperature will again begin to decline as heat is removed.

The reaction relation of forsterite and melt r also may be ascertained by considering the three-phase triangles which exist at this isobaric invariant temperature. Two high-temperature ones, forsterite–protoenstatite–melt r and forsterite–anorthite–melt r, are replaced at 1,260°C by two lower-temperature three-phase triangles, forsterite–protoenstatite–anorthite and protoenstatite–anorthite–melt r. Sketch these chemographic relations for yourself, noting that the reaction relation between Mg_2SiO_4 and liquid is required by the fact that melt r lies outside the three-phase triangle forsterite–protoenstatite–anorthite. In contrast, at the ternary eutectic, e', three high-temperature liquid-bearing three-phase triangles give way to a single subsolidus three-phase triangle which encloses the composition of the eutectic liquid, namely, protoenstatite–anorthite–tridymite.

Now consider the disequilibrium crystallization of the same initially molten bulk composition x, as illustrated in Figure 4.31b. On cooling, the first segment of the liquid descent path, xa, is the same as for equilibrium, because Mg_2SiO_4 again precipitates out. However, where the melt reaches the reaction curve pr at a, instead of reacting with the liquid, the early-formed forsterite crystals persist metastably. In this situation it is just as if the olivine had been physically removed from the system; so we are dealing with a new bulk composition, that of melt a. Since the forsterite does not maintain chemical communication with the liquid, the latter is not constrained to follow the reaction curve. Accordingly, as heat is removed, protoenstatite precipitates out and the melt moves directly away from $MgO \cdot SiO_2$, along path ab. At b the liquid has reached the cotectic curve re', and further cooling will cause the crystallization of both anorthite and additional crystalline $MgSiO_3$. Finally, the residual liquid attains the ternary eutectic composition, e', at 1,222°C; further loss of heat will result in the isothermal, simultaneous crystallization of protoenstatite, anorthite, and tridymite in the eutectic proportions until the melt is entirely consumed. Examination of the resultant assemblage would reveal the presence of early-formed forsterite grains surrounded by monomineralic protoenstatite and a peripheral zone of intergrown anorthite + protoenstatite, all set in an eutectic mesostasis of protoenstatite + anorthite + tridymite. Thus, a bulk composition which has neither normative forsterite nor normative silica polymorph completes solidification under disequilibrium conditions with both. In nature, where early-formed crystals may be mechanically separated from the magma, it is clear that this process is capable of giving rise to a chemically differentiated igneous-rock series.

On heating, the sequence of loss of crystalline phases and the compositional path followed by melt strongly depend on the initial phase assemblage as well as on the bulk composition, in this case, x, and on whether equilibrium melting or fractional fusion takes place. The equilibrium melting history of bulk composition x is exactly the reverse of the equilibrium crystallization path previously described (Figure 4.31a). On the other hand, if fractional fusion of our equilibrium assemblage of crystals occurs, melt r will be produced at 1,260°C, and as thermal energy is supplied at this temperature, all the anorthite disappears as well as some of the protoenstatite, whereas appreciable amounts of forsterite are generated. If all the reaction liquid r is removed, no further fusion takes place until the peritectic temperature, 1,557°C, is attained; at this stage all the remaining $MgSiO_3$ disappears with the production of additional Mg_2SiO_4 and melt p. If this liquid migrates away, the residual monomineralic forsterite assemblage remains stable up to its melting temperature of 1,890°C.

Where the initial crystalline assemblage represents chemical disequilibrium (Figure 4.31b), a more complicated history is recorded during heating even though fractional fusion may not be involved. The lowest-temperature liquid forms at 1,222°C and has the eutectic composition. At this point all the tridymite and some of the protoenstatite and anorthite are consumed. If equilibrium is maintained between melt and protoenstatite + anorthite (but not forsterite), the heating path will be the reverse of that illustrated in Figure 4.31b. Anorthite disappears at b and $MgSiO_3$ at a, where forsterite at last establishes chemical communication with the liquid; more heating results in a diminution of the forsterite and its eventual disappearance as the melt follows path ax.

If fractional fusion occurs during the heating of a disequilibrium assemblage, the melting sequence is even more complex. At e', all the tridymite, and some of the $MgSiO_3$ + $CaAl_2Si_2O_8$ will melt. On removal of the resultant liquid, we are left with the solid assemblage protoenstatite + anorthite (+ forsterite); with the continued supply of thermal energy, all the anorthite and most of the protoenstatite will melt at 1,260°C, generating reaction melt r and moderate amounts of crystalline Mg_2SiO_4. The migration away of this liquid leaves a residual solid assemblage consisting of forsterite and minor protoenstatite; on continued heating no additional fusion occurs until the $MgSiO_3$ disappears at 1,557°C, where liquid p and minor additional Mg_2SiO_4 are generated; with the draining off of this liquid, forsterite finally melts only at 1,890°C.

(4) Combinations of Limited or Complete Solid Solution with the Discontinuous-reaction Relation

Let us now study ternary phase relations in which extensive binary solid solution is combined with peritectic-type equilibrium. The system MgO–FeO–SiO_2 will be employed as a first example. This system actually displays quaternary behavior, because, even at relatively low oxygen fugacities where the charge is in equilibrium with native iron, small amounts of ferric iron are stable in the melt. We will ignore this problem for now, but will return to it later. Magnesio-wüstites, $(Mg, Fe^{2+})O$ solid solutions, are not important minerals in igneous rocks; so, although they are an important part of this system, they will not be considered in our discussion.

As is well-known, both olivine and hypersthene series exhibit extensive Fe^{2+}–Mg substitution, and we have previously studied the peritectic relation of the magnesian end-members. That this behavior carries over into the ternary system is clear from examination of Figure 4.32. However, unlike the pre-

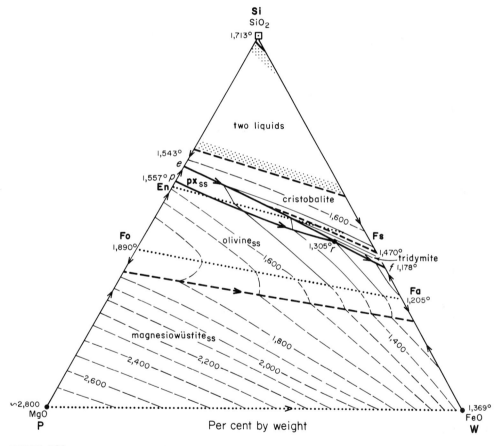

FIGURE 4.32.

One-atmosphere isobaric phase diagram for the ternary system MgO–FeO–SiO_2, after Bowen and Schairer (1935) and Muan and Osborn (1956). Charges were heated in a nitrogen atmosphere and equilibrated with an iron crucible. Proportions are in percentages by weight. Temperatures are in °C; isotherms on the liquidus are shown for 100°C intervals. The crystalline phase in equilibrium with melt at liquidus temperatures is also indicated. Maximum compositional ranges of the crystalline solid solutions, magnesiowüstites, olivines, (proto-)hypersthenes, are shown by dotted lines. Boundaries of the two-melt field are stippled. Abbreviation: px_{ss} = (proto-)hypersthene solid solution.

viously described systems, (proto-)hypersthenes of an intermediate Fe^{2+}–Mg
compositional range are stable on the liquidus, because the primary crystallization field of pyroxene solid solution encompasses a portion of the binary
solid solution in chemographic space. This primary crystallization field is situated between the binary peritectic p and the binary eutectic e along the side
line MgO–SiO_2 (see also Figure 4.10), but extends over the **En–Fs** join well
out into the ternary system. Owing to the instability of the ferrosilite end-
member at low pressures, the pyroxene solid-solution field must terminate
before reaching the bounding binary FeO–SiO_2, and does so at the ternary
reaction point r. The extent of (proto-)hypersthene solid solution appears to
range from Fs_{00} to a maximum of about Fs_{90} at one atmosphere total pressure.
The curve pr is a reaction curve in its more magnesian portions, but becomes a
cotectic curve near its termination at r; this change in character takes place at a
relatively iron-rich bulk composition—and a temperature of about 1,425°C—
where the composition of the pyroxene solid solution in equilibrium with a
melt lying on curve pr just emerges from beneath the primary crystallization
field of olivine (see Figure 4.33a). Curves er and rf are also cotectic curves.

Let us examine the behavior of three-phase triangles as a function of temperature in order to more fully understand the nature of the equilibria involving pyroxene solid solutions. Chemographic relations for the petrologically
important subsystem $2MgO{\cdot}SiO_2$–$2FeO{\cdot}SiO_2$–SiO_2 are presented schematically
in Figure 4.33 (note that this diagram, and Figures 4.34 and 4.35 as well, are
shown in terms of mole per cent rather than per cent by weight). Four sets of
three-phase triangles must be considered: (1) $olivine_{ss}$ + $pyroxene_{ss}$ + liquid;
(2) silica polymorph + $pyroxene_{ss}$ + liquid; (3) $olivine_{ss}$ + tridymite + liquid;
and (4) $olivine_{ss}$ + $pyroxene_{ss}$ + tridymite. The highest-temperature ones,
(1) and (2), open up on departure from the **Fo–Si** binary side line as small
amounts of ferrous iron are added to the system. With falling temperature,
these three-phase triangles sweep to the right (i.e., to more iron-rich phase
compositions), and have their lowest-temperature junction at the ternary reaction point. Only four pairs of these triangles are illustrated in Figure 4.33a, but
these are sufficient to show two things: first, olivine is always more ferrous than
a coexisting (proto-)hypersthene; second, although the pyroxene solid solution
in equilibrium with a silica polymorph at high temperatures is more magnesian
than pyroxene solid solution in equilibrium with olivine (compare $pyroxene_{ss}$
1 with $pyroxene_{ss}$ 2 at 1,527°C), these Fe^{2+}/Mg ratios are reversed at lower
temperatures (compare $pyroxene_{ss}$ 5 with $pyroxene_{ss}$ 4 at ~ 1,400°C). Such
relations require that, at an intermediate temperature (about 1,450°C), a specific intermediate solid solution ($pyroxene_{ss}$ 3) be stable both with silica-rich
cotectic liquid j + tridymite and, on the other hand, with the silica-poor reaction melt m + $olivine_{ss}$ b.

As is brought out in Figure 4.33b, the tie lines connecting cotectic melt (er)
with (proto-)hypersthene describe a twisted, fan-shaped array; a similar relation could equally well have been figured for tie lines connecting $pyroxene_{ss}$
and liquid pr.

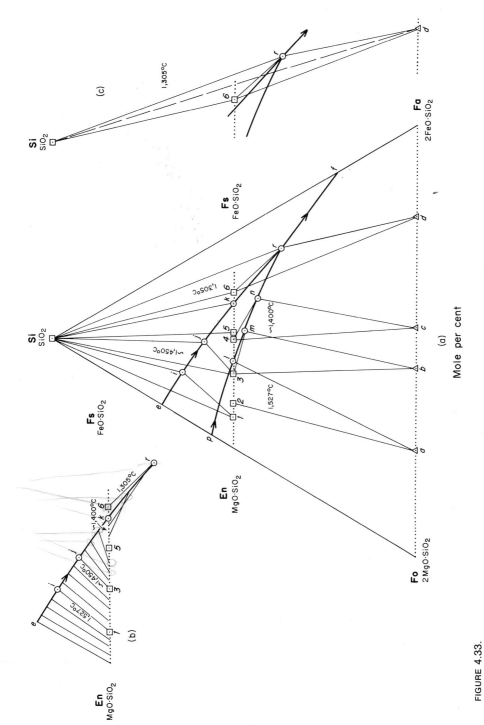

FIGURE 4.33.

Diagrammatic representation of the system **Fo–Fa–Si** at one atmosphere total pressure, from Figure 4.32 (ignoring the two-melt field for simplicity) and recast in mole proportions. Lettering corresponds to Figure 4.32. In (a), which is based on the work of Bowen and Schairer (1935), pairs of three-phase triangles are illustrated for the temperatures 1,527, 1,450, 1,400, and 1,305°C. For the 1,527°C isotherm, the pyroxene$_{ss}$ 2 – liquid *l* tie line, which by coincidence falls along a portion of the **En–Fs** compositional line, has been omitted for clarity. In (b), details of the tie lines connecting (proto-)hypersthene and cotectic melt *er* are shown. In (c) the nature of the three-phase triangles and tie-line shifts at the ternary reaction point are presented.

At 1,305°C, the two three-phase triangles, (1) olivine$_{ss}$ + pyroxene$_{ss}$ + melt and (2) tridymite + pyroxene$_{ss}$ + melt, possess a common tie line, from pyroxene$_{ss}$ *6* to liquid *r*. At this stage, withdrawal of heat causes the reaction of hypersthene *6* and liquid to produce iron-rich olivine$_{ss}$ *d* and additional tridymite, as shown by the dashed line in Figure 4.33c. When the reactants have been completely consumed, continued loss of heat will allow temperature to decline. For appropriate bulk compositions, two new three-phase triangles generated at 1,305°C must be considered: (3) the more iron-rich phase association olivine$_{ss}$ + tridymite + liquid; and (4) the more magnesian subsolidus assemblage olivine$_{ss}$ + pyroxene$_{ss}$ + tridymite. The extent of pyroxene solid solution appears to increase slightly with decreasing temperature (Bowen and Schairer, 1935, Figure 8).

A schematic isothermal section for the system **Fo–Fa–Si** at 1,527°C is presented in Figure 4.34. In this construction, a 1,527°C contour on the liquidus surface has been interpolated from Figure 4.32; also, the corresponding three-phase triangles, cristobalite + pyroxene$_{ss}$ *1* + melt *i*, and olivine$_{ss}$ *a* + pyroxene$_{ss}$ *2* + melt *l*, have been employed from Figure 4.33a. Details of phase relations in the neighborhood of the **En–Fs** join at a slightly lower temperature are also shown. Isothermal sections for other temperatures may be readily drawn by using available liquidus and three-phase-triangle information. It must be emphasized, however, that the disposition of tie lines linking phases of variable composition must be evaluated by experiment or thermodynamic calculation, or else must be reasonably approximated.

Now consider the line of liquid descent for some initially molten bulk composition *x* in the system **Fo–Fa–Si**, as shown in Figure 4.35. For brevity, we will discuss only the equilibrium path, in which crystals and melt maintain chemical communication. As heat is withdrawn from the melt, temperature falls until the liquidus surface is reached at, say, 1,600°C; here the first crystals of relatively magnesian olivine solid solution *g* precipitate out. As is evident from this initial two-phase tie line (shown in Figure 4.35 by short dashes) connecting melt *x* with crystal *g*, the subtraction of olivine causes the composition of liquid to migrate directly away from *g*. With falling temperature, the melt gradually becomes enriched in iron (and silica); hence the early-formed olivine solid solution continuously reequilibrates, becoming more fayalitic in the process, and accordingly the melt follows the curved path *xm*. The temperature at which melt reaches the curve *pr* at *m* is 1,450°C. Under these conditions, olivine$_{ss}$ *b*, the first crystals of (proto-)hypersthene$_{ss}$ *3*, and melt *m* stably coexist; this is the highest-temperature three-phase triangle appropriate for bulk composition *x*. With continued subtraction of heat, substantial amounts of pyroxene solid solution precipitate out from the melt, in small part because of reaction of olivine solid solution with liquid, and the melt moves along curve *pr* toward *q*. Concomitantly the (proto-)hypersthene and the olivine become enriched in ferrous iron. Where the tie line linking the composition of the pyroxene solid solution with melt first becomes tangent to the curve *pr* (at the

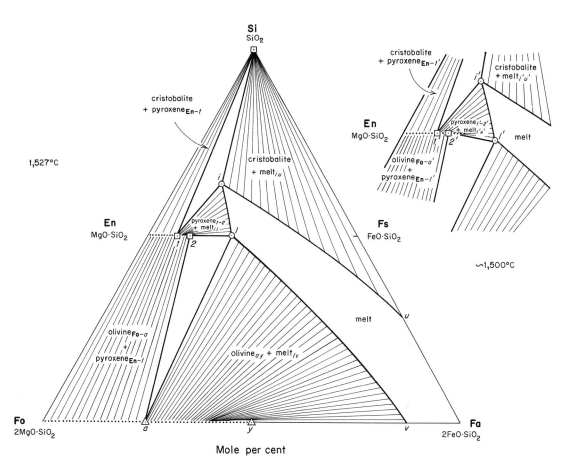

FIGURE 4.34.

Diagrammatic isothermal, isobaric section of the system **Fo–Fa–Si** (see Figure 4.32) at 1,527°C, at one atmosphere total pressure, and in mole proportions. The three-phase triangles, cristobalite + pyroxene$_{ss}$ l + melt i, and olivine$_{ss}$ a + pyroxene$_{ss}$ 2 + melt l, are taken from Figure 4.33. Phase relations at about 1,500°C in the vicinity of the MgO·SiO$_2$–FeO·SiO$_2$ join are illustrated at the right; note that here the maximum extent of (proto-)hypersthene solid solution occurs along the binary, and does not coincide with either of the two three-phase triangles.

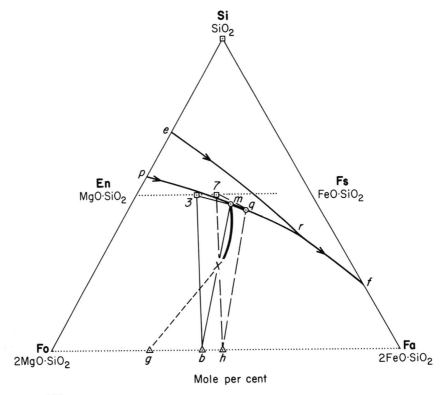

FIGURE 4.35.
Schematic one-atmosphere section of the system **Fo–Fa–Si**, from Figure 4.32, recast in mole proportions. The path of compositional migration on cooling under equilibrium conditions is *xmq*. The highest-temperature three-phase triangle for this composition (shown in solid lines) occurs at 1,450°C and corresponds to the triangle for olivine$_{ss}$ *b* + (proto-)hypersthene$_{ss}$ *3* + melt *m* from Figure 4.33a. The lowest-temperature three-phase triangle for this composition (shown in long dashes) occurs at about 1,420°C and is defined by the assemblage olivine$_{ss}$ *h* + (proto-)hypersthene$_{ss}$ *7* + melt *q*.

point where the $MgO \cdot SiO_2$–$FeO \cdot SiO_2$ join emerges from under the boundary of the olivine primary crystallization field), this curve changes its nature from a reaction relation to cotectic precipitation. Under equilibrium conditions, the last drop of melt, *q*, is consumed at a temperature of about 1,420°C, where it is stable with olivine solid solution *h* and pyroxene solid solution *7*.

When phase relations in the system MgO–FeO–SiO_2 were introduced at the beginning of this section, it was noted that some of the iron in the melt occurs as Fe^{3+} even where the charge equilibrates with an iron crucible in the presence of a nitrogen atmosphere (see Figure 4.32). Under such conditions, oxygen fugacity is maintained at a very low value, but this does not preclude the differential partitioning of electrons among the iron contained in the various solid phases and in the liquid (Bowen and Schairer, 1935). More recent one-atmosphere liquidus experiments (Muan and Osborn, 1956) have involved

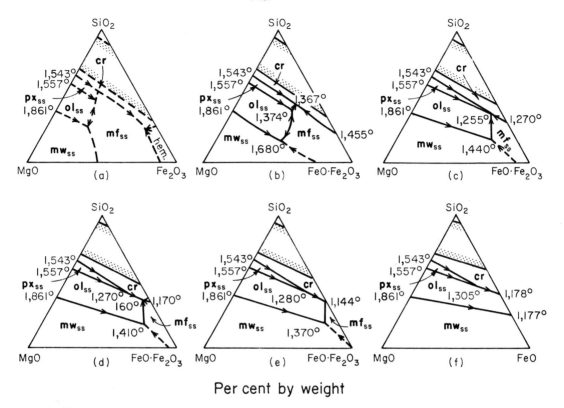

Per cent by weight

FIGURE 4.36.

Simplified, diagrammatic one-atmosphere liquidi for the quaternary system $MgO–FeO–Fe_2O_3–SiO_2$ at various oxidation states decreasing from (a) to (f), after Muan and Osborn (1956). (a) = sufficiently high f_{O_2} to keep most of the iron in the ferric state. (b) = in air. (c), (d) and (e) = CO_2/H_2 ratios of 40, 24, and 19, respectively. (f) = Figure 4.32 for comparison. Margins of the two-melt field are stippled. Abbreviations are as follows: **cr** = cristobalite (the lower temperature primary crystallization field of tridymite has been omitted for simplicity); hem = hematite; **mf**$_{ss}$ = magnesioferrite solid solution; **mw**$_{ss}$ = magnesiowüstite solid solution; **ol**$_{ss}$ = olivine solid solution; **px**$_{ss}$ = (proto-)hypersthene solid solution.

precise mixing of components in a gas phase flowing over the charge in order to maintain specific CO_2/H_2 ratios (hence oxygen proportions, thus providing f_{O_2} control). Simplified phase relations for various oxidation states are shown in Figure 4.36.

Several deductions emerge from an examination of this illustration. (1) The primary crystallization fields for the ferrous iron + magnesium solid-solution minerals (proto-)hypersthene, olivine, and magnesiowüstite are restricted to successively more magnesian compositions as oxidation state increases; whereas, understandably, the ferric iron-bearing minerals hematite and magnesioferrite, $(Fe^{2+}, Mg)Fe^{3+}_2O_4$, have correspondingly expanded ranges. (2) The reaction relation between magnesian olivine and melt is retained regardless of the relative oxidation, because phase relations along the MgO–SiO_2 binary are unaffected by variations in f_{O_2}. (3) In general, liquidus temperatures for iron-rich bulk compositions are elevated by increasing oxidation. (4) The basic thermal "topography" of this system is modified at high f_{O_2} values only to the extent that a pseudoternary eutectic is produced, replacing the relatively low f_{O_2} eutectic located on the binary side line (point f in Figure 4.32). In terms of application to igneous processes, perhaps the most important conclusion to be derived from Figure 4.36 is that, at relatively high oxidation states, and for appropriate bulk compositions, the early precipitation of magnesioferrite ($\approx$ magnetite) results in pronounced enrichment of the residual liquid in silica, but without a marked change in the Fe^{2+}/Mg ratio of the melt; in contrast, at low f_{O_2} values, strong concentration of ferrous iron with respect to magnesium occurs in residual liquids, which also become impoverished in silica during the crystallization sequence.

A second example of a system showing (limited) solid solution combined with the discontinuous-reaction relation is the ternary system $2MgO \cdot SiO_2$–$CaO \cdot MgO \cdot 2SiO_2$–$SiO_2$. Liquidus relations are presented in Figure 4.37. As is clear from this diagram, the primary crystallization field of forsterite overlaps all but the Ca-richest portions of the pseudobinary join $MgO \cdot SiO_2$–$CaO \cdot MgO \cdot 2SiO_2$, which means that the discontinuous-reaction relation between Mg_2SiO_4 and melt extends to **Di**-rich portions of the system. The **En**–**Di** join is further complicated by the fact that, at near-solidus temperatures in the $MgO \cdot SiO_2$-rich region, pigeonite (clinopyroxene of intermediate calcium content) has a small field of stability situated within the protoenstatite$_{ss}$–diopside$_{ss}$ solvus. At somewhat lower temperatures, pigeonite recrystallizes to **di**$_{ss}$ + **px**$_{ss}$. In spite of such intricacies, it is evident from this diagram that extensive solid solution along the **En**–**Di** join does not significantly modify the discontinuous-reaction relation between forsterite and liquid in the ternary system. For a diagram of a portion of the one-atmosphere **Di**-rich side of this pyroxene solvus, see Figure 5.7b; at these subsolidus temperatures, solid solution on the **En** limb is much less extensive.

QUATERNARY AND QUINARY DIAGRAMS

Most magmas of mafic and intermediate composition contain too many components to allow reasonably complete phase-equilibrium representation by the ternary (and quaternary) liquidus diagrams presented in the preceding sections. However, the so-called granite system is adequately defined by no more than five components, $KAlSi_3O_8$, $NaAlSi_3O_8$, $CaAl_2Si_2O_8$, SiO_2, and H_2O. As in Figure 4.15, we have used simple structural formulae for the ternary feldspars to preserve as closely as possible the atomic modal proportions, even though the figures will be given in terms of per cents by weight.

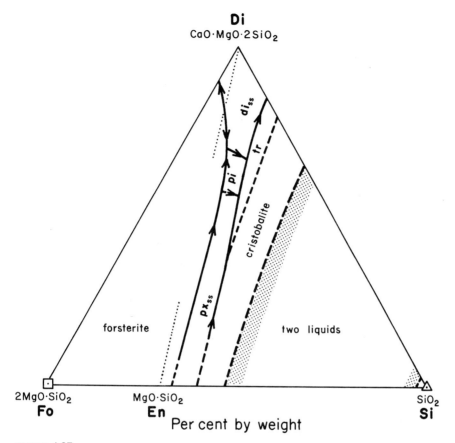

FIGURE 4.37.

One-atmosphere isobaric liquidus diagram for the ternary system 2MgO·SiO₂–
CaO·MgO·2SiO₂–SiO₂, after Bowen (1914) and Kushiro (1972a). The crystalline phase in
equilibrium with melt at liquidus temperatures is also indicated. The approximate maximum
extent of solid solution for pyroxenes lying along the pseudobinary join MgO·SiO₂–
CaO·MgO·2SiO₂ is also shown (Boyd and Schairer, 1964; Kushiro, 1972a). Abbreviations
are: di_{ss} = diopside solid solution; **pi** = pigeonite; px_{ss} = protoenstatite solid solution;
tr = tridymite.

Several important relationships in portions of the granite system have already been presented in Figures 4.12, 4.16, and 4.20, as well as discussed in the
text dealing with binary equilibria. At low pressures it is clear that K-feldspar
will melt incongruently to leucite + silica-rich liquid. However, the presence
of H₂O suppresses the discontinuous-reaction behavior of K-rich feldspar,
so that, above 2.6 kb P_{fluid}, even pure **Or** melts congruently. The alkali feldspar join is characterized by complete solid solution and an internal minimum either "dry" or at low H₂O pressures. In this system, the temperature of

the solidus and liquidus are decreased by elevated aqueous fluid pressures. A solvus occurs on this **Or–Ab** binary join. Increased total pressure ("dry or wet") causes a slight increase in the temperature of the solvus limbs; hence at P_{fluid} values approaching 5 kb, the solvus and solidus intersect along the pseudo-binary join **Or–Ab** (–H_2O).

However, before plunging into the more complicated H_2O-bearing systems, let us begin by studying the "dry" granite-nepheline syenite system. Ternary liquidus relations are shown for the system $KAlSiO_4$–$NaAlSiO_4$–SiO_2 at one atmosphere total pressure in Figure 4.38. Under such conditions, as in the **Or–Ab** binary, alkali feldspars exhibit a complete solid-solution series at near-solidus temperatures. Leucite shows considerable (synthetic) substitution of Na for K, and there is also extensive (although incomplete) solid solution along the $NaAlSiO_4$–$KAlSiO_4$ join. Two polymorphic transitions are also illustrated in Figure 4.38, tridymite–cristobalite and nepheline$_{ss}$–carnegieite$_{ss}$. Since the latter is complicated by solid solution, nepheline$_{ss}$ and carnegieite$_{ss}$ possessing differing Na/K ratios exist within a small but finite temperature interval. The extent of solid solution for all the subsilicic solid phases is imperfectly known. Much as for the **Or–Ab** solid solutions, the approximate maximum compositional ranges are represented as dotted lines in Figure 4.38. A further feature of this system at one atmosphere total pressure is that there are two ternary minima, *m* and *n*, in the quartz-normative and nepheline-normative subsystems, respectively. Silica-rich melts therefore fractionate toward *m*, the granite minimum, whereas subsilicic magmas differentiate toward *n*, the nepheline syenite minimum. Furthermore, these minima occur at the relatively low temperatures of about 990°C for *m* and approximately 1,010°C for *n*. The ternary reaction point *r* is an invariant point at 1,020°C, where leucite + melt *r* react to form alkali feldspar solid solution + nepheline solid solution.

Isothermal sections at 1,300 and 1,050°C for the one-atmosphere system **Ne–Ks–Si** are presented in Figures 4.39a and b, respectively. Relationships are rather straightforward in the extensively studied granite subsystem **Ab–Or–Si**. In contrast, the nepheline-normative subsystem is much more complicated, both because many more phases are present, and because of the uncertainty in the compositions of the various solid solutions; furthermore, much less experimental work has been done on this subsystem.

At 1,050°C in the silica-excess subsystem, cotectic melt *9* represents the leading apex of a three-phase triangle, tridymite + feldspar *i* + melt *9*; with falling temperature, melt moves to the right toward the minimum composition *m* (see Figure 4.38), and the coexisting alkali feldspar becomes more potassic. At temperatures slightly below 1,050°C, another three-phase triangle opens up away from the $KAlSi_3O_8$–SiO_2 side line, and this cotectic melt moves to the left toward *m*. At approximately 960°C, these two three-phase triangles collapse to a binary line along which tridymite, liquid *m*, and alkali feldspar are colinear in chemographic space, as is required by the melting behavior of the minimum composition.

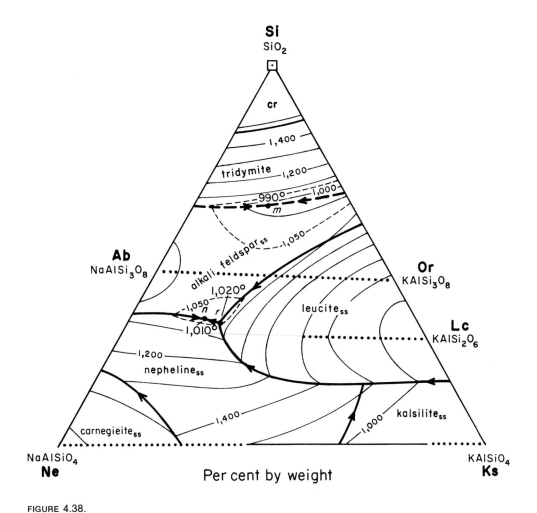

FIGURE 4.38.

One-atmosphere isobaric phase diagram for the ternary system KAlSiO₄–NaAlSiO₄–SiO₂ after Schairer (1957, Figure 29) and Fudali (1963). Isotherms on the liquidus are shown for 100°C intervals; extra detail is provided by the 1,050°C isotherm (dashed). The solid phase in equilibrium with melt is also indicated. Approximate maximum compositional ranges of crystalline solid solutions (alkali feldspars, leucites, nephelines, and kalsilites) are shown schematically by dotted lines. Abbreviations: **cr** = cristobalite; ss = solid solution.

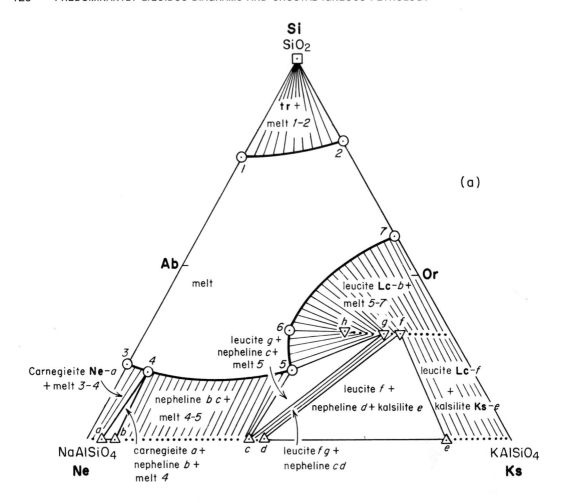

FIGURE 4.39.

One-atmosphere isobaric, isothermal phase diagrams for the ternary system **Ks–Ne–Si**, from Figure 4.38. Isothermal sections are at: (a) 1,300°C; and (b) 1,050°C. Approximate maximum compositional ranges of crystalline solid solutions are indicated schematically by dotted lines. Abbreviations: **tr** = tridymite.

The same general sort of geometric phase relations pertain to the nepheline-normative subsystem. One of the three-phase triangles, feldspar *j* + nepheline *p* + melt *12*, is seen in the 1,050°C isotherm of Figure 4.39b; with decreasing temperature, the compositions of all the phases become more potassic as the leading apex of the three-phase triangle moves toward minimum *n*. At temperatures just below 1,020°C, another melt-bearing three-phase triangle forms by the merging of two higher-temperature three-phase triangles (not shown in Figure 4.39) because of the complete reaction of leucite with melt; this new three-phase triangle is defined by the assemblage K-rich feldspar$_{ss}$ + K-rich nepheline$_{ss}$ + liquid, and it migrates toward the left as heat is sub-

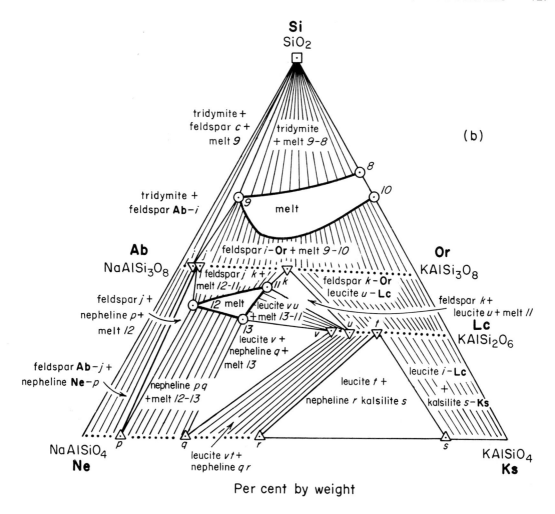

Si
SiO₂

(b)

tridymite +
feldspar *c* +
melt *9*

tridymite
+ melt *9–8*

8

10

tridymite +
feldspar **Ab**–*i*

9   melt

Ab
NaAlSi₃O₈

feldspar *i*–**Or** + melt *9–10*

i

Or
KAlSi₃O₈

feldspar *j* *k* +
melt *12–11*

k

feldspar *k*–**Or**
leucite *u*–**Lc**

feldspar *k* +
leucite *u* + melt *11*

Lc
KAlSi₂O₆

feldspar *j* +
nepheline *p* +
melt *12*

12 melt

leucite *v u*
+ melt *13–11*

feldspar **Ab**–*j* +
nepheline **Ne**–*p*

13

leucite *v* +
nepheline *q* +
melt *13*

v

u

t

leucite *i*–**Lc**
+
kalsilite *s*–**Ks**

nepheline *pq*
+melt *12–13*

leucite *t* +
nepheline *r* kalsilite *s*

NaAlSiO₄
Ne

p

q

r

leucite *vt* +
nepheline *qr*

s

KAlSiO₄
Ks

Per cent by weight

tracted. The two three-phase triangles degenerate to a single line along which alkali feldspar solid solution + liquid *n* + nepheline solid solution are colinear at the minimum temperature of about 1,010°C. Here there is no constraint that Na/K ratios in these three phases must be the same.

The dramatic decrease in liquidus and solidus temperatures at elevated aqueous fluid pressures in the granite system results from the fact that H₂O is highly soluble in the liquid, which fact also means that there is a large decrease in volume—especially at moderately low total pressures—accompanying the entropy increase on melting for the reaction: silica polymorph + alkali feldspar solid solution(s) + fluid → aqueous melt. (For a binary analogue, see Figure 4.5, the system SiO_2–H_2O.) Phase relations at 10 kilobars fluid pressure are illustrated in Figure 4.40. Comparison with the one-atmosphere isobaric liquidus diagram (Figure 4.38) reveals some interesting differences. Although the gross thermal "topography" is preserved at high values of P_{fluid}, liquidus temperatures are greatly lowered; the incongruent melting of K-rich alkali feldspar is eliminated; the pseudoternary eutectic, e, replaces the minimum, m (Figures 4.40 and 4.38, respectively); and there is incomplete binary solid solution between **Ab** and **Or**. These last two phenomena are related. Because of the depression of solidus temperatures and modest elevation of the crest of the solvus,

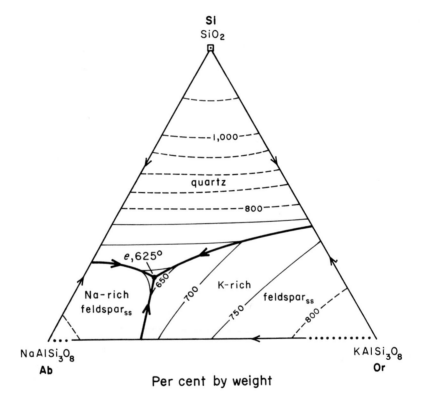

FIGURE 4.40.

Ten-kilobar P_{fluid} isobaric phase diagram for the pseudoternary system $KAlSi_3O_8$–$NaAlSi_3O_8$–$SiO_2(-H_2O)$, after Luth, Jahns, and Tuttle (1964, Figure 3), projected from the H_2O apex of the tetrahedron onto the anhydrous ternary base **Or–Ab–Si**. Isotherms on the liquidus are shown for 50°C intervals. The crystalline phases in equilibrium with the aqueous melt and an H_2O-rich fluid phase are also indicated. The approximate maximum compositional ranges of alkali feldspar solid solutions are represented by the dotted lines. For the treatment of a four-kilobar section of this system in which the condensed assemblages are not saturated with respect to H_2O, see Steiner, Jahns, and Luth (1975).

these two features intersect near the granite minimum composition in chemographic space at and above about 3.6 kilobars aqueous fluid pressure; for this reason, at values of P_{fluid} greater than 3.6 kilobars, two feldspar solid solutions coexist with a silica polymorph and liquid, resulting in eutectic-type subsolvus equilibria rather than minimum-type hypersolvus phase relations. At aqueous fluid pressures slightly exceeding 3.6 kilobars, say, about 4 kilobars, the cotectic curve along which two alkali feldspars are in equilibrium with melt terminates at the eutectic at one end, and at a critical end-point at the other. With more elevated values of P_{fluid}, solvus and solidus intersect over a broader compositional region, so that, at aqueous fluid pressures approaching 5 kilobars, the cotectic curve has intersected the **Ab–Or**(–H_2O) sideline, as shown in Figure 4.20. There is no requirement that the maximum on the solvus and the minimum on the solidus be the initial point of tangency between these two curves. Therefore, at pressures just above 3.6 kilobars, it is possible that the minimum is maintained along the alkali feldspar$_{ss}$ + silica polymorph cotectic curve near the center of the diagram, whereas the intersection of solvus and solidus provides a pseudoternary reaction point on the lefthand side of this curve (Stewart and Roseboom, 1962). This is, however, a comparatively minor point, and has not been investigated experimentally.

The chemical and *P–T* migrations of the lowest-melting compositions in the system $Na_2O \cdot Al_2O_3 \cdot 6SiO_2 - K_2O \cdot Al_2O_3 \cdot 6SiO_2 - SiO_2(-H_2O)$ are illustrated in Figures 4.41a and b, respectively. With increasing aqueous fluid pressure, the chemical composition of the most fusible portion of this system moves toward **Ab**. Concomitantly, the melts carry increasing amounts of dissolved H_2O. The most abrupt decline of the solidus temperature—on the order of 300°C—takes place within the first two kilobars P_{fluid}, with only a moderate decrease of 60–70°C occurring during the next 8 kilobars of pressure increment. Evidently the great compression in the relatively tenuous subsolidus aqueous fluid is chiefly accomplished at low pressures (e.g., see Appendix 5). It is apparent from Figure 4.41b that, for H_2O-saturated magmas, hypersolvus granites containing little or no normative Ca-plagioclase are confined to depths of intrusion shallower than about 13–14 km (= ~3.6 kilobars).

Up to this point we have been considering phase equilibria in the granite system *sensu stricto*, since the compositions of the melts have fallen within the quaternary system **Or–Ab–Si–H_2O**. Nearly all granitic rocks—certainly those which occur in batholithic proportions—contain substantial amounts of an additional constituent, **An**. For those masses which have passed through a molten or partially molten stage, such liquids solidify to "granites" only in the general sense; because plagioclase of compositional range **An**$_{10-45}$ is a major phase, and is often at least as abundant as the coexisting alkali feldspar, these granitic rocks are more appropriately termed monzonites, quartz monzonites, diorites, granodiorites, and quartz diorites (see Figure 4.2). Whatever they are called, it is necessary to include $CaAl_2Si_2O_8$ as an additional component in a discussion of the crystallization history of these igneous-rock types.

130

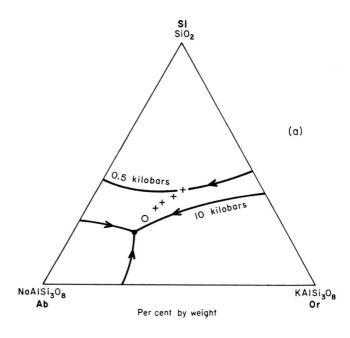

(a)

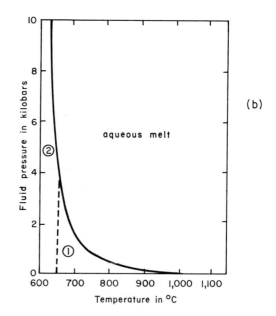

(b)

FIGURE 4.41.

(a) The progressive shift of the lowest-melting composition in the pseudoternary system $KAlSi_3O_8$–$NaAlSi_3O_2(-H_2O)$ as a function of increasing (total) fluid pressure, after Luth, Jahns, and Tuttle (1964, Figure 4), projected from the H_2O apex onto the anhydrous base **Or–Ab–Si**. Field boundaries at 0.5 and 10 kilobars P_{fluid} are shown. Crosses indicate the isobaric minima at 0.5, 1.0, 2.0, and 3.0 kilobars aqueous fluid pressure, whereas circles signify the isobaric eutectics at 5.0 and 10.0 kilobars. (b) Solidus of the pseudoternary system **Or–Ab–Si**$(-H_2O)$ as a function of aqueous fluid pressure, chiefly from Luth, Jahns, and Tuttle (1964, Figure 1). The dashed line represents the P–T migration of the critical point on the alkali feldspar solvus (Yoder, Stewart, and Smith, 1957; Orville, 1963). Fields are: 1 = one alkali feldspar solid-solution-type granite + H_2O; 2 = two alkali feldspar solid-solution-type granite + H_2O.

Unfortunately, there has not been enough experimental investigation of phase relations in the system $KAlSi_3O_8$–$NaAlSi_3O_8$–$CaAl_2Si_2O_8$–SiO_2–H_2O, mainly because of the sluggishness of the various important reactions, to provide detailed data. For instance, polymorphic transitions in the alkali and plagioclase feldspars, and the extent of solid solutions in the ternary feldspars, are imperfectly known. Moreover, liquidus phase relations in the quinary system have been only approximated.

To begin with, let us consider the compositions of the ternary feldspars with which we must deal. As can be seen from Figure 4.42, **Or** and **An** components show only limited miscibility in the analyzed natural phases; hence the plagioclase and alkali feldspar solid solutions may be regarded as essentially binary. The limbs of the ternary solvus extend virtually to the end-member compositions along the join $KAlSi_3O_8$–$CaAl_2Si_2O_8$; this solvus is also transected by the join $KAlSi_3O_8$–$NaAlSi_3O_8$ (see Figures 4.16 and 4.20 at three and five kilobars, respectively). Because of complete high-temperature solid solution in the plagioclase series (we will ignore the problem of plagioclase unmixing at low temperatures such as is implied, for instance, by the peristerite gap), the ternary solvus does not reach the binary join $NaAlSi_3O_8$–$CaAl_2Si_2O_8$. In the ternary

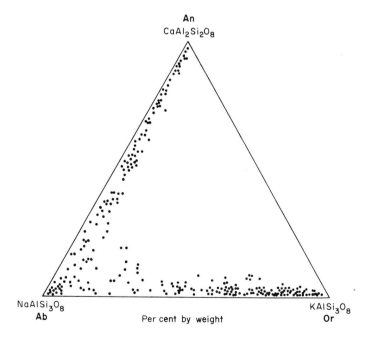

FIGURE 4.42.

Compositions of analyzed natural ternary feldspars, from Deer, Howie, and Zussman (1963, Figure 46).

system **Or–Ab–An**, the solvus evidently is a dome-shaped surface in T–x coordinates; bulk feldspar compositions projecting within this volume are characterized by the coexistence of two more-or-less binary solid solutions, a plagioclase and an alkali feldspar. The solvus dome for the ternary feldspars is illustrated in Figure 4.43. Aside from noting its general shape, which we were able to predict from a consideration of the bounding binary joins, we should pay attention to two additional features.

FIGURE 4.43.
Solvus surface for the pseudoternary system $KAlSi_3O_8$–$NaAlSi_3O_8$–$CaAl_2Si_2O_8(-H_2O)$ in percentages by weight, after Yoder, Stewart, and Smith (1957, Figures 45 and 46). The alkali-feldspar portion of this boundary surface is stippled, whereas the rest of it, which is more or less parallel to the plagioclase side line, is faced away from the reader, hence is obscured. Various isothermal sections are also indicated. The dome is truncated at high temperatures by intersection with the solidus. Diagram (a) is an isobaric section at two kilobars P_{fluid}; the dashed curve on the solvus surface is the trace of the critical composition (i.e., the isothermal isobaric point where two coexisting feldspars become identical). Feldspars a and b coexist with aqueous melt at temperature T_1. Diagram (b) is an isobaric section at five kilobars P_{fluid}. Feldspars c and d coexist with aqueous melt at temperature T_2 ($<T_1$).

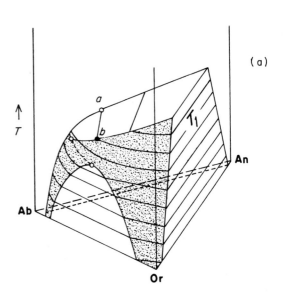

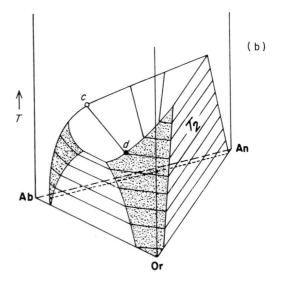

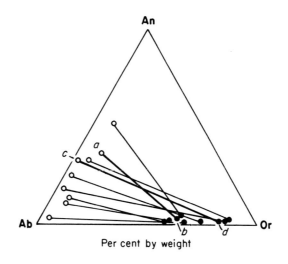

FIGURE 4.44.

Chemically analyzed natural pairs of ternary feldspars, after Yoder, Stewart, and Smith (1957, Figure 47). Coexisting a and b correspond roughly to those phases stable with H_2O-bearing melt at temperature T_1 in Figure 4.43a; coexisting feldspars c and d are in equilibrium with aqueous melt at the lower value of T_2 in Figure 4.43b.

First, tie lines linking coexisting feldspars at high temperatures and low aqueous fluid pressures, such as feldspars a and b in Figure 4.43a, show a smaller ratio of $Ab_{plagioclase}/Ab_{alkali feldspar}$ than at low temperatures and high values of P_{fluid}, where the component $NaAlSi_3O_8$ is strongly concentrated in the plagioclase c relative to the K-rich feldspar d (Figure 4.43b). This difference results from the fact that the partitioning of components among the coexisting phases is strongly a function of temperature, with disparate fractionation of Ab most extreme at low temperatures. That this holds true for analyzed natural pairs of ternary feldspars may be seen from Figure 4.44.

Second, the limbs of the solvus rise to slightly higher temperatures as pressure increases because of the excess volume of mixing for ternary feldspars of intermediate compositions. This phenomenon has already been described for the alkali feldspar binary solid solutions. As a consequence, with elevated total pressure, the extent of solid solution is diminished at constant temperature. Limits of solid solubility for the ternary system as a function of aqueous fluid pressure are depicted schematically in Figure 4.45. Relations are compatible with the observation that analyzed natural pairs of high-temperature volcanic feldspars exhibit considerably more extensive solid solution than those of hypabyssal and mesozonal intrusive igneous rocks, the feldspars of which nevertheless show more solid solution than the even lower-temperature coexisting alkali feldspar + plagioclase from pegmatites.

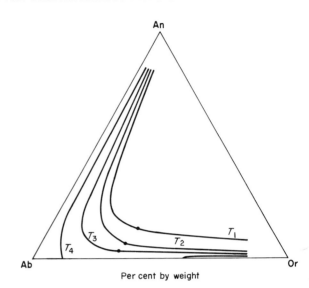

FIGURE 4.45.

Diagrammatic representation of feldspar solid solutions in equilibrium with aqueous melt in the pseudoternary system KAlSi$_3$O$_8$–NaAlSi$_3$O$_8$–CaAl$_2$Si$_2$O$_8$ (–H$_2$O) after Tuttle and Bowen (1958, Figure 66). T_1 indicates the highest temperature range presented (i.e., at relatively low values of P_{fluid}), whereas T_4 stands for the lowest temperature range. The location of the critical point on the solvus is also shown by a dot for each temperature range.

Schematic isobaric phase diagrams for the system **Or–Ab–An**–H$_2$O at 2 and 5 kilobars P_{fluid} are presented in Figures 4.46a and b, respectively. Laboratory data are available for the 5-kb section, but are lacking for the other. Since the solvus and solidus on the **Or–Ab**(–H$_2$O) bounding side line intersect at aqueous fluid pressures slightly less than 5 kilobars, the cotectic curve which begins at the eutectic f along the KAlSi$_3$O$_8$–CaAl$_2$Si$_2$O$_8$(–H$_2$O) pseudobinary side line passes through **Or**-rich portions of the pseudoternary system and terminates at eutectic e along the KAlSi$_3$O$_8$–NaAlSi$_3$O$_8$(–H$_2$O) pseudobinary side line, as shown in Figure 4.46b (see also Figure 4.20). At 2 kilobars P_{fluid}, this cotectic curve, which starts at eutectic d, terminates at a critical end point, cep, near the **Or–Ab**(–H$_2$O) side line where the relationship is that of a minimum, m, as indicated in Figure 4.46a (see also Figure 4.16). As in the **Or–Ab–Si**(–H$_2$O) system, there is no thermodynamic requirement for the crest of the solvus and the minimum on the solidus to provide the first point of tangency, although from geometric relationships it is obvious that the two curved surfaces initially must touch close to these loci (see Stewart and Roseboom, 1962).

Let us trace the 2-kb P_{fluid} equilibrium crystallization path on cooling of some arbitrary, initially molten bulk composition x in the system KAlSi$_3$O$_8$–NaAlSi$_3$O$_8$–CaAl$_2$Si$_2$O$_8$(H$_2$O). Phase relations are illustrated in Figure 4.47. On subtraction of heat, the temperature of the melt will decline until the plagioclase saturation surface is reached. At this stage, the first crystallites of very calcic plagioclase$_{\text{ss}}$ i appear; this feldspar is not saturated with respect to **Or**, because the coexisting liquid is similarly undersaturated. On further cooling, the melt migrates in chemographic space directly away from the crystals being removed; because these crystals continuously react with the liquid, the melt follows a curved path (x-1) to the cotectic curve (d-cep). On

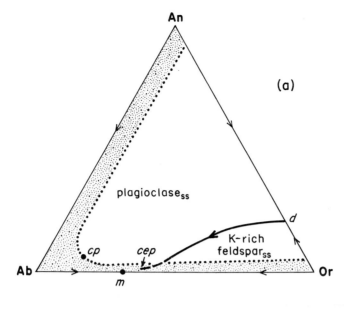

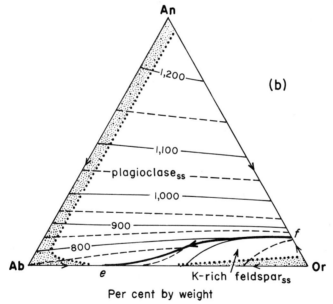

Per cent by weight

FIGURE 4.46.

Isobaric liquidus diagrams for the quaternary system $KAlSi_3O_8$–$NaAlSi_3O_8$–$CaAl_2Si_2O_8$–H_2O projected from the H_2O apex of the tetrahedron onto the anhydrous ternary base **Or–Ab–An** at: (a) 2,000 bars P_{fluid}, after Tuttle and Bowen (1958, Figure 64); and (b) 5,000 bars P_{fluid}, after Yoder, Stewart, and Smith (1957, Figure 41). The very small nonquaternary field for leucite + melt is ignored in (a). Isotherms are presented for 50°C intervals in (b). The solid phase in equilibrium with liquid is indicated; maximum solid solution limits are shown diagrammatically by the stippled pattern. Abbreviations are: cep = critical end point on the two feldspar + melt boundary curve; cp = critical point on the ternary feldspar solvus; ss = solid solution.

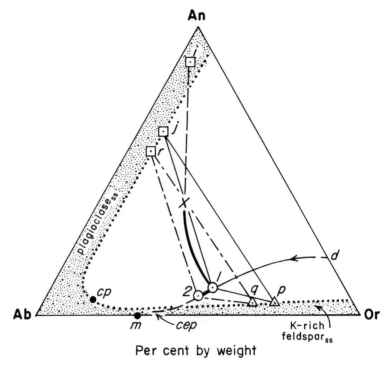

Per cent by weight

FIGURE 4.47.

Isobaric phase diagram for the quaternary system $KAlSi_3O_8$–$NaAlSi_3O_8$–$CaAl_2Si_2O_8$–H_2O projected onto the anhydrous base **Or**–**Ab**–**An** at 2,000 bars P_{fluid}. Symbols and abbreviations correspond to those of Figure 4.46a. The equilibrium line of liquid descent is x–1–2. Highest-temperature and lowest-temperature three-phase triangles, assuming equilibrium crystallization, are also shown.

reaching the K-rich feldspar$_{ss}$ saturation surface, melt has achieved the composition 1, and the homogeneous plagioclase solid solution is of composition j; here the first crystals of K-rich feldspar$_{ss}$ p also join the assemblage, defining the highest-temperature three-phase triangle for the chosen bulk composition (and for any other bulk composition lying on the line j-1). With further loss of heat, both plagioclase$_{ss}$ and K-rich alkali feldspar$_{ss}$ continuously react with the melt as they increase in amount and become more albitic; concomitantly the proportion of cotectic melt decreases as it, too, achieves an ever more sodic composition. The liquid is exhausted as it reaches the composition 2, where it is in equilibrium with K-rich alkali feldspar q and plagioclase r (the lowest-temperature three-phase triangle for bulk composition x). As temperature continues to fall in the subsolidus region, the **Ab** component will reproportion itself between the two competing feldspar solid solutions in such a way that it enriches the plagioclase$_{ss}$ in sodium relative to the potassium feldspar solid solution, as discussed previously (see Figure 4.44).

Employing Figure 4.47 again, we now will consider the line of liquid descent when crystal fractionation occurs. During the cooling of an initial bulk composition x, the path followed by melt through the primary crystallization field of plagioclase$_{ss}$ will be less strongly curved than that for the equilibrium situation. A very small amount of liquid will move down the cotectic curve at late stages in the cooling history towards—and perhaps even reaching—the critical end-point, cep. The two types of feldspar solid solution produced from this fractionating melt will be strongly zoned toward sodic rims, and trace amounts could attain identical compositions at the solvus critical point, cp. If any liquid of composition cep still remains at this point, it will move down a shallow thermal trough toward the minimum, m, on the pseudobinary side line **Or–Ab**($-H_2O$); concomitantly the single homogeneous alkali feldspar solid solution crystallizing from this melt leaves the solvus curve and moves towards an **Or/Ab** ratio identical to m.

It should also be pointed out that, for compositions of melt lying along the boundary between the plagioclase$_{ss}$ and K-rich feldspar$_{ss}$ fields, this curve is truly a cotectic curve only as long as its extension is disposed in the **Ab**-rich direction from the tie line linking K-rich feldspar with liquid. Clearly this curve shows cotectic behavior for conditions such as are described by the three-phase triangles adu and bev of Figure 4.48. However, as the composition of the K-rich feldspar$_{ss}$ in equilibrium with both melt and plagioclase$_{ss}$ migrates toward the critical point with decreasing temperature, it eventually achieves a composition lying on the plagioclase side of the intersection (in projection) of the solvus and field boundary curves—for instance, at physical conditions defined by the three-phase triangle cfw in Figure 4.48. In this region, early-formed plagioclase$_{ss}$ reacts with liquid to produce a homogeneous alkali feldspar solid solution; hence near its termination at the critical end-point, the two-feldspar$_{ss}$ boundary curve changes from a cotectic to a reaction curve.

We have just seen that, for fractionation in the ternary feldspar($-H_2O$) system, middle stages of crystallization are characterized by the coexistence of two feldspars; whereas, near the end, early-formed plagioclase may be armored by, or react with, liquid to form a single alkali feldspar$_{ss}$. In contrast, as discussed earlier (e.g., see Figures 4.38 and 4.40), the granite system *sensu stricto* is typified by the crystallization of just one alkali feldspar solid solution throughout the cooling history at low aqueous fluid pressures ($=$ hypersolvus granite) or by the coprecipitation of two alkali feldspars during the middle and late stages of crystallization at high values of P_{fluid} ($=$ subsolvus granite).

The complete quinary system $KAlSi_3O_8$–$NaAlSi_3O_8$–$CaAl_2Si_2O_8$–SiO_2–H_2O may be handled graphically employing the methods used by von Platen (1965) and Winkler (1974). As in the previous treatments of portions of this system (e.g., Figures 4.40, 4.41, 4.46, 4.47, and 4.48), we project from the H_2O apex of the n-component polyhedron onto an anhydrous base. However, because there are now four anhydrous components to consider, we make use of the fact that, at high temperatures, **Ab** $+$ **An** are miscible in all proportions, and examine pseudoternary sections through the pseudoquaternary system **Or–Ab–**

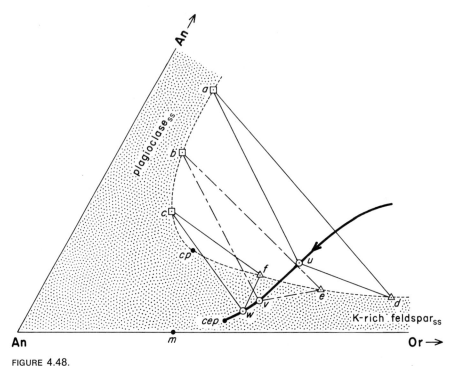

FIGURE 4.48.
Schematic isobaric **Ab**-rich portion of the system $KAlSi_3O_8$–$NaAlSi_3O_8$–$CaAl_2Si_2O_8$–H_2O projected onto the anhydrous base **Or–Ab–An** at 2,000 bars P_{fluid}, not drawn to scale. Abbreviations are the same as in Figure 4.46a. Three-phase triangles approaching the critical condition are illustrated. Where K-rich feldspar solid solution lies on the Na-rich side of the field boundary, the liquid in equilibrium with two feldspars is actually reacting with plagioclase; hence for such liquids the curve is a reaction curve rather than a cotectic.

An–Si(–H₂O) at fixed **Ab/An** ratios. For instance, the 2,000-bar isobaric P_{fluid} section shown as Figure 4.49 depicts primary crystallization fields for the system in which the **Ab/An** ratio is 3.8 (the normative plagioclase$_{ss}$ would be approximately **An₂₁**). For comparison, the same isobaric section is shown in projection for the pseudoternary system **Or–Ab–Si(–H₂O)**, in which the **Ab/An** ratio is infinite; for this system a single homogeneous alkali feldspar solid solution is stable, and a pseudoternary minimum, *m*, occurs at about 680°C. Obviously the presence of moderate amounts of $CaAl_2Si_2O_8$ component result in subsolvus granite-type behavior and the occurrence of an internal eutectic at *e*; in addition, the composition of the most fusible portion of the system shifts away from the **Ab** (+ **An**) apex, and the lowest temperature at which melting begins rises slightly, to 695°C. A difficulty with Figure 4.49 is that the compositions of the ternary feldspars cannot be plotted quantitatively on the anhydrous base, because it is only pseudoternary—nor can the composi-

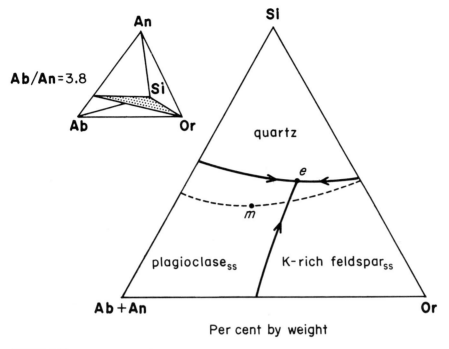

FIGURE 4.49.

Isobaric phase diagram at 2,000 bars P_{fluid} for the quinary system $KAlSi_3O_8$–$NaAlSi_3O_8$–$CaAl_2Si_2O_8$–SiO_2–H_2O projected from H_2O onto the anhydrous base Or–(Ab + An)–Si, after von Platen (1965, Figure 6); the value chosen for **Ab/An** is 3.8 (or about 21 per cent **An** in the normative plagioclase). For comparison, the field boundary at 2,000 bars P_{fluid} in the quaternary system **Or–Ab–Si–H₂O** is also illustrated, projected onto the anhydrous base **Or–Ab–Si**, after Tuttle and Bowen (1958, Figure 24). Abbreviations are: e = eutectic at 695°C; m = minimum at about 680°C; ss = solid solution.

tions of the melts with which they are in equilibrium be shown exactly. Nevertheless, this diagram does indicate clearly the influence of the **An** component on the disposition and thermal structure of the primary crystalline phase fields.

In passing, it should be remarked that, for the granite system in the broad sense, the effects of increasing aqueous fluid pressure and increasing **An** content somewhat oppose one another. At elevated values of P_{fluid}, the composition of the most fusible material migrates toward the **Ab** corner of the polyhedron and the minimum melting temperature declines, as shown in Figure 4.41; in contrast, as we have just seen in Figure 4.49, increased **An** content causes a small rise in the solidus temperature and an appreciable displacement of the composition of the most fusible mixture away from the **Ab** (+ **An**) apex. However, both increased P_{fluid} and increased **An** content result in the suppression of supersolvus, one-feldspar-type, crystal-melt equilibria and in the generation of a cotectic curve.

PETROGENESIS OF IGNEOUS ROCKS

Previous sections of this chapter have dealt with experimentally established crystal-melt equilibria in systems of two to five components. The latter are rather simplified analogues, at best, of natural magmas; except for the quinary **Or–Ab–An–Si–**H_2O, none closely approximate the chemical range of igneous-rock compositions which are thought to represent once largely molten silicate solutions. Nevertheless, we have seen that the phase-equilibrium behavior displayed in unary and binary systems carries over into the more complicated systems of three to five components. For this reason, we conclude that the phase relations as discovered in the laboratory for relatively simple synthetic systems may be applied confidently as models for the behavior of complex natural magmas—always bearing in mind, of course, the more or less predictable effect resulting from the addition of other components. Where refractory constituents such as CaO, TiO_2, MnO, and SiO_2 are involved, the multicomponent liquidus and solidus $T–x$ surfaces in general are lowered somewhat, except when crystalline solid solutions incorporate these components preferentially. Where less refractory, more volatile components such as ferrous iron, alkalis, and H_2O must be considered, a drastic lowering of the melting range is characteristic.

It is evident from the liquidus diagrams introduced in this chapter that crystal fractionation is a powerful and persuasive mechanism with which to explain magmatic differentiation. The course of partial melting and fractional crystallization is toward residual liquids of low melting temperature which are enriched in alkalis, ferrous iron (relative to MgO), silica, and H_2O. This course of differentiation is abundantly documented by chemical analyses of cogenetic lava series and epizonal plutons from various parts of the world. Several examples are illustrated in Figure 4.50. Although all intergradations exist, two principal series may be distinguished: (a) the oceanic tholeiite or Skaergaard trend (e.g., see Wager and Deer, 1939); and (b) the calc-alkaline or orogenic trend (e.g., see Daly, 1933). The former exhibits extreme iron enrichment (i.e., increment in the FeO/MgO ratio), typically contains pigeonite, and characterizes the differentiation of midoceanic ridge lavas, abyssal tholeiites, and fissure basalts of the continents. In contrast, the latter normally carries hypersthene and typifies the magma sequences of orogenic belts developed within or at the margins of continental crust, and shows pronounced alkali + silica enrichment and virtually constant FeO/MgO. Because of the differences in fractionation trends as seen from Figure 4.50, tholeiitic ferromagnesian mineral series exhibit a marked zonal development with Fe^{2+}/Mg enrichment in the more felsic and iron-rich differentiates, whereas individual mafic phases of calc-alkaline suites generally show more nearly constant phase compositions.

Whether K_2O + Na_2O + SiO_2 enrichment or iron enrichment dominates an igneous series of rocks (e.g., see Bowen, 1928, Chapter 7; Wager and Deer,

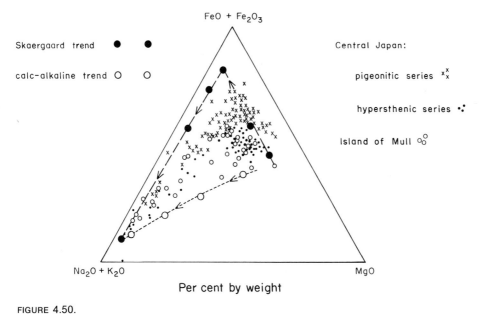

FeO + Fe₂O₃

Skaergaard trend ● ●

calc-alkaline trend O O

Central Japan:

pigeonitic series

hypersthenic series

Island of Mull

Na₂O + K₂O   MgO

Per cent by weight

FIGURE 4.50.

Proportions of magnesia, total iron oxides, and alkalis from lavas and fine-grained hypabyssal intrusions, including the Skaergaard, east Greenland (Wager and Deer, 1939), pigeonitic (≈ tholeiitic), and hypersthenic (≈ calc-alkaline) lavas from the Izu-Hakone area near Tokyo (Kuno, 1968), igneous rocks from the Island of Mull, Scotland (Bailey and others, 1924), and an idealized calc-alkaline lava trend (Daly, 1933). The transitional nature of melts between the oceanic tholeiitic and orogenic calc-alkaline differentiation trends is obvious. Arrows indicate the contrasting lines of liquid descent (i.e., differentiation on cooling of magmas).

1939) is very much a function of the relative oxygen fugacity range or oxidation state (Muan and Osborn, 1956). As is clear from an examination of Figure 4.36, liquids fractionate toward iron-rich, silica-depleted compositions at very low oxygen fugacities (Figure 4.36f), whereas high oxidation states promote liquid descent toward SiO_2-rich melts of intermediate or nearly constant FeO/MgO ratio (Figure 4.36b). The reason for this latter situation may be found in the stability of magnesioferrite solid solutions, $Fe^{3+}_2(Mg,Fe^{2+})O_4$, on the liquidus surface under oxidizing conditions; this early, high-temperature precipitation of a ferric oxide-bearing phase causes impoverishment of the melt in iron concomitant with silica enrichment.

Osborn (1959, 1962) used this relationship to argue that nonorogenic mafic magmas, displaying the iron-enrichment trend, evidently rise to upper levels of the Earth's crust without significant modification, thus reflecting the relatively low oxygen fugacities of the source region. In contrast, calc-alkaline melts, which show the trend toward silica + alkali enrichment, were thought to represent primary liquids which have been contaminated by crustal material, particularly H_2O; the molecular dissociation of the aqueous fluid would provide high values of f_{O_2} (especially if hydrogen escapes), thereby accounting for the nature of the differentiation. Although the process certainly is a viable one, calc-alkaline igneous terranes do not seem to exhibit the great volumes of oxidized early crystal cumulates which would be required to corroborate this genetic hypothesis; this relationship is also a strong argument against the proposed generation of abundant andesitic suites by the fractional crystallization of parental basaltic melts.

Since we will seek the ultimate origin of basaltic and andesitic magmas by a process of fractional fusion of the upper mantle—as seems to be required by their chemical and isotopic compositions (e.g., Sr^{87}/Sr^{86} ratios)—discussion of the generation of these melts will be deferred to Chapter 5. Suffice it to say here that, once produced, the course of both plutonic and near-surface differentiation is dictated by the kinds of processes and phase-equilibrium behaviors described in the present chapter.

The oceanic crust consists predominantly of tholeiitic basalt (Engel and Engel, 1964; Engel *et al.*, 1965) with minor amounts of alkali olivine basalt and partially serpentinized peridotites, whereas the aggregate bulk composition of the sialic crust appears to be andesite or granodiorite (Poldervaart, 1955; Mason, 1966, Chapter 3; Ronov and Yaroshevsky, 1969). For this reason, the origin and growth of the continents is linked directly with the generation of the calc-alkaline magma series (Ringwood, 1974); more on this subject will be presented in the next chapter. Chemical diversity of the continents seems to be provided in part by igneous fractionation and to a minor degree by metamorphic differentiation, but also to an important extent by both chemical and mechanical sedimentary processes (e.g., see Barth, 1961). The compositional reorganization of continental-type material therefore is complex and certainly multicycle. Consider the history of pristine andesite material subsequent to its addition to the sialic crust: (1) multistage erosion and chemical fractionation during weathering, transportation, and sedimentation; (2) metasomatism during metamorphism; and (3) partial, or nearly complete, melting. Stages (1), (2), and (3) may occur in variable order and frequency. Stage (1) is not a subject of this book however, and stage (2) will be discussed only briefly in Chapter 6. What we are concerned with here is (3), partial fusion (anatexis) of preexisting rocks under crustal conditions.

That melting can occur for silicic, alkalic bulk compositions has been demonstrated in Figures 4.40, 4.41, 4.46, and 4.49 (see especially Figure 4.41b), provided that an aqueous fluid is present. Since temperatures near the base of thickened portions of the continental crust are thought to lie in the range 500–700°C at depths approaching 40 km (Clark and Ringwood, 1964), the partial fusion of juicy portions of the sial is to be expected. This anatectic process may occur as equilibrium partial melting or as fractional fusion (see Presnall and Bateman, 1973, for a thorough discussion of both phenomena). The diapiric rise of such buoyant masses into middle and upper levels of the crust would provide a reasonable mechanism for devolatilizing the residual, basal sections of the continental crust, which would become more mafic and refractory as a consequence of the removal of the low melting-temperature aqueous silicate liquid. Migmatite terranes appear to be the product of partial fusion prior to phase separation and upward migration of the plastic, partly molten granitic magma; in contrast, subjacent granitic batholiths and mafic granulites + amphibolites of high metamorphic grade represent the lithologic assemblages after such a phase separation.

CHAPTER 5

Mantle Petrology

COMPOSITION OF THE MANTLE

Both the mineralogy of the mantle and its bulk rock chemistry remain subjects of speculation today, for the very good reason that few unambiguous or unmodified samples of mantle material have been recognized to have arrived at the Earth's surface. Of all the lithologies now available for study, only the ultramafic rocks seem to occur widely in a geologic context that suggests possible derivation from the mantle. That the over-all composition of the mantle is peridotitic seems assured, because of the relatively unique ability of such lithologies to provide the appropriate radiogenic heat production, which, coupled with crustal generation, provides the observed terrestrial heat flow (~ 1.0 to 2.0 μcal/cm^2/sec), the measured seismic-transmission velocities ($V_P = 7.8$ to 8.2 km/sec just beneath the M Discontinuity), the computed bulk densities ($\rho = 3.3$ to 5.5, grading from the top to the bottom of the mantle), and probably also the measured electric and magnetic properties. An analogy may also be drawn between the composition of the mantle and that of stony meteorites—the latter being assumed to represent chemically typical samples of a fragmented planetary body which, in an earlier, more coherent stage, underwent a somewhat Earth-like differentiation. Pyrolite is a term coined by Ringwood for a model upper-mantle composition (e.g., see Ringwood, 1966). In many scenarios for the evolution of the Earth, a relatively cold accretion of planetary material is postulated. The subsequent decay of radioactive isotopes (especially the relatively short-lived ones) would cause a rapid initial heating of the Earth and at least partial fusion. Pyrolite would represent the silicate slag formed during this very early stage of metal liquifaction and during subsequent accumulation of the iron-nickel core by infall, but before chemical fractionation had produced crustal rocks + differentiated (i.e., depleted) mantle material.

The degree and scale of chemical and mineralogic homogeneity of the mantle are also conjectural subjects. Lateral heterogeneities in bulk rock composition, from peridotitic to basaltic, undoubtedly exist locally, but judging from the relative chemical and isotopic uniformity of the mafic terrestrial lavas which have been extruded over the past 3.5 billion years, we would expect such horizontal gradations to be modest, at least for mantle phase compositions. Provided that such liquids represent the most fusible fraction of relatively refractory source peridotite, then, their uniformity indicates that the mineral compositions and the phases present in the protolith are essentially fixed, or at least not widely variable. However, according to principles discussed in Chapter 4, the relative proportions of the phases could vary significantly from area to area without appreciably affecting the composition of the minimum melt. Systematic variations in bulk composition of the mantle with depth seem plausible, considering the over-all density stratification of the Earth. The process of subduction might also be anticipated to produce both vertical and horizontal bulk chemical heterogeneities in the upper 600–700 kilometers of the mantle. Moreover, even assuming a mantle of constant composition, we would expect mineralogic transitions, from the relatively low $P-T$ mineral assemblages of the upper mantle to high-density phase compatibilities stable at much greater depths. Such transformations also seem to be required by the geophysical data. In summary, we may conclude that lateral inhomogeneities, resulting from variations in proportions of phases, but not including important changes in mineralogic composition, may occur in the mantle; furthermore, both chemical and mineralogic variations within the mantle are expected as a function of depth.

Estimated values for the chemical constitution of the upper mantle are presented in Table 5.1. Also included are chemical analyses of Archaean ultramafic lavas (komatiites) from South Africa and Australia, and of the average chondritic meteorite. Differences in composition between the peridotite inclusions and ultramafic melts on the one hand, and stony meteorites on the other, as seen in Table 5.1, probably reflect the more complete phase separation of the Earth's mantle and core during infall of the latter (i.e., migration of molten metal droplets down the gravity gradient to accumulate near the Earth's center) compared to that of the parental meteorite body. Moreover, terrestrial materials seem to have been subjected to a more subtle chemical differentiation subsequent to core formation, whereas most meteorites appear to represent an early, arrested stage of fractionation. However, all analyses of terrestrial rocks listed in the table show sufficient CaO so that Ca-rich clinopyroxene

TABLE 5.1.
Estimated compositions for the upper mantle compared with compositions of Archaean ultramafic extrusives from greenstone belts, largely compiled by Green (1972).

Component	1	2	3	4	5	6	7	8
SiO$_2$	45.16	45.0	43.8	45.1	42.86	46.63	44.80	38.04
TiO$_2$	0.71	0.07	0.02	0.5	0.33	0.34	0.23	0.11
Al$_2$O$_3$	3.54	3.01	1.45	4.1	6.99	3.02	5.26	2.50
Cr$_2$O$_3$	0.43	0.41	0.45	0.3	0.18	—	—	0.36
Fe$_2$O$_3$	0.46	1.28	1.61	2.0	0.36	1.00	1.00	—
FeO	8.04	6.70	6.75	7.9	8.97	9.63	9.46	12.45
MnO	0.14	0.11	0.12	0.2	0.14	0.18	0.18	0.25
NiO	0.20	0.25	0.29	0.2	0.20	—	—	—
MgO	37.47	39.7	44.0	36.7	35.07	34.23	34.34	23.84
CaO	3.08	3.15	1.38	2.3	4.37	4.79	4.35	1.95
Na$_2$O	0.57	0.24	0.15	0.6	0.45	0.15	0.35	0.98
K$_2$O	0.13	0.04	0.03	0.02	0.003	0.03	0.03	0.17
P$_2$O$_5$	0.06	—	—	0.1	—	—	—	0.21
Fe								11.76
Ni								1.34
FeS								5.73
Co								0.08

All numbers are percentages.
Column 1. Pyrolite model composition.
Column 2. Mean of analyses of 20 spinel lherzolite xenoliths from Rocher du Lion, Haute-Loire, France.
Column 3. Mean of analyses of 40 spinel lherzolite xenoliths from Dreiser Weiher, Eifel, W. Germany.
Column 4. Estimated upper-mantle composition from which volatiles (H$_2$O, CO$_2$ Cl) have been lost but without any basalt removal by partial fusion.
Column 5. Estimated upper-mantle composition, based on genetic relationship by partial-fusion, partial-crystallization model to account for ultramafic inclusions in basalt.
Column 6. Average peridotitc komatiite Barberton area, South Africa. Recalculated to 100 per cent anhydrous, and with Fe$_2$O$_3$ arbitrarily made 1.00 per cent.
Column 7. Peridotite with quench texture, Mt. Ida, western Australia. Recalculated to 100 per cent anhydrous, and with Fe$_2$O$_3$ arbitrarily made 1.00 per cent.
Column 8. Average of 94 superior analyses of chondritic meteorites (Urey and Craig, 1953).

would be an essential phase, along with the more abundant olivine + ortho-pyroxene, at pressures less than about 100–130 kilobars; such natural assemblages are termed lherzolites, and compare favorably with a model pyrolite composition. As is also apparent from Table 5.1, all rocks are moderately aluminous, hence carry an Al-rich phase in addition.

From the chemical analyses presented in Table 5.1, it seems plausible that magnesian olivine, and its very high-pressure equivalent assemblage, is the dominant constituent of the mantle; hence all subsolidus and fusion relations must involve equilibration with it. At least in the upper portions of the mantle, this means that all phases must be saturated with respect to $(Mg,Fe)_2SiO_4$, a fact which must be kept in mind when we consider mechanisms for the generation of basalts and lavas of the calc-alkaline suite.

More or less direct evidence bearing on the bulk chemistry and mineralogic constitution of the upper mantle may be obtained from a study of ultramafic inclusions brought to the surface in lavas and diatremes. These xenoliths occur chiefly in alkali olivine basalts and kimberlites (Boyd, 1973; Boyd and Nixon, 1973; MacGregor, 1974). In general, alkali olivine basalts are thought to have been derived from a parental pyrolite by partial melting at depths approximating or in excess of 100 kms (see pp. 161–162 on partial fusion, and pp. 170–171 on crust-mantle differentiation and plate tectonics). Xenoliths contained in them, presumably representing samples of conduit material ripped from the wall rocks during ascent of the magma, include olivine nodules, gabbros, pyroxenites, feldspathic peridotites, spinel lherzolites, and amphibole-bearing peridotites. In contrast, kimberlite diatremes—more or less solid mantle material fluidized by a $CO_2 + H_2O$ volatile phase—appear to have come from even greater depths, on the order of 200 kms. Such pipes contain fragments of eclogite and a variety of ultramafic inclusions, such as garnet and spinel lherzolites, the rarer phlogopite, and diamond-bearing peridotites. (For a phase diagram for the carbon polymorphs, see Figure 6.5.) Thus it would appear that contrasting phase assemblages have developed in solid mantle materials both of basaltic and of lherzolitic bulk compositions, and that these mineralogic differences are a function of depth. As we will see later on, such observations are compatible with experimental phase-equilibrium studies on model mantle (and crustal) materials at high temperatures and pressures.

In general, we can recognize two mineralogically distinct varieties of the uppermost mantle, depleted and undepleted. A somewhat similar classification may be based on trace elements, but will not be elaborated on here. Undepleted mantle material more or less corresponds to chondritic meteorites and pyrolite in composition, and carries essential amounts of Ca-pyroxene and an aluminous phase, such as plagioclase, spinel, or garnet, in addition to olivine and orthopyroxene (= lherzolite). In contrast, depleted mantle material consists largely of olivine ± hypersthene (= dunite ± harzburgite). As will be discussed farther on, small amounts of lherzolite fusion produce basaltic melt in the uppermost mantle; if the liquid is subsequently drained off, the residual olivine-rich solid assemblage is depleted in composition.

GROSS STRUCTURE OF THE UPPER MANTLE

The mantle is divisible into a series of nested shells or zones, which are distinguished by having relatively different propagation velocities of seismic waves.

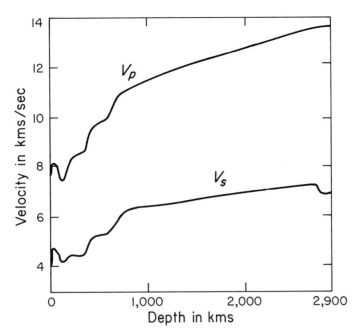

FIGURE 5.1.
Compressional (*P*) and shear (*S*) wave velocity distribution in the
mantle (Anderson and Kovach, 1969, Figure 1).

A velocity versus depth distribution is presented in Figure 5.1; the fine structure of the upper 50 ± 25 km of the mantle cannot be adequately represented
on such a scale, however. From the illustration it can be seen that, although
transmission speed in general increases with depth in the Earth, there is a pronounced low-velocity channel, especially for shear waves, about 100 km down;
it is shallower under the ocean basins and appears to deepen and become less
distinct under ancient, stable cratons. This seismic low-velocity region corresponds to the top of the asthenosphere, and the attenuation of earthquake
energy reaches a maximum here. Further down, the rate of increase of transmission velocity with depth changes at several places within the depth interval
400–750 km, the so-called transition zone which marks the base of the upper
mantle.

Although other explanations have been advanced, the seismic low-velocity,
high-attentuation zone seems to result from the onset of incipient melting in
the mantle, providing on the order of a few per cent of intergranular molten
material. Such an hypothesis readily accounts for the weakness of the asthenosphere, as well as for its seismic properties, in contrast to those of the overlying,
cooler, more rigid lithosphere. This idea is also supported by the contrasting
textures and compositions of xenoliths in kimberlite pipes (Boyd, 1974;
MacGregor and Basu, 1974; MacGregor, 1975). The apparently shallower

spinel- and garnet-bearing peridotites (see section on Subsolidus Reactions, pp. 154–155) carry phlogopite, possess granular, annealed fabrics, and show relatively depleted compositions—indicating that they may be the refractory residua of small degrees of fractional fusion. In contrast, the seemingly deeper garnet lherzolite inclusions are devoid of hydrous phases, display intensely deformed grains, and represent more primitive, undepleted mantle material; if melt were present here, phase separation has not occurred. By inference, the granular lherzolites may be samples of somewhat depleted subcontinental mantle; whereas the sheared peridotite xenoliths probably came from the underlying, compositionally more primitive asthenosphere (the deformation observed in this material may have attended cooling and ascent toward upper levels of the lithosphere).

The bands of discontinuity within the transition zone of the upper mantle shown in Figure 5.1 are thought to signal a two-stage polymorphic change in the crystal structure of $(Mg,Fe)_2SiO_4$ from olivine to spinel-type, at about 400–500 km depth, followed by heterogeneous reactions of the sort

$$\underset{\text{(enstatite)}}{MgSiO_3} = \underset{\text{(periclase)}}{MgO} + \underset{\text{(stishovite)}}{SiO_2},$$

and

$$\underset{\text{(olivine-spinel)}}{Mg_2SiO_4} = \underset{\text{(periclase)}}{2\,MgO} + \underset{\text{(stishovite)}}{SiO_2}$$

about 650–750 km down. No detectable transitions appear to take place at greater depths within the lower mantle, according to the seismic data presented in Figure 5.1. Of course, there may well be other, higher-pressure transformations in the deeper portions of the mantle, but discussion of their possible mineralogic configurations would be quite conjectural, since pertinent geophysical data are lacking at this time.

SUBSOLIDUS REACTIONS

Basaltic Compositions

Because many segments of the Earth's crust are basaltic in composition, and because rocks of this composition are thought by many geophysicists and petrologists to occur at least locally in the upper mantle, high-pressure subsolidus phase-equilibrium experiments on natural basalts have been conducted by a number of investigators. Although the various sets of laboratory data are largely compatible at near-solidus temperatures, there is a divergence of opinion over phase relations that have been extrapolated to lower temperatures. Some of these discrepancies undoubtedly result from the different compositions of the basalts employed as starting materials, as well as from contrasting experimental techniques (e.g., iron loss or gain in the charge, redox reactions, and hydration, as functions of the composition of the enclosing capsule or of the furnace assembly). Other discrepancies may result from the exceedingly difficult problem of demonstrating chemical equilibrium and of establishing the exact nature of the phase assemblage for such compositionally complex systems. However, the biggest single source of discrepancies is probably the great magnitude of the $P-T$ extrapolation from the data presented by the various working groups.

Three phase diagrams for the conversion of essentially dry natural basalt to eclogite are presented, along with the critical run data, in Figures 5.2, 5.3, and 5.4. These studies are compared in Figure 5.5. Also presented in Figure 5.2 is

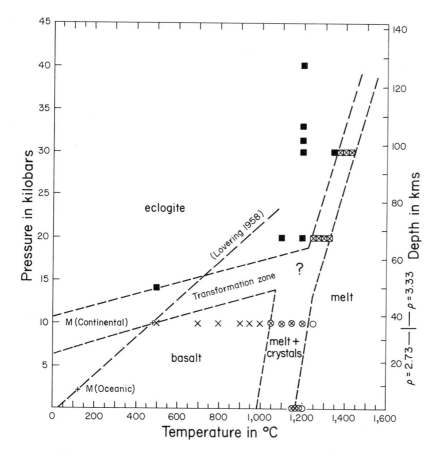

FIGURE 5.2.

Transformation of Glenelg, Scotland, eclogite to basalt, and fusion relations in the absence of H_2O, largely by Yoder and Tilley (1962, Figure 43). Experiments were performed in the piston-cylinder apparatus (20 kilobars and above) and internally heated gas apparatus (10 kilobars). The basalt-eclogite transition and suboceanic and subcontinental M Discontinuities (plus signs) after Lovering (1958) are also illustrated.

FIGURE 5.3.

Phase diagram for two dry quartz tholeiites (*A*) and (*B*), after Ringwood and Green (1966) and Green and Ringwood (1967a, Figure 7). Experiments used piston-cylinder equipment. Quartz tholeiite *B* is richer in silica and alkalis, whereas quartz tholeiite *A* contains more CaO and MgO. The first appearance of garnet in quartz tholeiites with $100 \text{ Mg}/(\text{Mg} + \text{Fe}^{2+}) = 61$ is on the high-pressure side of *AB*. The line *G* marks the first appearance of garnet in quartz tholeiite with $100 \text{ Mg}/(\text{Mg} + \text{Fe}^{2+}) = 10$, and the line *F* marks the first appearance of garnet in quartz tholeiite with $100 \text{ Mg}/(\text{Mg} + \text{Fe}^{2+}) = 90$. Plagioclase is absent on the high pressure side of the line *CD* in quartz tholeiite (*B*) composition, but is absent on the high-pressure side of the line *E* in quartz tholeiite (*A*) composition. The line *BD* is the approximate solidus of the dry quartz tholeiite (*B*) composition. Open symbols indicate runs on quartz tholeiite (*A*) only.

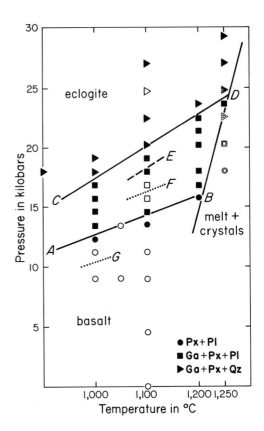

Lovering's (1958) estimation of the basalt-eclogite transition, which assumes that it represents the M Discontinuity in both oceanic and continental areas. At crustal temperatures, the calculated geothermal gradients for the suboceanic and subcontinental shield regions lie between the Yoder and Tilley (1962) and Ringwood and Green (1966) basalt–eclogite transition zones illustrated in the summary diagram of Figure 5.5. As is clear from Figures 5.2, 5.3, and 5.4, experiments on the transition are essentially in accord around the temperature of 1,000°C and above.

Ignoring the melt region for the time being, we can see from the various experimental studies that the relatively low-pressure, high-temperature field for the assemblage labradorite + Ca-clinopyroxene + olivine ± hypersthene (± sphene) is separated from the relatively high-pressure, low-temperature associaton Na-Ca-clinopyroxene + garnet (± rutile ± quartz) by a *P–T* zone that is about 3–5 kilobars or more in width, consisting of intermediate phase compatibilities. With the exceptions of rutile and quartz, all the participating phases show extensive solid solution. Within the transition zone of garnet granulite + plagioclase eclogite, as pressure is elevated at constant tempera-

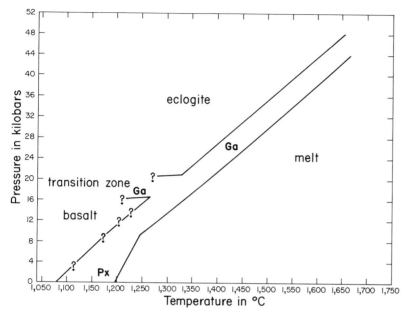

FIGURE 5.4.

Phase relations of anhydrous olivine tholeiite within and adjacent to the basalt and eclogite melting interval, after Cohen, Ito, and Kennedy (1967, Figure 3); as demonstrated by later work (Ito and Kennedy, 1971), small amounts of quartz are present in the eclogite field, and the solidus lies at lower temperatures than shown in the earlier study. Except for one-atmosphere runs, experiments were performed in the piston-cylinder apparatus.

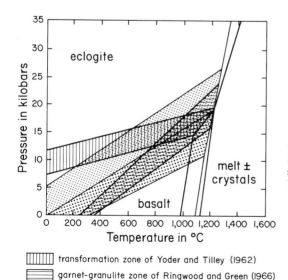

FIGURE 5.5.

Summary of dry phase-equilibrium studies on several different basaltic compositions

	transformation zone of Yoder and Tilley (1962)
	garnet-granulite zone of Ringwood and Green (1966)
	plagioclase-eclogite zone of Ito and Kennedy (1971)
	garnet-granulite zone of Ito and Kennedy (1971)

ture, plagioclase decreases in amount and becomes more sodic, olivine and hypersthene diminish in abundance (for some bulk compositions, one or both eventually disappear), clinopyroxene increases and changes to a more jadeitic composition, and almandine appears and becomes slightly enriched in pyrope component as it increases in amount. The garnet/clinopyroxene ratio also increases with elevated pressure. It is still not clear whether these mineralogic transformations are a continuous function of increasing pressure or take place in a series of relatively abrupt steps, but it is certain that the direction of phase compositional change with isothermal pressure increment is as described.

Since natural basalts must be described in terms of 8 to 10 components, but only 5 ± 2 phases occur in the subsolidus region, several degrees of compositional freedom are present, even at fixed temperature and pressure. Thus variations in the bulk chemistry of the rock may cause sympathetic changes in the compositions of the phases and in the $P-T$ zones of their stabilities, but need not cause the production of a new mineral species. For instance, laboratory experiments have demonstrated that for basaltic compositons, an increase in the Fe^{2+}/Mg ratio favors the appearance of garnet at lower pressures, whereas an increase in silica content delays the formation of garnet to higher pressures. Increase in the proportion of normative plagioclase extends the stability field of garnet granulite + plagioclase eclogite to higher pressures.

Peridotite Compositions

As described in the section dealing with the chemical and mineralogic constitution of the mantle, a peridotitic composition approaching aluminous lherzolite— or pyrolite—is expected to be typical for at least large portions of the upper mantle. Rocks possessing such compositions exhibit several important heterogeneous reactions in the laboratory, which allow us to establish a petrogenetic grid for the upper 100 km of the mantle. In contrast, the volumetrically less common harzburgite and dunite lithologies sustain important mineralogic transitions only at considerably greater depths; since lherzolites, of course, contain essential orthopyroxene and olivine, this rock type will show the same deeper transformations too, perhaps displaced slightly in $P-T$ space because of the presence of minor amounts of lime and alumina.

Piston-cylinder experiments on synthetic pyrolite compositions and on natural lherzolites have been performed by several groups (e.g., Green and Ringwood, 1967b; O'Hara, 1967; Ito and Kennedy, 1967; MacGregor, 1968). For these Ca- and Al-bearing peridotites, it seems clear that, at depths less than 20 to 40 kms, the stable assemblage is olivine + Al-poor orthopyroxene + Ca-clinopyroxene + plagioclase, succeeded then by olivine + Al-rich orthopyroxene + Ca-clinopyroxene + spinel compatibilities to depths approaching 60 to 80 kms, whereas at even greater depths the stable association

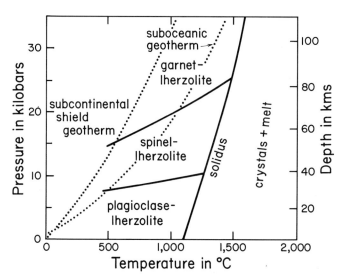

FIGURE 5.6.

Experimentally established petrogenetic grid for synthetic and natural Al-bearing anhydrous lherzolite bulk compositions (from Green and Ringwood, 1967b, Figure 11; O'Hara, 1967; Ito and Kennedy, 1967, Figure 3; MacGregor, 1968), summarized by Wyllie (1970, Figure 9). Geothermal gradients shown as dotted lines, were computed by Clark and Ringwood (1964).

is olivine + Al-poor orthopyroxene + Ca-clinopyroxene + garnet. A summary diagram for the bulk composition of dry lherzolite is presented in Figure 5.6; the geothermal gradients for mantle beneath the continental shields and beneath the ocean basins are also illustrated. It is evident from this figure that, within the first 100 km or more, the Earth's apparently normal geothermal gradient does not intersect the anhydrous peridotite solidus. If the asthenosphere contains trace amounts of interstitial melt, as is inferred by many petrologists and geophysicists, small amounts of H_2O are probably required to lower the mantle solidus temperature enough to account for partial melting at depths of about 100 kms and greater. Partial fusion of the mantle will be considered in the next section.

The solubility of $MgSiO_3$ in $CaMgSi_2O_6$ at 30 kilobars total pressure is illustrated in Figure 5.7a, and is compared with the one-atmosphere solvus relations in Figure 5.7b (Boyd and Schairer, 1964; Davis and Boyd, 1966; Warner and Luth, 1974). Within this range of physical conditions, pressure scarcely seems to influence the **En** content of iron-free synthetic Ca-clinopyroxene. For this reason, the equilibration temperature of enstatite

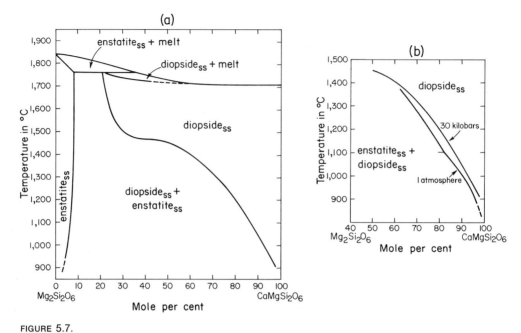

FIGURE 5.7.

(a) Experimentally established anhydrous phase relations along the join MgO·SiO$_2$–CaO·MgO·2SiO$_2$ at 30 kilobars pressure, after Davis and Boyd (1966, Figure 1). (b) Comparison of the **Di**-rich portion of the join MgO·SiO$_2$–CaO·MgO·2SiO$_2$ at one atmosphere (Boyd and Schairer, 1964) and at 30 kilobars pressure (Davis and Boyd, 1966); similar results were obtained by Warner and Luth (1974).

+ diopside-bearing peridotites (lherzolites) may be calculated by analyzing the Ca-clinopyroxene and comparing with the experimental data. In addition, if the peridotite carries an aluminous phase, such as spinel or garnet, the pressure of equilibration may be evaluated from the Al-content of the enstatite (once the temperature is known by employing the Ca-clinopyroxene composition), as can be see from Figure 5.8 (MacGregor, 1974). These piston-cylinder experiments have all been carried out in portions of the systems MgO–CaO–SiO$_2$ (Figure 5.7) and MgO–Al$_2$O$_3$–SiO$_2$ (Figure 5.8); so where small amounts of other constituents, such as Na$^+$, Cr^{+3}, Fe^{2+} or Fe^{3+}, are present in the natural peridotites, there will probably be departures from the *P–T* values that can be deduced from the described synthetic phase equilibria. However, to a first approximation, the laboratory results are appropriate for most natural lherzolites, which contain negligible quantities of these other components.

Olivine transforms to a denser, spinel-type or related structure at somewhat higher pressures; because of Fe–Mg solid-solution relations, this transition takes place over a *P–T–x* interval, which accounts for the finite width of the ~400 km discontinuity shown in Figure 5.1. Isothermal *P–x* sections for the system 2MgO·SiO$_2$–2FeO·SiO$_2$ by Akimoto and Fujisawa (1968) and by

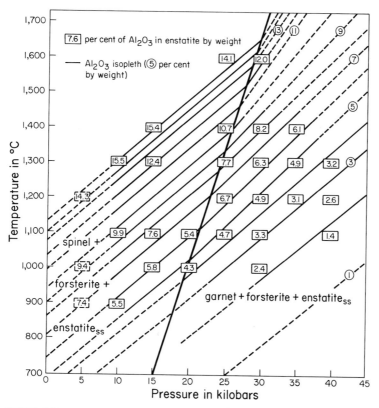

FIGURE 5.8.

Experimentally established Al_2O_3 contents of enstatites coexisting with forsterite and spinel or pyrope in the system $MgO-Al_2O_3-SiO_2$, after MacGregor (1974, Figure 2). Note that temperature is plotted on the ordinate axis and pressure on the abscissa.

Ringwood and Major (1970) are illustrated in Figure 5.9. An additional complication for Mg-rich bulk compositions was discovered by the latter workers, who showed that forsterite inverts to an as yet incompletely characterized orthorhombic spinel-like phase, β, rather than to the normal spinel polymorph, γ, of cubic symmetry. Nevertheless, it is evident from Figure 5.9b that, at 1,000°C, fayalite is converted to the denser γ phase at about 55 kilobars, whereas more elevated pressures are required to cause the transition to take place in intermediate olivines, and pure forsterite reacts to form the β phase only at pressures approaching 120 kilobars. The recrystallization of intermediate solid solutions is characterized by the coexistence of $olivine_{ss} + \gamma_{ss}$ or $olivine_{ss} + \beta_{ss}$ throughout a pressure interval whose magnitude depends on the normative olivine composition. As can be seen from Figure 5.9a, increasing tempera-

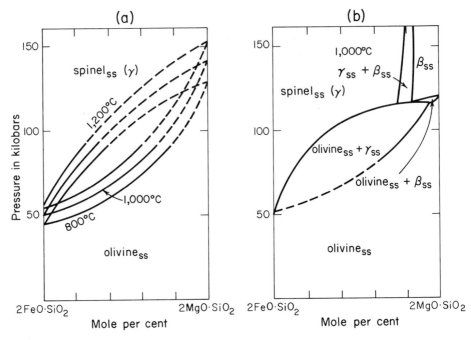

FIGURE 5.9.

High-pressure anhydrous experiments in the system $2MgO \cdot SiO_2$–$2FeO \cdot SiO_2$, after: (a) Akimoto and Fujisawa (1968); and (b) Ringwood and Major (1970). Field boundaries are dashed where inferred; β is an orthorhombic spinel-like phase, γ is true cubic spinel.

ture raises the pressures necessary to produce the denser γ or β assemblage by about 30°C per kilobar.

Mantle olivines are thought to have a composition close to Fo_{90}, judging from peridotite inclusions and nodules in alkali olivine basalts and kimberlite pipes. Such a phase would begin the transformation to β_{ss} at about 108 kilobars at 1,000°C. However, at 400 kilometers depth—the top of the transition zone in the mantle—the confining pressure is approximately 130 kilobars. Hence the temperature attending the onset of transformation for Fo_{90} should be on the order of 1,600°C. This value is compatible with computed geothermal gradients, being essentially the predicted temperature for the ~400-km discontinuity.

At slightly shallower depths, pyroxene components may transform to a more dense garnet solid solution, although where this occurs is by no means certain. (Note that the over-all conversion of pyroxene to a garnet-type structure requires that a portion of the silicon assume octahedral coordination in the high-pressure assemblage.) Because pyroxene is much less abundant than olivine, the transformation would not be expected to produce as marked a change in mantle properties as the two-stage conversion of $(Mg,Fe^{2+})_2SiO_4$ to the β form, then to the olivine–spinel, γ. Hence the velocity profile illustrated in Figure 5.1 would not be strongly influenced by the hypothetical pyroxene$_{ss}$ → garnet$_{ss}$ reaction.

But what is the nature of the reaction, or reactions, which occur at mantle depths of 650 kilometers and somewhat greater? Few experimental and calorimetric data are available to provide guidance here. By analogy with shallower transformation, however, it is thought that small-volume assemblages characterized by the closest possible packing of anions, chiefly oxygen, predominate. For instance, stishovite—in which silicon is totally six-fold coordinated, as is Ti in rutile—and periclase probably replace olivine–spinel and orthopyroxene. It is also possible that $(Mg,Fe)SiO_3$ inverts to a corundum- or ilmenite-type crystal structure and ultimately to a perovskite-type configuration under such conditions (Ringwood, 1975). In any case, reactions are expected to be of the sort illustrated in Figure 5.9, since iron and magnesium are both present in solid solutions, at least on the low-pressure sides of the hypothesized transition loops.

PARTIAL FUSION

Anhydrous Systems

Experiments dealing with small degrees of melting under mantle $P–T$ conditions have been carried out in the absence or virtual absence of H_2O on two principal compositional types of material: (1) natural basalt + more silicic differentiates; and (2) both natural and simplified synthetic peridotite. However, partial fusion experiments at high pressures have also been performed on calc-alkaline igneous rocks of the basaltic andesite–dacite–rhyolite series. $P–T$ diagrams for the melting relations of eclogite, quartz tholeiite, olivine tholeiite, and aluminous lherzolite have been presented already as Figures 5.2, 5.3, 5.4, and 5.6, respectively. Tholeiites carry normative hypersthene and either normative olivine or normative quartz; alkali olivine basalts, in contrast, have normative nepheline and olivine but lack normative hypersthene, and never contain normative quartz. Of course, from these figures, it is not possible for the reader to ascertain the precise chemical nature of the liquids produced at near-solidus conditions. However, electron microprobe analyses of quenched glasses in several of these studies have shown that, in most cases, the melts are somewhat more silicic and alkalic than the basaltic source material, especially if the proportion of melt is low.

Results of experiments for a natural andesite are shown in Figure 5.10, and isobaric high-pressure melting intervals for a number of calc-alkaline and tholeiitic lava types are shown in Figure 5.11. It is clear from this latter study that andesitic liquid occupies a pronounced thermal minimum. In contrast, even more silicic, alkalic melts, such as dacite and rhyolite, evidently have higher solidus and liquidus temperatures than andesite, at ($\sim$ dry) pressures

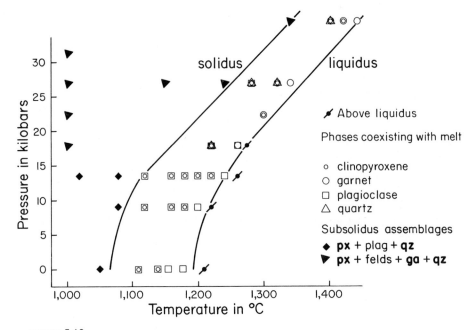

FIGURE 5.10.

Results of phase-equilibrium experiments on a natural andesite, in the virtual absence of H_2O, after Green and Ringwood (1968, Figure 4).

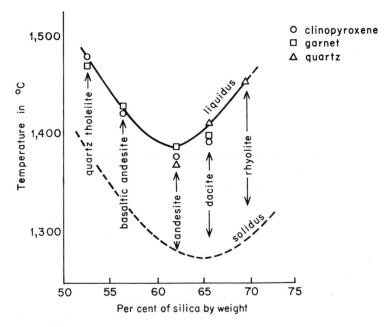

FIGURE 5.11.

Isobaric T–X projection of sensibly anhydrous liquidus and solidus temperatures for various crustal lava types at 30 kilobars total pressure, after Green and Ringwood (1968, Figure 8). Note that the thermal range of melt + crystals regularly increases as a function of silica content.

of about 30 kilobars; furthermore, quartz occurs as a near-liquidus phase for these silicic, alkalic bulk compositions; hence such magmas fractionate toward andesite, not rhyolite, and are in equilibrium with quartz rather than olivine. Thus, among possible liquids in the basalt–rhyolite spectrum, andesitic melts represent the most fusible material at moderate depths (on the order of 100 kms) in the mantle. Another point of interest in this diagram is that basalts possess the smallest melting interval; whereas liquidus and solidus temperatures diverge with increasing SiO_2 content of the source rock. The significance of melting interval is discussed in the following paragraph. It should be evident already, however, that these phase relationships do not apply directly to the deep generation of calc–alkaline melts, because such silicic protoliths apparently do not occur in abundance in the predominantly peridotitic mantle.

Provided that basaltic melt represents the initial liquid generated at the peridotite solidus, then if the melt is physically separated from the enclosing parental lithology, it should (being totally molten) be at liquidus conditions for its own bulk composition. In other words, the mantle solidus must coincide with the basalt liquidus at the P and T of generation and phase separation. Furthermore, the crystalline phase or phases in equilibrium with the liquid must be those characteristic of the now depleted source region of the mantle, namely, olivine $\pm$ orthopyroxene. It can be seen from the various anhydrous basaltic and andesitic systems illustrated in Figures 5.2, 5.3, 5.4, and 5.10 (see also Figure 5.11) that at elevated pressures, none of these homogeneous magmas is saturated with respect to olivine. Even at pressures less than about 8 kilobars, only the olivine tholeiite of Figure 5.4 could be in equilibrium with a hypothetical parental peridotite source. Such a conclusion has been arrived at by Ito and Kennedy (1967), who demonstrated that the fraction of the lherzolite which melts at the lowest temperature at mantle pressures is olivine-rich picritic basalt. Finally, because basaltic liquid represents a minimum melting composition for the source peridotite, the thermal range between liquidus and solidus, *at the P–T conditions of generation,* should be small—as is suggested by examination of Figure 5.11. Presumably olivine tholeiite would have an even smaller fusion interval than the quartz tholeiite illustrated (e.g., at 30 kilobars, the olivine tholeiite of Cohen, Ito, and Kennedy, 1967, possesses a melting range of 60–65°C as shown by Figure 5.4).

Evidently, very small degrees of (dry) partial fusion deep within the asthenosphere produce silica-undersaturated liquids, the so-called alkali olivine basalts, which are typically nepheline normative. They are nepheline normative because the alkalis are strongly concentrated in the very small amounts of liquid; such subsilicic melts seem to carry so little silica that it cannot, when combined with alkalis and alumina during crystallization, form feldspars without associated nepheline. At shallower levels, olivine tholeiites can be formed by partial melting of peridotite, or by the sustained precipitation and

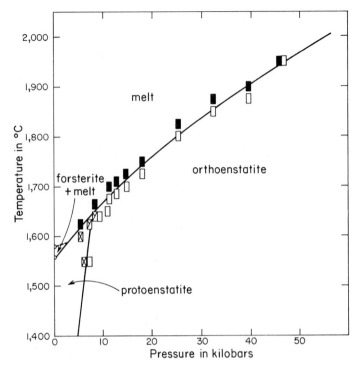

FIGURE 5.12.

Piston-cylinder experiments on polymorphic transitions and dry melting relations for the bulk composition MgO·SiO$_2$, after Boyd, England, and Davis (1964, Figure 1). Note that temperature is plotted on the ordinate axis and pressure on the abscissa.

separation of early-formed olivine crystals during magma ascent and cooling. Finally, at shallow depths, quartz tholeiites can be generated by this process. The fundamental reason for this change in partial melting relations appears to be, at least in part, that the anhydrous incongruent melting of MgSiO$_3$ is suppressed by pressures in excess of approximately 5 kilobars; phase relations are shown in Figure 5.12. Thus, at shallow levels, olivine is stable with a quartz normative liquid (e.g., see Figure 4.10); whereas, at high pressures, only silica-undersaturated melts can be generated by the partial fusion of peridotite. This experimentally established relationship is compatible with the observed shallow production of more or less silica-saturated oceanic and island arc tholeiites in regions where the asthenosphere rises to shallow levels. In contrast, alkali olivine basalt fields characteristically are situated on stable, continental-crust-capping lithospheric slabs of great thickness (the lavas presumably representing derivation from a deep asthenospheric magma source).

The essentially dry partial melting of a rather silicic basaltic eclogite at 27–36 kilobars gives rise to andesitic liquid, according to experiments by Green and Ringwood (1968). Calculations performed by Marsh and Carmichael (1974) indicate the production a similar calc-alkaline liquid by partial fusion of eclogite in the presence of K-feldspar. This andesitic composition of melt also appears to represent the most fusible composition in the calc-alkaline series at high pressures, as is obvious from Figure 5.11. In contrast, Ito and Kennedy (1974) have shown that incipient melting of a more mafic eclogite at pressures in the range of about 11–25 kilobars provides a nepheline normative liquid.

We have seen that the partial fusion of peridotite in the absence of H_2O provides melts of picritic to basaltic composition, depending on the P–T conditions of generation. However, anhydrous liquids more silicic than basalt clearly are not in equilibrium with an olivine-bearing solid assemblage, even in the uppermost portions of the mantle. Therefore, although basalts probably are derived from typical, essentially or nearly dry pyrolite, laboratory experiments suggest that the protoliths for andesites and the related calc-alkaline magma suite apparently do not include anhydrous, olivine-rich materials. A plausible but less firmly established conclusion from the H_2O-absent phase-equilibrium studies is that liquids of the andesitic clan possibly may be generated by the high-pressure partial fusion of basalt which has been previously transformed to quartz eclogite.

Hydrous Systems

Certain peridotite inclusions in alkali olivine basalts contain hydrous phases such as hornblende (pargasite, kaersutite, or tremolite–richterite) or phlogopite; hence H_2O may be an important minor constituent in the upper mantle. Such hydrous minerals appear to be restricted to the lithosphere, judging from the mantle xenoliths described from kimberlite diatremes. As discussed in Chapter 4 (e.g., see Figure 4.5), solution of H_2O in a melt lowers its solidus temperature relative to an anhydrous subsolidus assemblage (+ fluid); moreover, adding H_2O to a system causes the appearance of hydrous minerals at low temperatures. Thus, for a particular bulk composition, we expect to encounter melting at lower temperatures where H_2O is present than where it is not; furthermore, the subsolidus association of hydrous solids is likely in the former situation.

For hydrous, chiefly tholeiitic basalt compositions, amphibole-bearing mineral assemblages are stable at pressures up to approximately 25 kilobars, according to numerous experimental studies (e.g., Yoder and Tilley, 1962; Green and Ringwood, 1968; Lambert and Wyllie, 1972; Allen *et al.*, 1972; Holloway and Burnham, 1972; Helz, 1973). Generalized phase relations for basaltic compositions $\pm H_2O$ are illustrated in Figure 5.13. H_2O saturation is shown in Figure 5.13a; whereas only minor amounts of H_2O (i.e., low a_{H_2O}) characterize the equilibria illustrated in Figure 5.13b. Except at pressures less than a few hundred bars, the silicate liquid is more hydrous than hornblende; hence melting of the H_2O-bearing mineral is completed at a higher temperature under conditions where $a_{H_2O} < 1$ than where $a_{H_2O} \approx 1$. Why is this? The answer lies in the fact that, as was discussed in Chapter 2 (p. 34), the H_2O-rich condensed assemblage (melt, in this case) is favored by high activities of H_2O; whereas less hydrous condensed assemblages have expanded $P-T$ stability fields under conditions of reduced a_{H_2O}. The H_2O contents of amphiboles and phlogopite, which have at least limited occurrences in the mantle, of about 2 and 4 per cent by weight, respectively, are less than that of aqueous peridotite melts, which contain about 5–10 per cent of H_2O by weight under elevated confining pressures. Therefore, high values of P_{H_2O} promote melting at the expense of the thermal stability ranges of these hydrous minerals.

At total pressures exceeding about 25 kilobars, anhydrous eclogitic phase associations + aqueous fluid are favored over amphibolite; the reason for this phenomenon is that the assemblage omphacite$_{ss}$ (i.e., Na–Ca–clinopyroxene) + garnet$_{ss}$ + H_2O occupies a smaller volume than does the chemically equivalent amphibolite. Thus, if partial fusion of basaltic amphibolite is to happen at all, it must occur within the upper 60–80 kilometers of the mantle. Similar to the near-solidus melts derived from the higher-pressure eclogitic lithologies of equivalent bulk composition, liquids generated from the incongruent melting of amphibolites are andesitic. For natural alkali olivine basalts and simplified synthetic analogous systems, biotite and phlogopite are stable to much higher pressures than are amphiboles in the presence of H_2O (Essene et al., 1970; Boettcher, 1973). Accordingly, partial fusion of alkali olivine basalts and mica peridotites at depths of 100–150 kilometers should produce more potassic melts than the shallower partial melting of amphibolites of oceanic tholeiite derivation.

In the presence of H_2O, the initial partial-fusion products of natural and synthetic lherzolites are also andesitic, according to some laboratory investigations (Yoder, 1969; Kushiro, 1972b, 1973b, and Kushiro et al., 1968; see also O'Hara, 1965). That these results are still a matter of contention is clear from the discussion by Mysen et al., (1974) and Ringwood (1974). The hydrous melts are silicic because enstatite melts incongruently to forsterite + hydrous liquid at least up to aqueous fluid pressures of 30 kilobars, as Figure 5.14 shows, in contrast to the dry situation (compare Figures 5.14 and 5.12). Kushiro (1972b) showed that, in the H_2O-saturated quaternary system $CaO \cdot MgO \cdot 2SiO_2$–$2MgO \cdot SiO_2$–$SiO_2$–$H_2O$, the primary crystallization field of forsterite is more extensive at 20 kilobars (see Figure 5.15) than in the one-atmosphere anhydrous ternary system **Di–Fo–Si**; this indicates that substantially quartz-normative hydrous liquids are in equilibrium with olivine + orthopyroxene solid solution + Ca-clinopyroxene solid solution crystalline assemblages (= lherzolite). In contrast, experiments by Nicholls (1974) indicate that peri-

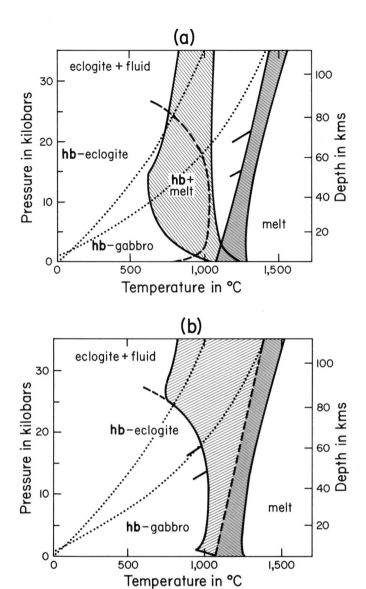

FIGURE 5.13.

Composite phase relations for the basaltic bulk composition, with special reference to the presence of H_2O and hornblende (**hb**), after Wyllie (1970, Figures 11 and 12). The lower-temperature subcontinental, and higher-temperature suboceanic, geothermal gradients are also indicated. Figure (a) illustrates the anhydrous liquidus-solidus region (heavy shading) plus the basalt-eclogite transition, and both the thermal stability limit of hornblende (dashed line) and the fusion interval (light shading) under conditions in which H_2O is present in excess; the unlabeled P–T region between the two melt regions consists of anhydrous crystalline assemblages where aqueous fluid is absent, or homogeneous H_2O-saturated melt where excess fluid is present. Figure (b) illustrates the dry melting zone (heavy shading) plus the basalt-eclogite transition, and both the stability of hornblende and the fusion interval (light shading) where only trace amounts of H_2O are present.

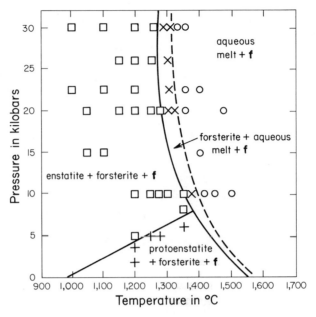

FIGURE 5.14.

Piston-cylinder experiments on the bulk composition $MgO \cdot SiO_2$ + excess H_2O, after Kushiro, Yoder, and Nishikawa (1968, Figure 2). Small amounts of forsterite occur at subsolidus temperatures because the supercritical aqueous fluid, f, contains more silica than magnesia.

FIGURE 5.15.

Liquidus diagram for the system $CaO \cdot MgO \cdot 2SiO_2 - 2MgO \cdot SiO_2 - SiO_2 - H_2O$ at 20 kilobars total (aqueous fluid) pressure, and projected onto the anhydrous **Di–Fo–Si** base, after Kushiro (1927b, Figure 1). All phases are saturated with respect to H_2O. Point x is the quaternary reaction point; quartz-normative liquids of this composition are in equilibrium with fosterite, diopside solid solution, and enstatite solid solution. Temperatures are in °C. Compare equilibria presented here with the one-atmosphere dry phase equilibria (Figure 4.37).

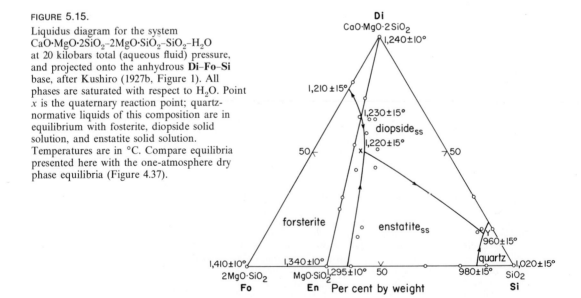

dotite can be in equilibrium with H_2O-rich andesitic melt only at values of $P_{fluid} \leq 10$ kilobars; at higher aqueous-fluid pressures, magmas saturated with olivine are less silicic than andesite. Evidently, although the process is still disputed, liquids of the calc-alkaline magma suite might be produced by the incipient fusion of upper (or uppermost) mantle peridotite under hydrous conditions as well as by partial melting of amphibolite and of quartz eclogite. H_2O probably does not occur as a separate, pure phase in the mantle, to judge from compositional considerations, but more likely is present along hydrated grain boundaries, and to some extent is diluted by other soluble components, especially CO_2. In any case, it is clear from the laboratory studies that as a_{H_2O} increases at fixed P and T, the amount of silica in the initial melt also increases.

High-pressure experiments involving the equilibration of natural and synthetic peridotites with a binary CO_2–H_2O fluid phase have demonstrated the moderate solubility of CO_2—in addition to H_2O—in the melt (Boettcher, Mysen, and Modreski, 1975; Mysen and Boettcher. 1975; Eggler, 1974). At a given P and T, where the mole fraction of CO_2 is high in the peridotite system, nepheline-normative liquids are produced by partial fusion; whereas high a_{H_2O} values promote the generation of silica-saturated melts. Moreover, Holloway (1973) has shown that, at fluid pressures of 4 kilobars and above, the incongruent melting of the calcic amphibole pargasite, $NaCa_2Mg_4AlSi_6$ $Al_2O_{22}(OH)_2$, is delayed to higher temperatures where the aqueous volatile phase is diluted with moderate amounts of CO_2, compared to the pseudobinary system **Par**–H_2O. This phenomenon results from the fact that, at high X_{H_2O}, H_2O is more soluble in the melt than in the amphibole—in which it is constrained to be present in the stoichiometric proportions dictated by the mineral formula.

Finally, in our discussion of the role of H_2O in ultramafic rocks subjected to high pressure, it should be noted that phlogopite may be stable to pressures greater than 35 kilobars in rocks of appropriately potassic bulk composition (Modreski and Boettcher, 1973); similar to our conclusion regarding the beginning of melting for alkali olivine basalts, the deeper is the zone of hydrous partial melting of lherzolite, the more K-rich should be the product magma. The experimentally established greater thermal stability range of phlogopite relative to Ca-amphibole is also in harmony with its occurrence in the deeper garnet lherzolites, compared to the apparent restriction of amphiboles to the shallower plagioclase- and spinel-bearing peridotites.

Calculation of Crystal-Melt Equilibria in the Mantle

Thus far, we have referred chiefly to high-pressure phase-equilibrium experiments which bear on the generation of magmas by partial fusion of peridotite, and of other, more silicic, alkalic compositions as well. By employing suitable simplifications, we can also calculate the activities of constituents in the liquid, the chemical potentials of which are established by equilibration with the solid assemblage from which the melt has been derived. Such computations have been pioneered by Carmichael (e.g., see Carmichael, Turner, and Verhoogen, 1974, p. 50–59, 78–123), and will be only very briefly outlined here.

The fundamental relationship between activities of participating species and the Gibbs free energy of a partial-fusion reaction is given by the Van't Hoff reaction isotherm. This relationship was discussed in Chapter 1, pp. 18–21, and will be elaborated for solid-state partition reactions in Chapter 6, pp. 238–249. At equilibrium, for any specified P and T, the expression for a generalized reaction of the sort $aA + bB = lL + mM$ reduces to equation (1.72),

$$\Delta G^0 = -RT \ln \frac{a_L{}^l a_M{}^m}{a_A{}^a a_B{}^b} = -RT \ln K,$$

where $a_M{}^m$ stands for the activity of species M raised to the stoichiometric coefficient m, and K is the distribution constant. For a simple reaction such as

$$\underset{\text{(forsterite)}}{\text{Mg}_2\text{SiO}_4} + \underset{\text{(in melt)}}{\text{SiO}_2} = \underset{\text{(enstatite)}}{2\text{MgSiO}_3},$$

the equilibrium activity of silica in the liquid phase is defined by equation (1.72). An alternative form of the reaction probably more closely approximates the actual equilibrium, and may be readily envisioned for a more complex system in which, say, ferrous iron is present, and Fe^{2+} and Mg exchange among octahedral sites on an ion-for-ion basis:

$$\underset{\text{(forsterite)}}{2\text{MgSi}_{05}\text{O}_2} + \underset{\text{(in melt)}}{\text{SiO}_2} = \underset{\text{(enstatite)}}{2\text{MgSiO}_3}.$$

In any case, the chemical potential of silica is likewise specified at constant temperature and pressure by equation (1.66):

$$\mu_{SiO_2} = \mu^0{}_{SiO_2} + RT \ln a_{SiO_2}.$$

For simplicity we have chosen the pure magnesian end-members for both olivine and orthopyroxene; so the activities of forsterite and enstatite are of unit value, at least, at one atmosphere total pressure, and thus

$$-\Delta G^0 = RT \ln K = RT \ln a_{SiO_2(melt)}.$$

The standard state selected for SiO_2 is as pure silica glass. If ideal solutions are assumed, the calculated activity of SiO_2 in the multicomponent liquid may be equated with its concentration, or mole fraction, in the melt. How one specifies X_{SiO_2} in melt is a nontrivial problem involving structure of the liquid. Of course, most mantle protoliths do not contain pure Mg_2SiO_4 or pure

$MgSiO_3$, but rather olivine and hypersthene with $Mg/(Fe^{2+} + Mg)$ ratios of about 0.9. Again assuming ideal solutions, for any specific protolith we could substitute the observed mole fractions of **Fo** and **En** in equation (1.72), which may then be written,

$$-\Delta G^0 = RT \ln K$$

$$= RT \ln X_{SiO_2(melt)} + RT \ln (X_{Fo})^2 - RT \ln (X_{En})^2.$$

Because the values of X_{Fo} and X_{En} are nearly identical here (see Ramberg and deVore, 1951), the computed silica mole fraction would not change appreciably from the pure Mg end-member system, even though the mole-fraction term for orthopyroxene and, (as written, that of the olivine as well) for this particular reaction is squared.

Numerous other liquid fractionation reactions may be envisioned which relate the activities (or mole fractions) of various other crystalline constituents, such as pyroxene and feldspar end-members and particular oxide components in liquid solution, thereby permitting an estimation of the activities—or mole fractions—of the latter in the melt. One such example is the following:

$$\overset{\text{(diopside)}}{CaMgSi_2O_6} + \overset{\text{(both in melt)}}{SiO_2 + Al_2O_3} = \overset{\text{(anorthite)}}{CaAl_2Si_2O_8} + \overset{\text{(enstatite)}}{MgSiO_3}.$$

All calculations of this type suffer from the uncertainties inherent in this method, including the assumption of ideal solution behavior, and the considerable difficulty of evaluating the mole fraction of a constituent such as SiO_2 or Na_2O in a multicomponent melt.

The influence of changing temperature and pressure on the equilibrium constant, K, was discussed in Chapter 1, pp. 20–21 (see also Chapter 6, p. 242), and was given by equations (1.77) and (1.79):

$$\left[\frac{\partial \ln K}{\partial \left(\frac{1}{T} \right)} \right]_P = -\frac{\Delta H^0}{R};$$

and

$$\left(\frac{\partial \ln K}{\partial P} \right)_T = -\frac{\Delta V^0}{RT}.$$

At crustal pressures, the isothermal compressibilities of condensed phases are small and, with an increase in both P and T, tend to offset the isobaric thermal expansions—hence as a first approximation both terms may be ignored. Because of the greater length of the extrapolations to mantle pressures, however, such an approximation no longer seems to be justified, and both thermal expansion and compressibility of all participating species must be taken into account. Unfortunately, not all the appropriate data are available, especially for glassy constituents. Quantitatively, then, the reliability of such computations is only modest. Their value lies in the fact that trends in component activities (mole fractions, i.e., compositions) of melts generated from chemically complex protoliths as a function of physical conditions can be estimated.

For instance, Marsh and Carmichael (1974) have demonstrated a theoretical basis for the observed phenomenon (Kuno, 1966; Hatherton and Dickinson, 1969) of increased K_2O content with apparent depth for calc-alkaline magmas (see Figures 5.18 and 5.19) produced through the postulated anatexis of a sanidine-bearing quartz eclogite, representing subducted oceanic crust. Inasmuch as the computations are based on the proviso that K–feldspar is present in excess, so that derived liquids are saturated with respect to this constituent, $KAlSi_3O_8$ must be rather insoluble in the melts or very low degrees of eclogite partial melting must occur. Furthermore, eclogites characteristically do not carry a potassium-rich feldspar. Nevertheless, the computations of Marsh and Carmichael do provide the K_2O variation recognized by Kuno, Hatherton, and Dickinson.

CRUST-MANTLE DIFFERENTIATION AND PLATE TECTONICS

In previous sections we have discussed mechanisms whereby magmas of basaltic and calc-alkaline type may have been generated in the past, as well as in the present. Let us summarize that discussion.

Basalts are produced by small degrees of partial melting of relatively undepleted mantle material (i.e., Al-bearing lherzolite = pyrolite). Sometimes this protolith may contain hydrous minerals, such as calcic amphibole and/or phlogopite. The zone of incipient fusion coincides with the Earth's asthenosphere, and is overlain by a cooler mantle segment attended by subsolidus $P–T$ conditions and surmounted by crust. The depth and degree of partial melting determine the nature of the basalt: in general, shallower levels of generation, small proportions of fusion, and greater amounts of H_2O favor tholeiitic magmas (the more extensive the partial melting is, the more olivine-rich the resultant magma series will be); whereas partial melting at more profound depths, very slight degrees of melting, and more nearly anhydrous or CO_2-rich conditions promote the production of alkali olivine basalts. Three experimentally verified mechanisms have been proposed by which andesites and related igneous rocks, ranging from basaltic andesites through dacites, may be produced: (1) by the incongruent melting of basaltic amphibolite at depths less than about 60–80 kilometers; (2) by the essentially anhydrous partial fusion of quartz eclogite at somewhat greater depths; and (3) by the incipient melting of wet (or at least damp) peridotite. Subsequent high-level differentiation of these magma types on ascent toward the surface may be responsible for the more silicic rhyolitic lavas; it is certain that unless such SiO_2-rich liquids are nearly saturated with respect to H_2O, they cannot possibly have been equilibrated with olivine-bearing mantle at considerable depths within the Earth, because a silica polymorph, not olivine, is stable in these siliceous melts at ($\sim$ dry) liquidus conditions.

Primary, undepleted mantle material is subjected to a process of liquid phase separation, or fractional fusion, in all these models of magma generation. On crystallization, the melt which has been distilled off the asthenosphere becomes basaltic oceanic crust or basaltic + calc-alkaline continental crust. The residual mantle consists chiefly of olivine (= dunite) or olivine + hypersthene (= harzburgite), and seems to be incapable of further significant differentiation.

A thermally-driven gravitative instability in the mantle presumably provides the over-all mechanism whereby sea-floor spreading is sustained (e.g., Elsasser, 1971). During this process, asthenospheric pyrolite rising beneath the spreading centers is apparently converted to depleted lithosphere by the process of partial fusion, and new oceanic crust is formed on solidification of the buoyant basaltic liquid. Subsequently, this oceanic crust, in part locally metamorphosed near the ridge, is conducted laterally away from the oceanic ridge during sea-floor spreading, and is largely consumed at convergent lithospheric plate junctions. Most of the subducted basaltic material, transformed either to amphibolite or to eclogite, appears to have undergone a second stage of partial melting at depth, thereby providing the andesitic suite of plutonic and extrusive magmas characteristic of island arcs and continental margins. These liquids, of course, are less dense than the surrounding crystalline assemblages; so, on coagulation, gravitational body forces drive the plutons toward the surface of the Earth.

Because aqueous fluids contained in sedimentary rocks and hydrated oceanic crust are carried to considerable depths in subduction zones, the hanging-wall (or nonsubducted) mantle lithosphere and underlying asthenosphere may be subjected to an upward stream of volatiles (e.g., see Best, 1975); the reaction of peridotite with H_2O would result in incipient fusion and the formation of additional calc-alkaline and/or basaltic igneous masses. Thus it is likely that, with the passage of time, the subcontinental mantle would tend toward a depleted composition, provided there is negligible differential motion between lithosphere and asthenosphere. Such a condition is suggested also by the occurrence of chemically depleted granular lherzolite inclusions in what are interpreted to be samples of subcontinental lithosphere caught up in kimberlite pipes. On the other hand, Nicolas and Jackson (1972) have called attention to contrasting occurrences of serpentinized peridotites in the Alps: although the ultramafic portions of ophiolite suites (regarded as oceanic crust and mantle underpinnings) are characterized by harzburgites in the eastern Mediterranean, the western Alps typically contain lherzolites thought to represent the mantle directly beneath the continental crust. Evidently in some instances sial rests on relatively undepleted peridotite.

The calc-alkaline rocks have aggregate densities too low for them to be permanently lost by subduction processes. Therefore, by-and-large, once continental crust forms, it appears to persist at or near the Earth's surface—or is underplated near the base of the continental crust—unlike oceanic crust, which is destroyed almost as rapidly as it is generated. Segments of oceanic crust appear to be incorporated in the continents by tectonic processes from time to time, and, of course, basaltic melts are intruded into and extruded on continental crust. Sea-floor spreading processes continually sweep masses of sial together into major accumulations, although rifting of large, continental-crust-capped lithospheric plates also occurs episodically. The maximum thickness of the continental crust may perhaps be controlled by the relatively low temperatures of melting for such silicic and alkalic materials in the presence of an aqueous fluid (e.g., see Figure 4.41b); the subcontinental geothermal gradient seems to intersect the solidus for sialic compositions at about 10–15 kilobars fluid pressure, corresponding to the observed continental crust thickness of 35–50 kilometers. Thus sialic material is subjected to further stages of partial melting, and to the processes of metamorphic and sedimentary differentiation.

The complex process of crust-mantle interaction may perhaps account for the observation by Miyashiro (1974) that in island arcs and continental margins (that is, convergent plate junctions), the proportion of calc-alkaline volcanic rocks increases relative to basalts with the passage of time; this trend reflects the increasing development of mature continental-type crust. Although the tholeiitic series is produced within the oceanic basins at a relatively constant rate, the andesitic magmas would be produced near convergent plate margins by both primary and secondary processes. The primary process is partial fusion of subducted ocean crust and/or partial melting of "wet" nonsubducted mantle; the secondary process is refusion of previously accumulated calc-alkaline volcanic rocks. Like that of oceanic tholeiite, the primary production of andesitic melts should be relatively constant with time; whereas the secondary generation would depend on how much calc-alkaline material had accumulated in the past, and would be expected to increase with time as the continental crust thickened.

However, no matter how much reworking takes place, the total bulk composition of the continental crust must remain nearly constant, since it is determined ultimately by partial fusion, which has been taking place in the upper mantle for the entire recorded lithologic history of the Earth (e.g., see Ringwood, 1974). A final implication of these relationships is that the Earth is continuing to differentiate according to density and refractory characteristics in such a way that the mantle is progressively depleted in the large- and low-atomic number elements, with the fusible constituents continually being added to the continental crust.

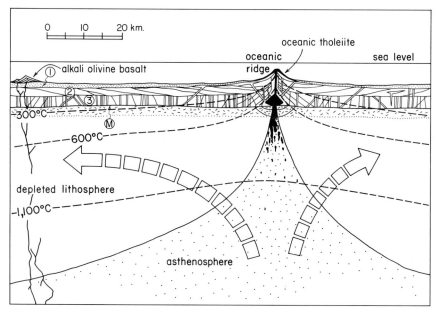

FIGURE 5.16.

Schematic diagram of an accreting plate margin or spreading center. Increasing degrees of anhydrous partial melting of the asthenosphere during ascent and depressurization provides tholeiitic ridge magmas shown in black. Alkali olivine basalts occur as more or less isolated masses which apparently have their source in deeper portions of the asthenosphere. Oceanic layers are: (1) deep-sea pelagic sediments; (2) pillow basalts and breccias; and (3) gabbros + sheeted diabase complexes. (M) is the Mohorovicic Discontinuity. The basal region of cumulate ultramafic material (random dash pattern) occurs as a near-surface fractional crystallization product of tholeiitic magma. The thermal structure shown is after Oxburgh and Turcotte (1968, Figures 10, 11). Large arrows indicate directions of material transport.

The spatial distribution of crustal rocks representing the various magma types—and metamorphic facies and sedimentary regimes as well—evidently is a function of lithospheric plate dynamics. This seemingly is so because motions of the several plates appear to influence the local and regional thermal structures, which in turn affect the generation of melting in the upper mantle—and the metamorphic phase assemblages and sediments developed in and on the crust. Figures 5.16 and 5.17 provide generalized thermal structures for accreting and consumptive plate junctions, respectively. Geologic relationships, with emphasis on igneous rock series, are also illustrated. It must be emphasized that these figures are *models,* and reflect a synthesis of petrologic + geophysical data, liberally interpreted.

174

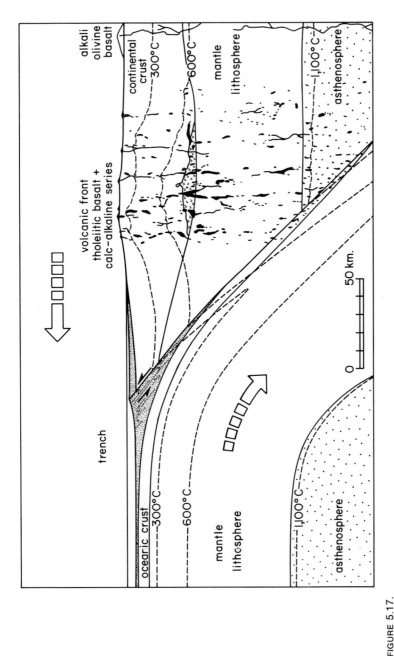

FIGURE 5.17.

Schematic diagram of a destructive (i.e., consumptive) plate margin. Partial melting of preexisting rocks occurs in three distinct regions: (a) the metabasaltic oceanic crust at the top of the downgoing plate, consisting of amphibolite at shallow levels and eclogite at deeper levels; (b) the stable, nonsubducted (hanging wall) asthenosphere through which volatiles are presumed to be migrating above the adjacent descending plate; and (c) the basal, thickened portions of the sialic, H$_2$O-rich crust. The first two magma types are shown in black, and include both tholeiitic and calc-alkaline compositions; whereas the third, indicated by a checked pattern, represents a relatively silicic, largely plutonic series. The width of the zone of active igneous activity at any one time is somewhat less than the composite field shown. Fields of alkali olivine basalt lie toward the interior of the stable, nonsubducted continental-crust-capped lithospheric plate and have been derived from great depths by very minute degrees of partial fusion of the mantle. The thermal structure is after Hasebe, Fujii, and Uyeda (1970). Oxburgh and Turcotte (1970), Turcotte and Oxburgh (1972), and Griggs (1972). Large arrows indicate directions of material transport.

These diagrams illustrate the observation that tholeiitic basalts characterize the oceanic basins and marginal seas, whereas island arcs and continental margin volcanic-plutonic belts contain both hypersthene normative basaltic and andesitic-dacitic igneous series. The alkali olivine basalts in general are located toward the interiors of the stable, continental-crust-capped lithospheric plates, although midoceanic volcanism can apparently sometimes tap deep, nepheline-normative liquids. Such a distribution is exhibited by Figure 5.18 for the Japanese arc and its environs. Here it is apparent that the alkali olivine

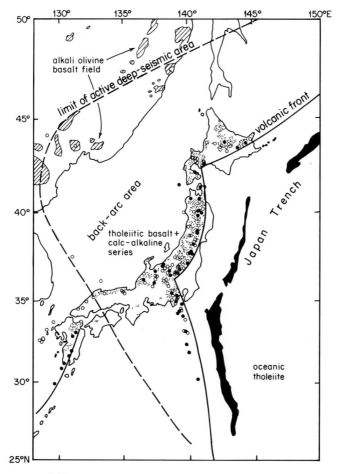

FIGURE 5.18.

Young volcanic rocks in the vicinity of the Japanese arc, modified after Matsuda and Uyeda (1970). Symbols for the tholeiitic and calc-alkaline magma types are: solid circles = active volcanoes; open circles = Quaternary volcanoes; dotted regions = Upper Tertiary volcanic rocks. The diagonal-lined area represents Pliocene-recent alkali olivine basalt fields.

basalt fields of the Asiatic mainland are sited on a terrane characterized by deep-focus earthquakes (i.e., ones at a great depth, locating the inclined lithospheric plate junction). In general, for a given silica content, there is a marked correlation between the per cent of K_2O by weight of recent lavas and the depth to the seismic zone, as shown in Figure 5.19. This remarkable correlation suggests that melts are generated along, or adjacent to, the Benioff zone. This phenomenon possibly is due to dissipative shear heating or to the progressive release of volatiles from the amphibole- and phlogopite-bearing descending slab; at least in some way the systematic increase in potash content of magmas is a function of the depth of partial fusion, as shown by the high-pressure

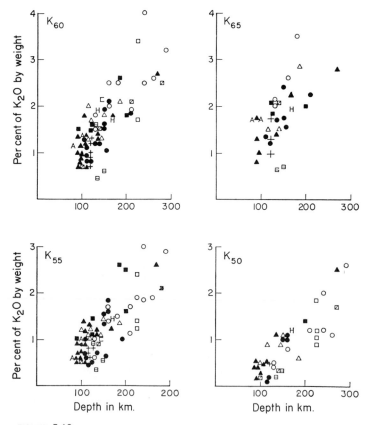

FIGURE 5.19.

Relationship between the K_2O contents of circumpacific Quaternary lavas for specific SiO_2 contents, and depths to the seismic zone, after Dickinson (1970). The correlation is shown for volcanics possessing 65(K_{65}), 60(K_{60}), 55(K_{55}), and 50(K_{50}) per cent silica by weight. Symbols designate different geographic lava suites.

laboratory investigations and calculations, and by proximity to the convergent plate junction.

We may conclude this chapter by observing that experimental phase-equilibrium studies—mostly employing piston-cylinder apparatus—have provided provisional explanations for the contrasting compositions of observed magma suites. The various types seem to be complex functions of the temperatures and pressures of partial melting, the activities of the volatile constituents, where present, and the bulk compositions of the source materials. In the same way, plate tectonic models and inferred thermal structures seem to indicate why the various magma types are sited where they are. Movements of the lithospheric slabs are related to the temperature regime of the upper mantle, and would provide access of the different protoliths to the conditions appropriate for fractional fusion.

CHAPTER 6

Predominantly Subsolidus Diagrams and Crustal Metamorphic Petrology

MINERALOGIC DIVERSITY AND THE CLASSIFICATION OF METAMORPHIC ROCKS

Metamorphism may be defined as the process whereby a preexisting rock is converted to a new texture or mineral assemblage by being subjected to physical—and often chemical—conditions which contrast with those under which it was initially formed. Such P–T regimes conventionally are restricted to physical conditions ranging between those appropriate to sedimentary environments on the one hand and those appropriate to igneous environments on the other. Thus, the processes of diagenesis and partial fusion (anatexis) are excluded by definition as being at the limits of metamorphism. Metamorphism is an essentially subsolidus but not surficial phenomenon, which may or may not involve the participation of a volatile phase. Large portions of at least the uppermost mantle (the lithosphere, for example) have recrystallized in the predominantly solid state, and hence in a real sense represent metamorphic terranes; however, this is a moot point, and such lithologies and their origins have already been discussed in Chapter 5. We will therefore confine our attention in this chapter to crustal recrystallization.

Granulation and mylonitization are mechanical processes which impart new textures to the milled-down rocks, and which thus represent a kind of metamorphism. The effects of stress strongly depend on the crystallographic and dimensional orientations of minerals, and on the aggregate physical properties of the constituting lithologies. This complex subject is beyond the scope of this book, which considers the chemical features of rocks but not their mechanical properties.

Although we will focus our attention on the chemical principles which govern heterogeneous equilibrium attending metamorphic recrystallization, it is useful to distinguish among several varieties, based on the tectonic environment. Broadly, these intergradational geologic regimes involve: (a) contact metamorphism; (b) regional metamorphism; and (c) burial metamorphism.

(a) Contact-metamorphic rocks are produced by an abnormally high thermal gradient and heat flow surrounding a hot mass, such as a high-temperature peridotite plug or a granitic stock. Marked thermal gradients generally are confined to the upper 10 to 20 kilometers of the Earth's crust because, at deeper levels, the rocks are already rather hot and soft (plastic)—hence marked thermal gradients are not produced.

(b) Regional, or dynamothermal, metamorphism is characteristic of orogenic belts where deformation accompanies recrystallization. Such metamorphic lithologies in general exhibit the effects of pervasive stress and resultant strain, including penetrative fabrics typified by preferred orientation of mineral grains.

(c) The term burial metamorphism is applied to large tracts of rock which have been feebly recrystallized by the gradually increasing pressure from superincumbant (lithostatic) load; this process characteristically involves neither important differential stress nor a thermal structure occasioned by the emplacement of a hot pluton. It is therefore typical of the medial and deeper portions of stable, nonorogenic crust.

Metamorphic terranes have been studied systematically for more than eighty years. It was apparent from the outset that systematic changes in mineralogy accompany progressive metamorphism. These alterations range from very feeble low-grade stages, where the original features of the protoliths are well-preserved, to intensely recrystallized, high-grade stages, in which the chemical and mineralogic nature of the preexisting rocks have been partially obscured or even totally obliterated. During progressive metamorphism, whether it is of the contact, regional, or burial variety, in general there is a gradual, over-all increment in grain size and progressive devolatilization proceeding toward the higher-grade rocks. Metamorphic grade, then, is a qualitative indicator of the physical conditions that have been operating, with elevated P–T conditions being characteristic of higher grade. Let us now briefly consider some of the advances in understanding of metamorphic mineral parageneses.

Working in the Scottish Highlands on pelitic schists, Barrow (1893, 1912) demonstrated the systematic entrance of new minerals proceeding upgrade. The most feebly recrystallized rocks contain clastic micas and the index-phase chlorite; these units give way spatially to progressively more thoroughly reconstituted lithologies carrying, in order of appearance, the index minerals biotite, almandine, staurolite, kyanite, and sillimanite. The individual minerals are systematically distributed in distinct regional zones in the field. Goldschmidt (1911) showed a topologically similar sequence of phase changes in contact-metamorphic aureoles that had developed in argillaceous and quartzofeldspathic hornfelses surrounding epizonal silicic igneous stocks in the Oslo area.

The line on the surface of the Earth defined by the lowest-grade appearance of an index mineral in any particular region was termed an isograd by Tilley (1925), since it should indicate metamorphic conditions of virtually constant grade. In three dimensions, of course, the locus of points defined by the first appearance of an index phase is a surface, and in general this surface would not be expected to coincide with either an isotherm or an isobar. As we shall see, this lack of coincidence in turn results from the fact that P–T curves for the production of individual minerals or phase assemblages generally, but not invariably, have $\infty > dP/dT > 0$.

Another concept, as useful as the isogradic concept, is that of metamorphic mineral facies, enunciated by Eskola (1914, 1939). Eskola studied recrystallized rocks of basaltic composition, chiefly in Norway and Finland, and demonstrated that the sequence of assemblages developed as a function of metamorphic grade. Relations are shown in a schematic metamorphic petrogenetic grid in Figure 6.1. Eskola remained unconvinced that certain very feebly recrystallized zeolite-bearing and allied rocks represented equilibrium-phase assemblages; hence he chose not to establish separate metamorphic mineral facies to designate them. However, some more recent workers (e.g., see Coombs, 1961) have noted the ubiquitous, systematic occurrence of such lithologies, and accordingly have proposed the recognition of two additional, very low-grade metamorphic facies, the zeolite and prehnite–pumpellyite

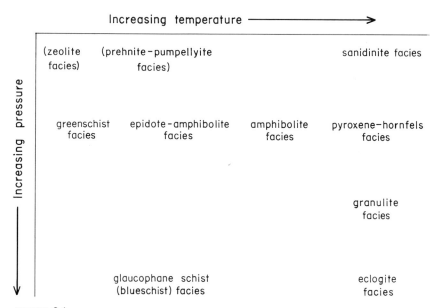

FIGURE 6.1.

Qualitative petrogenetic grid of metamorphic mineral facies, chiefly after Eskola (1939, Table II). Additional low-grade metamorphic facies, indicated by parantheses, have been recognized by Coombs and others (1959) and Coombs (1960, 1961).

TABLE 6.1.

Typical phases from the various metamorphic mineral facies.

Facies	Distinctive phases
zeolite	laumontite, heulandite clinoptilolite, wairakite, analcime
prehnite-pumpellyite	prehnite, pumpellyite, chlorite, albite
sanidinite	cordierite, sanidine, sillimanite, hypersthene, augite
greenschist	epidote, chlorite, actinolite, albite
epidote amphibolite	epidote, sodic plagioclase, blue-green hornblende
amphibolite	green-brown hornblende, intermediate plagioclase, garnet
pyroxene hornfels	hypersthene, augite, calcic plagioclase, andalusite
granulite	garnet, clinopyroxene, kyanite, intermediate plagioclase
blueschist	lawsonite, glaucophane, jadeitic pyroxene, aragonite
eclogite	omphacite, Mg-bearing garnet

facies. Some characteristic minerals for rocks of chiefly basaltic composition and representative of the different metamorphic facies are listed in Table 6.1 (see also the assemblages of Figures 6.2, 6.3, and 6.4).

The concept of metamorphic facies is of fundamental value to petrologists, because it allows us to obtain a semiquantitative understanding of the physical conditions which attend metamorphism. The thermodynamic principles underlying this concept and governing the phase equilibria have been elucidated by several investigators, especially Thompson (1955, 1959), Fyfe, Turner, and Verhoogen (1958), and Korzhinskii (1959). These workers have also provided extensive treatment of the process of changing bulk composition, or metasomatism. A working definition for rocks composing each metamorphic mineral facies is as follows: To a specific metamorphic facies belong all mineral assemblages which formed at chemical equilibrium—or in a state closely approaching chemical equilibrium—under a common range of physical conditions; variations in mineralogy of the several rock types constituting one metamorphic facies are strictly a function of their bulk composition. It is, of course, true that the mineral parageneses are influenced by the chemical potentials of the mobile components. These latter represent controlling intensive variables just like T and P. Accordingly, for completeness, the definition of a metamorphic facies should include specific μ limits for the mobile components, but typically these are not known in detail.

Clearly, the mineral assemblages which are produced during metamorphism depend on the chemistry of the protolith as well as on the P–T conditions. Furthermore, if a strong departure from chemical equilibrium occurs, various different phase assemblages may be produced under a common set of physical conditions, even from the same bulk composition. However, all reactions drive towards the equilibrium configuration, as discussed in Chapter 1; so petrographic relations, such as compositionally zoned minerals and replacement textures, may be employed to approximate the equilibrium assemblage.

Miyashiro (1961) drew attention to the fact that metamorphic mineral parageneses in different orogenic belts exhibit very marked differences in the sequences of observed mineral associations; hence they bear record to contrasting $P-T$ trajectories during recrystallization. He recognized three principal types: a low-pressure, high-temperature type characterized by andalusite and sillimanite; an intermediate-pressure, intermediate-temperature type, in which kyanite and sillimanite are formed; and a high-pressure, low-temperature type, signaled by the occurrence of jadeitic pyroxene and glaucophane. Diagrammatic parageneses for metabasaltic and metapelitic bulk compositions are illustrated in Figures 6.2, 6.3, and 6.4, for the low, intermediate, and high P/T types, respectively. The mineralogic effects of metabasaltic versus metashaley

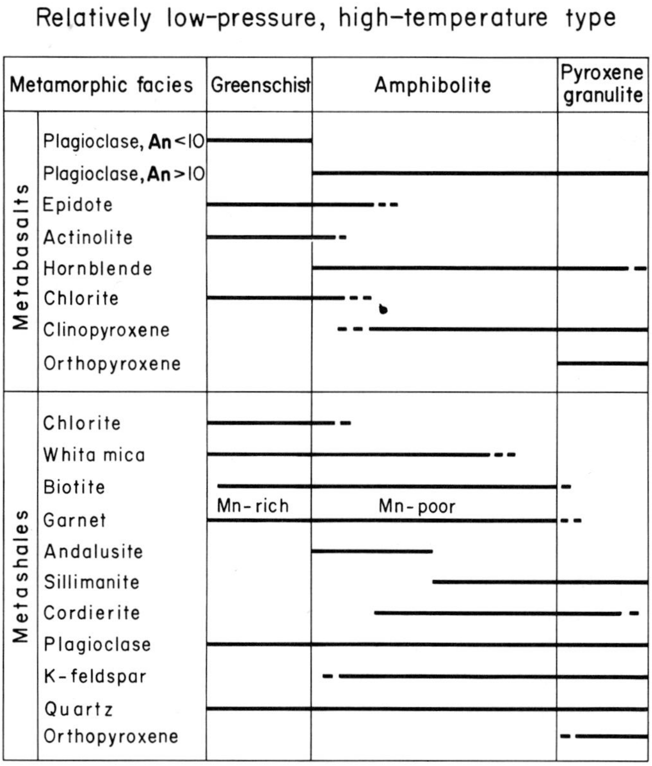

Relatively low-pressure, high–temperature type

FIGURE 6.2.

Mineral parageneses in the andalusite–sillimanite type of matamorphic facies series, after Miyashiro (1961, Figure 2). The example illustrated is from the Abukuma Plateau, Japan.

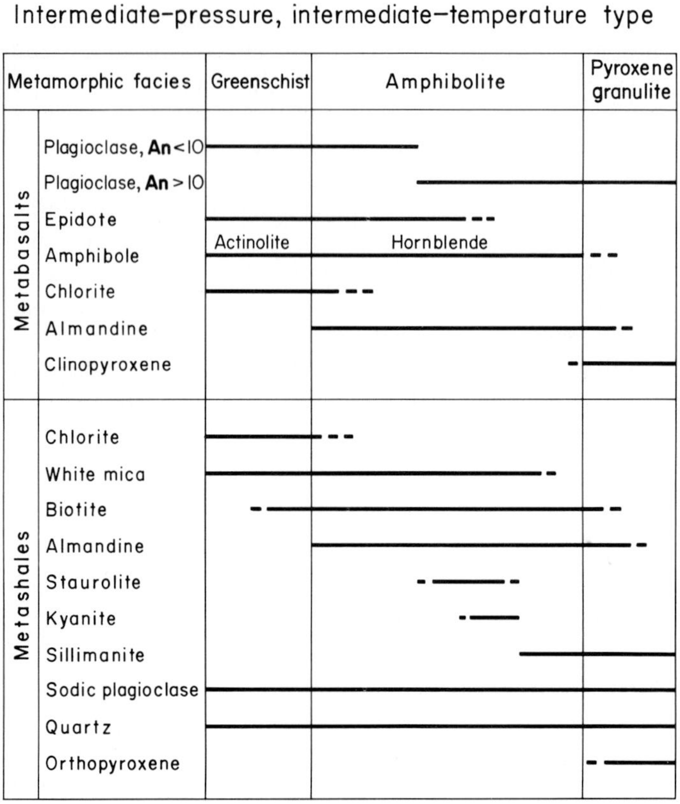

Intermediate–pressure, intermediate–temperature type

Metamorphic facies		Greenschist	Amphibolite	Pyroxene granulite
Metabasalts	Plagioclase, **An** < 10			
	Plagioclase, **An** > 10			
	Epidote			
	Amphibole	Actinolite	Hornblende	
	Chlorite			
	Almandine			
	Clinopyroxene			
Metashales	Chlorite			
	White mica			
	Biotite			
	Almandine			
	Staurolite			
	Kyanite			
	Sillimanite			
	Sodic plagioclase			
	Quartz			
	Orthopyroxene			

FIGURE 6.3.

Mineral parageneses in the kyanite–sillimanite type of metamorphic facies series, after Miyashiro (1961, Figure 1). The example illustrated is from the Scottish Highlands.

bulk composition is obvious in these figures. Miyashiro also noted the existence of metamorphic facies series intermediate to the types illustrated. All intergradations are possible, of course, and result from the fact that each metamorphic belt has been subjected to its own unique set of physical conditions. We tend to focus on mineral parageneses developed across a linear metamorphic belt, but in fact, variations in the observed mineral parageneses occur along the strike as well. From this phenomenon it may be concluded that differences in the controlling $P–T$ trajectories exist even within a single terrane.

Further efforts to quantify our knowledge of the physicochemical conditions of recrystallization have centered around: (1) recognition of progressive min-

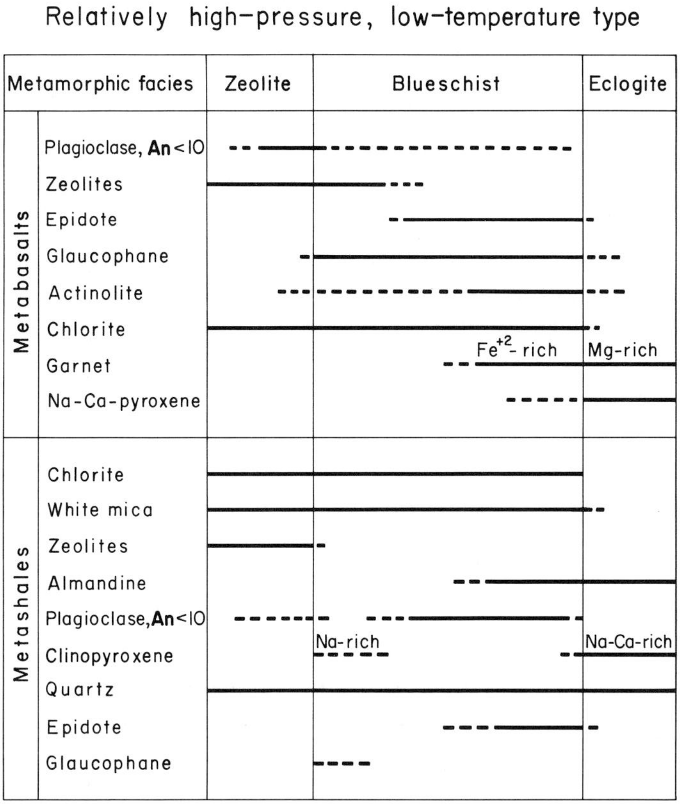

Relatively high-pressure, low-temperature type

Metamorphic facies	Zeolite	Blueschist	Eclogite

Metabasalts

- Plagioclase, **An** < 10
- Zeolites
- Epidote
- Glaucophane
- Actinolite
- Chlorite
- Garnet — Fe^{+2}-rich | Mg-rich
- Na-Ca-pyroxene

Metashales

- Chlorite
- White mica
- Zeolites
- Almandine
- Plagioclase, **An** < 10
- Clinopyroxene — Na-rich | Na-Ca-rich
- Quartz
- Epidote
- Glaucophane

FIGURE 6.4.
Mineral parageneses in the jadeitic pyroxene–glaucophane type of metamorphic facies series, after Miyashiro (1961, Figure 3). The example illustrated is from the Kanto Mountains, Japan. The zeolites here include the hydrous Ca–Al–(Fe) silicates lawsonite and pumpellyite.

eralogic and chemical (including isotopic) changes in individual phases and in coexisting mineral assemblages; (2) correlation of observed mineral parageneses with the results of phase-equilibrium experiments; and (3) thermochemical calculations of phase equilibria. All three methods are used in conjunction, but usually one or another is emphasized. Most of the rest of this chapter will be devoted to an application of experimentally established phase equilibria to the natural occurrences, although occasionally we will resort to the computational method, as outlined in Chapter 3. Finally, we will very briefly discuss the relationship of metamorphism to plate tectonics.

SOLID-SOLID REACTIONS

In metamorphic petrology there are only a few heterogeneous equilibria (other than certain reactions that partition elements and isotopes) that do not involve the evolution or consumption of one or more volatile species. The P–T locations of solid-state equilibria do not depend on values of P_{H_2O}, P_{CO_2}, P_{O_2}, etc., because volatile components do not participate in the reactions—hence fluid is an indifferent phase. Solid-solid equilibria are of three general types: (1) polymorphic transitions; (2) multiphase or coupled reactions; and (3) exsolution or solvus reactions. To the extent that the participating phases are pure, stoichiometric end-members and do not contain other components, the reactions are independent of the bulk composition of the rock and the chemical potentials of the various mobile components. In such cases, the only important chemical consideration is that the bulk chemistry of the rock be appropriate for the crystallization of the requisite minerals.

(1) Important examples of simple polymorphic conversions include orthoenstatite–protoenstatite, calcite–aragonite, kyanite–andalusite–sillimanite, the various structural modifications of ice and of silica, and diamond–graphite. An experimental phase diagram for the system $MgSiO_3$ was presented as Figures 5.12 and 5.14. Laboratory determinations and a thermochemical calculation of P–T values for reactions relating the $CaCO_3$ polymorphs were illustrated in Figure 3.2. The effect of strain on this equilibrium has been well-documented by Newton, Goldsmith, and Smith (1969). Phase relations among the aluminosilicate polymorphs were presented in Figure 2.7.

Let us briefly consider the variation of the Gibbs free energy with pressure for the Al_2SiO_5 polymorphs. At constant temperature, the change of molar G as a function of P for each polymorph is given by equation (1.33), $(\partial G/\partial P)_T = V$. A diagrammatic plot of Gibbs free energy against pressure for this unary system is presented in Figure 6.5 for an isotherm at 500°C. You can see that at low pressures andalusite is stable; whereas at a, at a pressure of 4.2 kilobars, andalusite + kyanite coexist, and at greater pressures, kyanite is the stable polymorph. At this temperature, sillimanite is not an equilibrium phase at any pressure. However, metastable equilibrium between a lower-pressure assemblage, sillimanite, and a higher-pressure assemblage, kyanite, occurs at b at a pressure of 2.5 kilobars, provided the formation of andalusite is inhibited; likewise, if the crystallization of kyanite does not take place, andalusite persists to c at 8.5 kilobars, above which pressure it metastably inverts to sillimanite.

Phase diagrams for the systems H_2O and SiO_2 were shown in Figures 4.3 and 4.4, respectively. All these transitions presented above have important applications to metamorphic rocks.

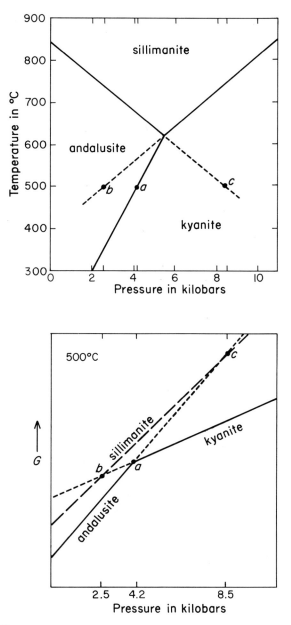

FIGURE 6.5.

P–T diagram after Richardson, Gilbert, and Bell (1969) and schematic *G–P* diagram for the system Al_2SiO_5. Note that temperature is plotted along the ordinate, pressure along the abscissa in the upper diagram. At 500°C, andalusite and kyanite are in chemical equilibrium at 4.2 kilobars pressure. At 500°C, metastable sillimanite forms from metastable kyanite (in the absence of the stable polymorph andalusite) at pressures less than 2.5 kilobars, and from metastable andalusite at pressures exceeding 8.5 kilobars (in the absence of the stable polymorph kyanite).

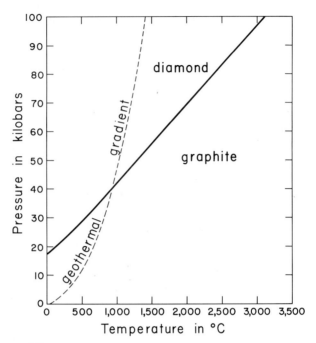

FIGURE 6.6.

Polymorphic transformations in the carbon system, after Bundy *et al.*, (1961) and Berman (1962). The subcontinental geothermal gradient computed by Clark and Ringwood (1964) is also shown.

The *P–T* stability ranges of diamond and graphite are illustrated in Figure 6.6. As is clear from this diagram, the Earth's subcontinental geothermal gradient passes into the diamond field only at lithostatic pressures approaching 40 kilobars. This pressure corresponds to a depth of about 120–130 kilometers—well within the upper mantle. Graphite represents the stable form of crystalline carbon under all crustal metamorphic conditions. Therefore, the persistance of diamonds in your girlfriend's necklace reflects the phenomenon of metastability—if they are really diamonds.

Using the diamond–graphite reaction as an example, recall from equation (1.73) that

$$\Delta G_{graphite-diamond} = -RT \ln K + RT \ln a_{graphite}/a_{diamond}.$$

The reader may be wondering how, in a situation that involves only pure phases with fixed activities, the equilibrium constant, K, can be balanced by the activity quotient, $a_{graphite}/a_{diamond}$. The answer lies in the fact that although the activities of the participating species are constant at any given P and T

independent of the phase proportions (and bulk composition in a multicomponent system), the activity values are unity only at one atmosphere, and change gradually as pressure increases. Combining equations (1.33), $(\partial G/\partial P)_T$ $= V$, and (1.64), $dG = RT\, d \ln a$, results in the expression

$$\left(\frac{\partial \ln a}{\partial P} \right)_T = \frac{V}{RT}, \tag{6.1}$$

which shows the effect of pressure on activity. At one atmosphere and the equilibrium temperature for the reaction diamond–graphite (the one-atmosphere terminus for the univariant diamond–graphite P–T curve appears to be located at temperatures below absolute zero, judging from Figure 6.6), the value of K would be unity, since at equilibrium it must balance the activity quotient. Now consider how the change of physical conditions away from the one-atmosphere equilibrium temperature influences K. Even if we choose to define G^0 in a way that precludes a pressure dependence (but see discussion on pp. 20–21, and equation 1.79), the equilibrium constant clearly departs from unit value as temperature changes; this is evidenced from equation (1.77), $[\partial \ln K/\partial(1/T)]_P$ $= -\Delta H^0/R$. Obviously, only where the activity quotient exactly equals K will the molar Gibbs free energies of reactant and product assemblages be the same; this situation defines the P–T locus of the diamond–graphite univariant curve. If physical conditions are displaced from those required for equilibrium between diamond and graphite, the activity quotient will be either too large or too small compared to K, and the product or reactant will begin to disappear. Such a change in phase proportions fails to influence the activity quotient, however, because the activities of pure, stoichiometric phases are independent of their amounts. For this reason, reactions such as diamond–graphite run to completion, provided chemical equilibrium is achieved, whenever physical conditions are displaced from the univariant P–T curve.

(2) Multiphase reactions which neither evolve nor consume volatile species do not appear to be very common in metamorphic processes. Several examples of such subsolidus reactions, actually more appropriate to mantle petrogenesis, were illustrated in P–T diagrams in the last chapter (e.g., see Figures 5.2, 5.3, 5.4, 5.6, and 5.8); the general natures of these reactions are clear, but because of their chemical complexity—hence multivariancy of the equilibria—it has usually not been possible to write demonstrably correct, balanced equations. Examples which are easily balanced stoichiometrically include the following equilibria of possible crustal metamorphic significance:

(a) $\quad \underset{\text{(talc)}}{Mg_3Si_4O_{10}(OH)_2} + \underset{\text{(enstatite)}}{4MgSiO_3} = \underset{\text{(anthophyllite)}}{Mg_7Si_8O_{22}(OH)_2};$

(b) $\quad \underset{\text{(talc)}}{Mg_3Si_4O_{10}(OH)_2} + \underset{\text{(diopside)}}{2CaMgSi_2O_6} = \underset{\text{(tremolite)}}{Ca_2Mg_5Si_8O_{22}(OH)_2};$

(c)   (lawsonite)   (glaucophane)
$$2CaAl_2Si_2O_7(OH)_2 \cdot H_2O + 5Na_2Mg_3Al_2Si_8O_{22}(OH)_2$$

(albite)   (tremolite)   (chlorite)
$$= 10NaAlSi_3O_8 + Ca_2Mg_5Si_8O_{22}(OH)_2 + 2Mg_5AlSi_3AlO_{10}(OH)_8;$$

(d)   (lawsonite)   (quartz)   (wairakite)
$$CaAl_2Si_2O_7(OH)_2 \cdot H_2O + 2SiO_2 = CaAl_2Si_4O_{12} \cdot 2H_2O;$$

(e)   (jadeite)   (quartz)   (albite)
$$NaAlSi_2O_6 + SiO_2 = NaAlSi_3O_8;$$

((f)   (grossular)   (quartz)   (anorthite)   (wollastonite)
$$Ca_3Al_2Si_3O_{12} + SiO_2 = CaAl_2Si_2O_8 + 2CaSiO_3;$$

(g)   (iron cordierite)   (almandine)   (sillimanite)   (quartz)
$$3Al_3Fe_2^{2+}Si_5AlO_{18} = 2Fe_3^{2+}Al_2Si_3O_{12} + 4Al_2SiO_5 + 5SiO_2.$$

The physical conditions for equilibrium (a) have been computed by Greenwood (1963, 1970) and by Zen (1970) employing slightly different thermochemical data. Their resultant $P-T$ curves are presented in Figure 6.7. Obviously, rather large disparities in the computed univariant curves attend small differences in the calorimetric data. However, it is evident from Figure 6.7 that, topologically, the anthophyllite field of stability is replaced at low temperatures and high pressures by the chemically equivalent assemblage of talc + enstatite. It appears that this reaction takes place under crustal conditions, as indicated, for instance, by occurrences of both assemblages in amphibolite facies metaperidotites of the Lepontine Alps (Evans and Trommsdorff, 1974).

The $P-T$ conditions defined by reaction (b) have been calculated using the method of Chapter 3 (specifically equation 3.7) and calorimetric data from Appendix 2. The computed equilibrium pressure at 25°C is

$$P - 1 = \frac{3083 \cdot 41.86}{4.49} = 28.7 \text{ kilobars.}$$

The value for the field boundary slope is

$$\frac{dP}{dT} = \frac{0.45 \cdot 41.86}{4.49} = 4.19 \text{ bars/°C.}$$

The univariant curve thus determined is illustrated in Figure 6.8. Quite clearly, tremolite should be stable under all crustal metamorphic conditions relative to the talc + diopside assemblage. However, it must be noted that the sum of the uncertainties in the entropies of tremolite, talc, and 2 diopside is ±0.85 cal/deg/mole of tremolite, hence there is *very* little control on the $P-T$ curve slope; likewise, the aggregate uncertainties in standard Gibbs free energies of formation of the participating phases far exceeds the ΔG^0 for the reaction.

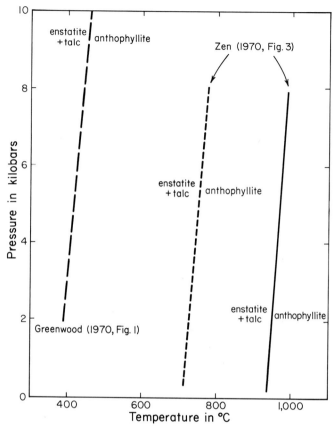

FIGURE 6.7.

Solid state reaction (a), $Mg_3Si_4O_{10}(OH)_2 + 4MgSiO_3$
$= Mg_7Si_8O_{22}(OH)_2$, computed from various, slightly different
thermochemical data by Greenwood (1963, 1970, and personal
communication) and Zen (1970).

Nevertheless, tremolite is a common mineral throughout a wide range of
crustal metamorphic conditions, whereas the assemblage talc + diopside is
not (e.g., see parageneses described by Trommsdorff and Evans, 1974) in harmony with the computed phase diagram.

Reaction (c) cannot be computed directly, because calorimetric data for
several of the participating phases have not been obtained. The volume change
of the reaction, as written, is about 10 per cent. Employing a method described
by Fyfe, Turner, and Verhoogen (1958, p. 34), we can approximate entropies
of the various minerals by adding the available entropy values for the constituent (crystalline) oxides in appropriate proportions, and making a correction
where the phase of interest has a molar volume different from that of the oxide
sum. Given such an estimation, the entropy change at 300°C and 3,000 bars

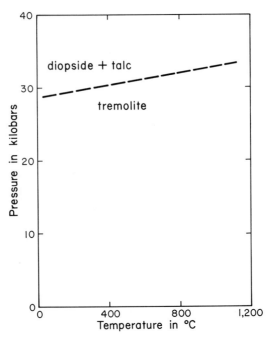

FIGURE 6.8.

Solid-state reaction (b), $Mg_3Si_4O_{10}(OH)_2$ + $2CaMgSi_2O_6$ = $Ca_2Mg_5Si_8O_{22}(OH)_2$, computed from thermo-chemical data of Appendix 2 and employing equations (3.7) and (1.49). The P–T curve locations coincide with that computed by Ahrens and Schubert (1975, p. 391 and Figure 10); see also Essene, Hensen, and Green (1970, Figure 3).

pressure for the reaction is about 5 per cent (Ernst, 1963a, Table 6). Hence the assemblage glaucophane + lawsonite is the relatively high-pressure, low-temperature analogue of the greenschist assemblage tremolite + albite + chlorite.

Results from phase-equilibrium experiments for reactions (d)–(g) are assembled in Figure 6.9; of necessity, several different bulk compositions are projected onto a single P–T plot in order to save space and to facilitate comparison. As with the polymorphic transitions and the multiphase or coupled reactions just discussed, the activities of the volatile components do not affect the equilibria, because volatiles do not take part in these reactions. However, except for the various structural modifications of SiO_2, $H_2O(?)$, Al_2SiO_5, and C—and perhaps to a first approximation, talc and lawsonite—all the participating phases in these reactions exhibit solid solution with other constituents. Because mineral groups including the garnets, pyroxenes, amphiboles, micas, feldspars, and zeolites are complex solid solutions which reflect rock bulk chemistry, extensive variables, such as the normative **Ab/An**, **En/Fs**, and

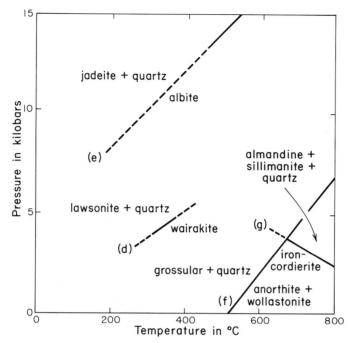

FIGURE 6.9.

Experimentally established solid-solid phase equilibria, dashed where extrapolated:
(d) $CaAl_2Si_2O_7(OH)_2 \cdot H_2O + 2SiO_2 = CaAl_2Si_4O_{12} \cdot 2H_2O$ (Liou, 1971a);
(e) $NaAlSi_2O_6 + SiO_2 = NaAlSi_3O_8$ (Birch and LeComte, 1960; Newton and Smith, 1967; Boettcher and Wyllie, 1968b);
(f) $Ca_3Al_2Si_3O_{12} + SiO_2 = CaAl_2Si_2O_8 + 2CaSiO_3$ (Newton, 1966b); and
(g) $3Al_3Fe_2^{2+}Si_5AlO_{18} = 2Fe_3^{2+}Al_2Si_3O_{12} + 4Al_2SiO_5 + 5SiO_2$ (Richardson, 1968).

Di/Hd ratios, will influence the variance and P–T locations of most of the equilibria we have treated. This effect has been verified in the laboratory, for instance, by Newton and Smith (1967, Figure 7), who presented different P–T curves for reaction (e), jadeite + quartz = albite, as a pure binary system on the one hand, and with minor clinopyroxene solid solution on the other.

To find out why this is so, let us begin by considering the pure binary reaction (e) as presented in Figure 6.9. The stability relations are illustrated in Figure 6.10 in terms of an isobaric G–T diagram at about 10 kilobars (see also Figure 4.7 and discussion on pp. 77–81). In this binary system, the heavy curves representing both albite and jadeite + quartz assemblages have negative slopes and are slightly concave to the temperature axis, as required by equation (1.32), $(\partial G/\partial T)_P = -S$. Now suppose that we add a small amount of another constituent which is soluble in the clinopyroxene but not in the albite— for instance, $CaO \cdot MgO \cdot 2SiO_2$. Regardless of its molar Gibbs free energy, because the same amount of the ternary component **Di** must be added to both

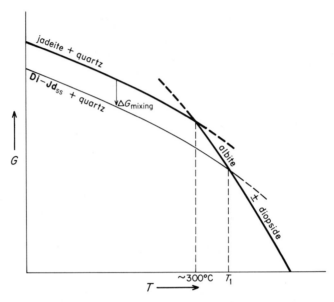

FIGURE 6.10.

Ten-kilobar isobaric diagram of molar Gibbs free energy versus temperature for the composition $Na_2O \cdot Al_2O_3 \cdot 6SiO_2$ (heavy lines) and for the composition $Na_2O \cdot Al_2O_3 \cdot 6SiO_2$ + minor $CaO \cdot MgO \cdot 2SiO_2$ (light line). For convenience, the Gibbs free energy of diopside is taken as zero.

reactant and product assemblages, there will be no net displacement of the isobaric temperature of intersection (= the equilibrium temperature, approximately 300°C) of the G–T curves for the mechanical mixtures of crystalline jadeite + diopside + quartz on the one hand, and of albite + diopside on the other. However, although the latter assemblage represents a stable compatibility, diopside and jadeite will spontaneously mix, yielding a concomitant decrease in Gibbs free energy; the effect can be seen from equation (1.29), $\Delta G_{mixing} = \Delta H_{mixing} - T \Delta S_{mixing}$. This miscibility results in a higher temperature of intersection—hence equilibrium at temperature T_1—between reactant and product assemblages. The magnitude of the temperature increment depends on the proportion of $CaO \cdot MgO \cdot 2SiO_2$ added, and on the values of the heat and entropy of mixing. Thus addition of **Di** component to the system causes the climopyroxene$_{ss}$ field of stability to expand to higher temperatures and lower pressures at the expense of albite.

Of course, similar arguments hold for other compositional variables, such as Na/K and Fe^{3+}/Al substitutions. The multiphase reactions we have considered, with the exception of (d), were all presented in terms of magnesian end-members, whereas we know that most rock bulk compositions are characterized by more or less similar proportions of MgO and FeO. How the P–T coordinates and variance of these equilibria—and polymorphic-type transitions among the iron- and magnesium-bearing carbonates—will be displaced with the change in Fe^{2+}/Mg depends on whether the solid solution causes a greater decrease in Gibbs free energy in the reactant assemblage or in the product assemblage.

Let us now consider another aspect of the influence of rock bulk chemistry on phase equilibria. Thus far, the stability range of sodic pyroxene in the presence of quartz has been discussed and illustrated in Figures 6.9 and 6.10. However, as was pointed out in Chapter 4 (p. 78), a mineral possesses its maximum P–T stability range in a rock of its own bulk composition. We will therefore investigate the stability of jadeite itself, and of various phase combinations, in the binary system $\frac{1}{2}Na_2O \cdot Al_2O_3 \cdot 2SiO_2$–$SiO_2$ (i.e., $NaAlSiO_4$–SiO_2). Phase relations are illustrated in Figure 6.11. For the bulk composition of jadeite, $NaAlSi_2O_6$, sodic pyroxene is stable on the high-pressure, low-temperature side of the univariant P–T curve (h), which is defined by the reaction 2 jadeite = nepheline + albite. Within the field bounded by curves (e) and (h), jadeite and quartz are incompatible, as are nepheline and albite. It would be instructive for the student to show, by means of a G–T or G–P diagram, how the presence or absence of excess silica influences the stability ranges of jadeite and albite in this simple binary system.

For your interest, an isothermal G–x diagram for the Ne–Si system at 727°C (= 1,000°K) and at isobaric values of one atmosphere, 15, and 30 kilobars pressure has been constructed from the thermochemical data of Appendixes 2 and 3, and is presented in Figure 6.12. The values for molar Gibbs free energy at high pressures were computed using equation (1.33), $(\partial G/\partial P)_T = V$. As a first approximation, because the compressibilities and thermal expansions of crystalline solids are small terms, the molar volumes at one atmosphere and 298°K were taken as constant throughout the P–T range of interest. To construct this simple binary system, for which $NaAlSiO_4$ and SiO_2 have been selected as components, note that $\frac{1}{2}$ jadeite = 50Ne,50Si (i.e., $\frac{1}{2}NaAlSiO_4$ + $\frac{1}{2}SiO_2$), and that $\frac{1}{3}$ albite = 33Ne,67Si (i.e., $\frac{1}{3}NaAlSiO_4$ + $\frac{2}{3}SiO_2$); hence, for G values employed, $\frac{1}{2}$ the value of the molar Gibbs free energy for jadeite and $\frac{1}{3}$ the value for albite at the various pressures were plotted. From the P–T and G–x diagrams of Figures 6.11 and 6.12, respectively, we can see that jadeite is metastable with respect to the assemblage albite + nepheline at low pressures; that the associations nepheline + jadeite, jadeite + albite, and albite + quartz are all stable for different rock bulk compositions under the P–T conditions limited by curves (e) and (h) of Figure 6.11; and that albite is metastable relative to jadeite + quartz at elevated pressures.

(3) Exsolution reactions are as important to metamorphic petrology as are the previous types of subsolidus reaction. This is because, provided solvus pairs have formed in chemical equilibrium with one another, and have not

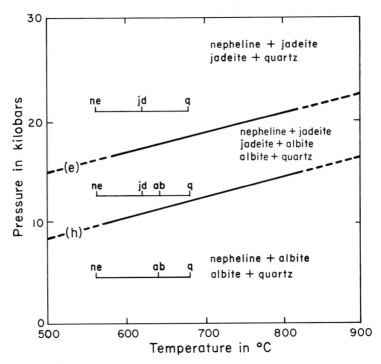

FIGURE 6.11.

Phase diagram for the binary system $Na_2O \cdot Al_2O_3 \cdot 2SiO_2 - SiO_2$, after Boettcher and Wyllie (1968b). Curve (e) corresponds to the reaction

$$\underset{\text{(jadeite)}}{NaAlSi_2O_6} + \underset{\text{(quartz)}}{SiO_2} = \underset{\text{(albite)}}{NaAlSi_3O_8}$$

presented in Figure 6.8. Curve (h) is the reaction

$$\underset{\text{(jadeite)}}{2NaAlSi_2O_6} = \underset{\text{(nepheline)}}{NaAlSiO_4} + \underset{\text{(albite)}}{NaAlSi_3O_8},$$

and compares well with the earlier laboratory investigation by Robertson, Birch, and MacDonald (1957).

suffered compositional change since the unmixing process, their compositions may be used to establish the attending physical conditions of metamorphism. Of course, since recrystallization takes place over both prograde and retrograde *P–T* trajectories sustained by the enclosing rock, any given assemblage must show the effects not just of the metamorphic culmination, but also of the subsequent history. However, because rates of reaction diminish rapidly with falling temperature, most equilibria are regarded as having been "quenched in," or immobilized, near the *P–T* peak of metamorphic conditions.

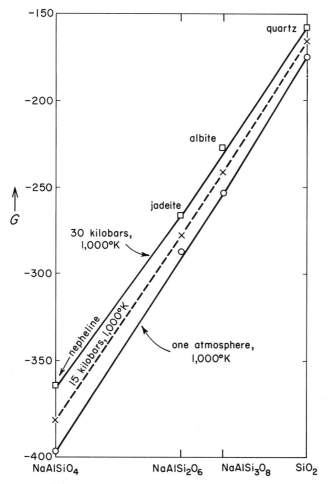

FIGURE 6.12.

Diagram of molar Gibbs free energy against composition for the
system $NaAlSiO_4$–SiO_2 at 727°C and pressures of one atmosphere,
15 kilobars, and 30 kilobars; thermochemical data are from
Appendixes 2 and 3. In terms of the end members $NaAlSiO_4$
and SiO_2, G values for $\frac{1}{2}$ mole jadeite and $\frac{1}{3}$ mole albite have
been plotted.

Solvus P–T relations were treated for the alkali and ternary feldspars in
Figures 4.16, 4.20, and 4.41 to 4.44, and for the diopside–enstatite solid-solution
series in Figure 5.7. The Gibbs-free-energy relationships and discussions of the
general phenomena of equilibrium and spinodal solvi were presented in Chapter 4 (pp. 92–94) and in Figures 4.18, 4.19, and 4.21.

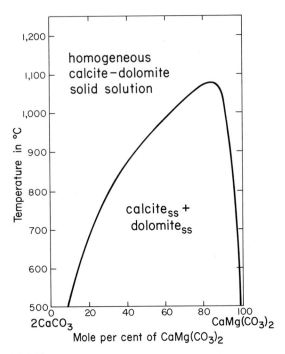

FIGURE 6.13.

Isobaric temperature-composition phase relations for the system $2CaCO_3$–$CaMg(CO_3)_2$, after Goldsmith (1959, Figure 1); the CO_2 pressure is low (e.g., $\approx$ 50 bars), but is sufficient to prevent decarbonation. The isothermal pressure effect on $MgCO_3$ solubility in calcite$_{ss}$ (in equilibrium with dolomite$_{ss}$) between 500 and 800°C and from 1 to 25 kilobars is about 0.12 mole per cent $MgCO_3$ per kilobar, according to Goldsmith and Newton (1969).

The isobaric calcite–dolomite solid-solution series and unmixing region are illustrated in the *T–x* diagram of Figure 6.13. Dolomite, a binary compound between $CaCO_3$ and $MgCO_3$, exhibits only minor solid solution toward calcite under geologically reasonable metamorphic conditions—reflecting the fact that the crystal structure involves a rather rigorous ordering of Ca^{2+} and Mg^{2+} in nonequivalent sites which alternate along the *c* axis. Calcite does incorporate moderately large amounts of magnesium with increasing temperature (naturally, the degree of disorder, under equilibrium conditions, also increases with rising temperature); hence this limb of the solvus is the most useful for establishing the physical conditions under which the coexisting carbonate pair will crystallize. Unfortunately, on cooling, calcite$_{ss}$ and dolomite$_{ss}$ anneal rather easily, resulting in continued unmixing as the material returns toward surface conditions. But what can be said about the P–T conditions of origin for a metamorphic rock which contains only a single carbonate phase, for example, a magnesian calcite? Evidently, for a given pressure of formation, the composition of the calcite provides only a minimum temperature of crystallization, because it is not saturated with Mg.

Before concluding this section dealing with various kinds of solid-state equilibria, we should briefly consider the order-disorder phenomenon. We have seen that any factor which differentially changes the Gibbs free energy of reactant or product assemblage will *displace* the set of physical conditions within which equilibrium is attained. Since different degrees of cation (or anion) ordering among participating phases affect the entropy change of a reaction, it is clear from equation (1.29), $\Delta G = \Delta H - T \Delta S$, that this parameter will influence the P–T stability ranges of the mineral assemblages. For instance, the degree of Si–Al ordering in albite (Hlabse and Kleppa, 1968) and in sillimanite (Zen, 1969; Greenwood, 1972) influences the P–T location of the jadeite + quartz = albite, jadeite = nepheline + albite, and aluminosilicate polymorphism transitions. Recalling previous G–T and G–P diagrams, you should be able to predict how these equilibria will be displaced in the contrasting cases of perfect order and marked disorder.

REACTIONS INVOLVING CO₂

Subsolidus equilibria involving decarbonation have already been discussed for the reactions calcite + quartz + anatase = sphene + CO_2, and calcite + quartz = wollastonite + CO_2. The first reaction was shown, as a plot of $\log f_{CO_2}$ against $1/T_K$, in Figure 1.1; the second was illustrated as a plot of P_{CO_2} against T in Figure 3.5. The physical parameters defined by such univariant equilibria can be calculated from the thermochemical and volumetric data of Appendixes 2, 3, and 6, as discussed in previous sections. We use equation (3.7) to compute the value of P_{CO_2} ($= P_{total}$) at some specified temperature; then choose several other values of temperature and compute the CO_2 pressure at equilibrium between reactants and products. Alternatively, we could assume that the ΔH of the reaction is constant throughout the temperature interval of concern, and employ equation (1.56) after having determined one P–T point on the univariant curve. Or we could use the method described for the brucite-periclase + H_2O equilibrium on pp. 48–50. In all of these approaches, it is important to remember that our equations are exact only where fugacities of the volatile constituents are used—in this case, CO_2. Such values may be calculated at any conditions for which PVT data are available (e.g., see Appendix 6), since we know from equations (1.58) and (1.60) that, at constant temperature, $dG = V_i \, dP_i$, and $dG = RT \, d \ln f_i$. Hence, we have the relationship

$$\left(\frac{\partial \ln f_i}{\partial P_i} \right)_T = \frac{V_i}{RT}. \tag{6.2}$$

The influence of bulk composition on stable phase assemblages involving, for instance, simple decarbonation reactions is illustrated in Figure 6.14. Here we investigate subsolidus equilibria in the ternary system $CaO-SiO_2-CO_2$. The five phases of concern here are: lime, a silica polymorph, carbon dioxide gas, and two binary compounds, calcite and wollastonite. In Figure 6.14a, curve (1) defines the $T-P_{CO_2}$ locus of the reaction whereby the lower-grade compatibility of calcite + quartz is replaced by the higher-grade association of wollastonite + CO_2. Curve (2) gives the breakdown of calcite itself to lime and carbon dioxide. Now consider the contrasting bulk compositions w, x, y, and z in our ternary system. The various phase assemblages for such "rocks" are listed as a function of physical conditions in Table 6.2. It is easy to see from a perusal of this tabulation that bulk chemistry exerts a powerful control on the equilibrium assemblage. Moreover, as $P-T$ conditions and hence phase compatibilities change, whether or not an individual "rock" will exhibit the effects of a specific univariant equilibrium curve will also depend on its bulk composition.

The importance of the presence or absence of a silica polymorph on calcite stability is evident in Figure 6.14a. The isobaric Gibbs-free-energy relations are shown diagrammatically in Figure 6.14b. Although the mechanical addition of SiO_2 does not change the relative Gibbs free energies of either the calcite-bearing or the lime-bearing assemblage, the latter is silica-undersaturated, and SiO_2 + CaO spontaneously react to produce wollastonite; this results in a net lowering of the total G for the new product assemblage relative to that of the reactant assemblage. Therefore at constant pressure, in the presence of silica, calcite decomposes at a lower temperature than in its absence.

In a classic paper, Bowen (1940) treated the progressive decarbonation of siliceous dolomitic marbles from a chemographic point of view. He recognized a series of devolatilization steps, based on natural assemblages. Each is characterized by the disappearance of a lower-grade phase and/or by the appearance of a higher-grade phase. Most reactions evolve carbon dioxide, but a few produce or consume H_2O. With rising temperature and roughly constant P_{fluid}, these index minerals, in order of appearance, were postulated to be tremolite, forsterite, diopside, periclase, wollastonite, monticellite, akermanite, spurrite, merwinite, and larnite. Although the number and exact sequence of phases has been revised subsequently, and is still under investigation (in fact, the univariant curves for some of the critical reactions probably intersect in $P_{total}-P_{CO_2}-P_{H_2O}-T$ space), the general phenomenon of progressive decarbonation is well-understood. A few of the experimentally established curves of T

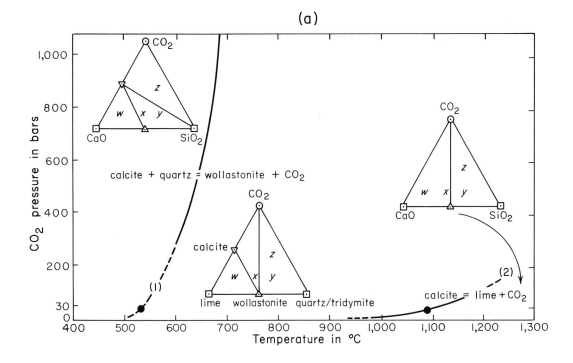

(a)

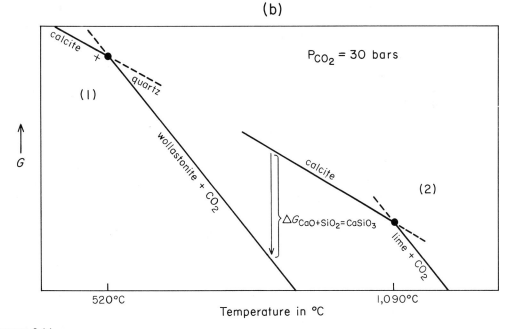

(b)

FIGURE 6.14.

(a) Phase equilibria in the system $CaO-SiO_2-CO_2$ after Smyth and Adams (1923); Harker and Tuttle (1956); and Greenwood (1967a). Curve (1) has been calculated by Danielsson (1950). (b) Isobaric Gibbs free energy against temperature plot at 30 bars CO_2 pressure for the equilibria: (1) $CaCO_3 + SiO_2 = CaSiO_3 + CO_2$; and (2) $CaCO_3 = CaO + CO_2$. Slight concavity to the temperature axis of G–T curves is not shown.

TABLE 6.2.

Phase assemblages for contrasting bulk compositions in the system $CaO-SiO_2-H_2O$, from Figure 6.14.

P–T region	Bulk composition			
	w	x	y	z
T < curve (1)	lime + calcite + wollastonite	calcite + wollastonite + quartz	calcite + wollastonite + quartz	calcite + quartz + CO_2
curve (1) < T < curve (2)	lime + calcite + wollastonite	calcite + wollastonite + CO_2	wollastonite + quartz/tridymite + CO_2	wollastonite + quartz/tridymite + CO_2
T > curve (2)	lime + wollastonite + CO_2	lime + wollastonite + CO_2	wollastonite + tridymite + CO_2	wollastonite + tridymite + CO_2

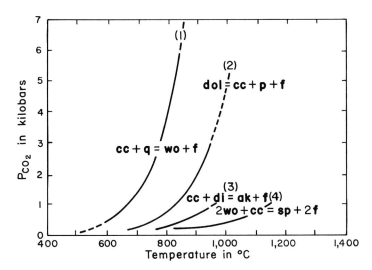

FIGURE 6.15.

Some simple decarbonation reactions, after the summary by Turner (1968, Figure 4.8). The reactions are as follows: (1) $CaCO_3 + SiO_2 = CaSiO_3 + CO_2$; (2) $CaMg(CO_3)_2 = CaCO_3 + MgO + CO_2$; (3) $CaCO_3 + CaMgSi_2O_6 = CaMgSi_2O_7 + CO_2$; and (4) $2CaSiO_3 + 3CaCO_3 = 2Ca_2SiO_4{\cdot}CaCO_3 + 2CO_2$. Abbreviations are: **ak** = akermanite; **cc** = calcite; **di** = diopside; **dol** = dolomite; **f** = CO_2; **p** = periclase; **q** = quartz; **sp** = spurrite; and **wo** = wollastonite.

against P_{CO_2} are summarized in Figure 6.15; at high temperatures and low carbon-dioxide pressures, other reactions (not illustrated) involve tilleyite and rankinite.

REACTIONS INVOLVING H$_2$O

Progressive metamorphism results in an over-all devolatilization of the pre-existing rocks, except at very low metamorphic grades, where certain initially relatively dry protoliths become extensively hydrated. The chief volatile constituent consumed or generated during recrystallization is H$_2$O, although, as we have seen in the last section, CO$_2$ may predominate in carbonate-bearing lithologies; moreover, for many rock bulk compositions, the expulsion or addition of hydrogen, methane, H$_2$S, ammonia, oxygen, or the halogens may exert important controls over the mineral paragenesis. Nevertheless, because of its greater abundance in crustal rocks, H$_2$O is the predominant volatile species to be considered.

The general aspect of experimentally established dehydration curves has been presented already in Figures 2.6 and 3.4 for the reaction brucite = periclase + H$_2$O. Stability relations of hornblende in the compositionally complex system basalt + H$_2$O were illustrated in Figures 5.13a (excess H$_2$O) and

5.13b (trace amounts of H_2O). In addition, chemographic relations for the occurrence of muscovite in the quaternary system $K_2O \cdot Al_2O_3 \cdot 6SiO_2 - Al_2O_3 - SiO_2 - H_2O$, and of analcime in the ternary system

$$Na_2O \cdot Al_2O_3 \cdot 2SiO_2 - Na_2O \cdot Al_2O_3 \cdot 6SiO_2 - H_2O,$$

were presented schematically in Figures 3.7 and 3.8, respectively. In these and other diagrams, the characteristic devolatilization $P-T$ curve is concave to the pressure axis, because, at low confining pressures, a reaction such as hydrous mineral(s) = anhydrous mineral(s) + H_2O involves a very large ΔV because of the large specific volume of H_2O. However, since the tenuous fluid is quite compressible, ΔV for the reaction diminishes substantially at elevated pressures. Thus, from relationship (1.49), the Clapeyron equation, it is evident that the quotient dP/dT is a small positive number at low pressures, but tends toward infinity as pressure increases.

In most conventional hydrothermal syntheses, $P_{fluid} = P_{total} \approx P_{H_2O}$ for experimental convenience (see Chapter 2, pp. 27–29); but as discussed in Chapter 3 (pp. 51–53), given such univariant $P_{fluid}-T$ curves, there are methods of computing the equilibria where $P_{fluid} < P_{total}$. As we also noted in the section in Chapter 2 dealing with control of volatile components (e.g., see Figure 2.6), and as we shall discover in the later section of this chapter concerning a multicomponent fluid phase, laboratory techniques allow for yet another way of performing experiments in which $P_{H_2O} < P_{total}$, namely, by dilution or contamination of the fluid phase. In this section, however, we will focus on reactions in which the pressure of an aqueous fluid, essentially pure H_2O in composition, is equal to the confining pressure.

First we examine equilibria in the ternary (or pseudoternary) system $CaO \cdot Al_2O_3 \cdot 2SiO_2 - SiO_2 - H_2O$. This is a system which to some extent models the transitions among zeolite, prehnite–pumpellyite, greenschist, and blueschist facies mineral assemblages and their $P-T$ conditions, although the attending plagioclase in nature much more closely approaches the **Ab** rather than the **An** component in composition. A phase diagram in which the aqueous ($\approx H_2O$) fluid pressure equals total pressure is presented in Figure 6.16. In these experiments, only ternary equilibria were observed for the bulk composition **An** + 2**Si** + excess H_2O, and the stability fields for the minerals laumontite and wairakite (zeolite facies), lawsonite + quartz (blueschist facies) and anorthite + quartz or zoisite + kyanite + quartz ($\approx$ greenschist facies?) were elucidated. Typical dehydration curves are represented by the equilibria (a), laumontite = wairakite + $2H_2O$, (b), wairakite = anorthite + 2 quartz + $2H_2O$, and (e), lawsonite = anorthite + $2H_2O$; whereas reaction (c), lawsonite + 2 quartz + $2H_2O$ = laumontite, is a hydration boundary, and (d), lawsonite + 2 quartz = wairakite, is a solid-solid equilibrium. Only curve (c) is unusual because, for this reaction, H_2O is consumed on the isobaric increase of temperature; the curve slope is positive but gentle in $P-T$ coordinates

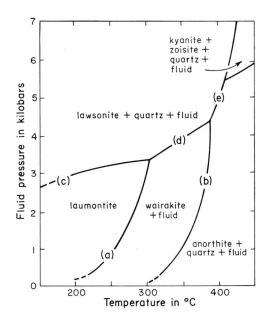

FIGURE 6.16.
Conventional $P_{fluid}-T$ diagram for the bulk composition $CaO \cdot Al_2O_3 \cdot 4SiO_2$ + excess H_2O (after Liou, 1971a, Figure 5; see also Nitsch, 1968; Thompson, 1970). High-pressure phase equilibria involving zoisite and kyanite are taken from an investigation by Newton and Kennedy (1963). Compositions of the hydrous calcium–aluminosilicates are: lawsonite = $CaAl_2Si_2O_7(OH)_2 \cdot H_2O$; laumontite = $CaAl_2Si_4O_{12} \cdot 4H_2O$; and wairakite = $CaAl_2Si_4O_{12} \cdot 2H_2O$.

and is slightly concave to the temperature axis. Remembering the Clapeyron equation, we can explain the $P-T$ disposition of this equilibrium as follows, keeping in mind the fact that temperatures are below the critical point for H_2O (see Figure 4.3); thus the fluid behaves like liquid water, and of course is characterized by a small molar volume. In contrast to the very dense, highly ordered lawsonite, the zeolite laumontite, being an open-framework structure, possesses both a large molar volume and a relatively high entropy value; thus the reaction as written involves a negligible entropy increase and a relatively large volume increase, even though H_2O is being consumed. With increasing temperature, the entropy increase diminishes as the participating fluid is heated, resulting in a negative sign for d^2P/dT^2.

The chemographic relationship of bulk composition to phase assemblage is illustrated in Figure 6.17, a schematic version of the system $CaO \cdot Al_2O_3 \cdot 2SiO_2$–$SiO_2$–$H_2O$. Three of the reactions are binary rather than ternary, as is evident from the compositional degeneracy (i.e., nongenerality of phase compositions) as well as from phase-rule considerations. For instance, there are only three participating phases in the binary equilibria: (a) laumontite = wairakite + H_2O (lawsonite and quartz being nonparticipating or indifferent phases); (d) lawsonite + 2 quartz = wairakite (laumontite and fluid being indifferent phases); and (e) lawsonite = anorthite + H_2O (indifferent phases wairakite and quartz). The other reactions presented are truly ternary, however. Compositional degeneracy is also reflected in the number of univariant curves which radiate out from the ternary invariant points (see pp. 53–55).

Suppose now that, by some mechanism such as dilution of the fluid phase,

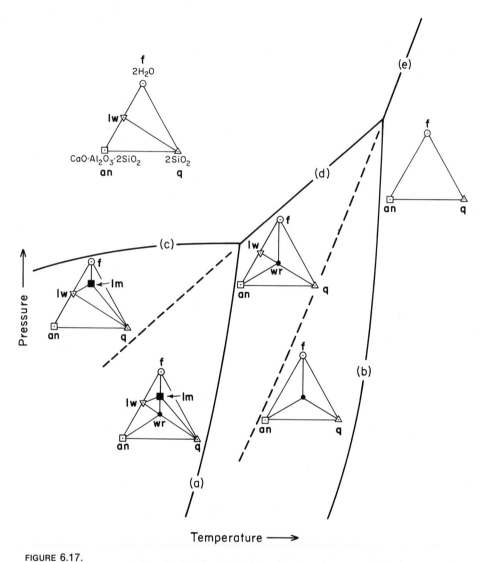

FIGURE 6.17.

Diagrammatic phase relations in the system CaO·Al$_2$O$_3$·2SiO$_2$–2SiO$_2$–2H$_2$O, after Figure 6.16. Abbreviations are: **an** = anorthite; **f** = fluid; **lm** = laumontite; **lw** = lawsonite; **q** = quartz; **wr** = wairakite. Phases individually stable at the ternary invariant point defined by the intersection of curves (a), (c), and (d) are : **f, lm, lw, q, wr**. Phases individually stable at the ternary invariant point defined by the intersection of curves (b), (d), and (e) are: **an, f, lw, q, wr**.

or fluid communication to a lower-pressure regime through a fissure system, or because there is not enough volatile constituent for phase saturation, the activity, partial pressure, fugacity, or chemical potential of H$_2$O were reduced. How would the stability fields illustrated in Figure 6.16 be displaced? Evidently the *P–T* location of curve (d) would be unaffected because it represents a solid-solid reaction, and fluid is not a participating phase. Curves (a), (b), and (e) would decline to lower temperatures for a given total pressure, because the more hydrous condensed assemblages are disfavored compared to the less hydrous condensed assemblages (+ aqueous fluid) of equivalent bulk composition as the activity of H$_2$O diminishes (e.g., see Figure 2.6). From equation (1.64), $dG = RT \, d \ln a$, it is clear that, as the activity of the free (or uncombined) volatile species decreases, the Gibbs free energy of the fluid-bearing assemblage sympathetically declines. Therefore, similar to the isobaric G–T diagram shown in Figure 4.7 (see also Figure 6.10), diminution in H$_2$O activity causes a contraction of the thermal-stability range of the hydrous condensed assemblage. It should be easy for you to verify for yourself the qualitative effect of reduced P_{H_2O} at constant total pressure on the univariant P–T curves (a), (b), and (e) employing G–T or G–P diagrams.

Curve (c), the hydration reaction, obeys the same principle, but because the condensed low-temperature assemblage of equivalent bulk composition is less hydrous than the higher-temperature laumontite assemblage, values of $a_{H_2O} < 1.0$ will cause curve (c) to move toward higher temperatures at constant total pressure relative to the phase relations depicted in Figure 6.16. An isobaric G–T diagram at 3,000 bars pressure which explains this phenomenon is pre-

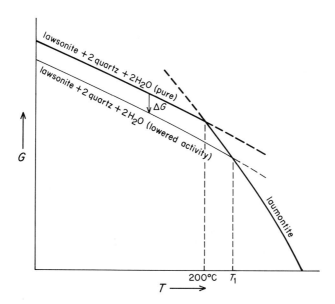

FIGURE 6.18.

Three-kilobar isobaric diagram of molar Gibbs free energy versus temperature for the composition CaO·Al$_2$O$_3$·4SiO$_2$·2H$_2$O and the equilibrium (c), lawsonite + 2 quartz + H$_2$O = laumontite, where the H$_2$O is present as a pure phase, $a_{H_2O} \approx 1$ (heavy lines), and where the activity of H$_2$O is much less (light lines). The decrease in Gibbs free energy on lowered activity of H$_2$O reflects the relationship $dG = RT \, d \ln a$.

sented as Figure 6.18. Where $P_{H_2O} = P_{total}$, and $a_{H_2O} \approx 1.0$, products and reactants are at equilibrium at about 200°C, but lowered activity of H_2O at 3 kilobars P_{total} raises the thermal-stability limit of the lawsonite-bearing assemblage to the temperature T_1.

Experimental studies in the quaternary system $CaO \cdot SiO_2$–Al_2O_3–SiO_2–H_2O demonstrate that the phase relations presented above for the bulk composition $CaO \cdot Al_2O_3 \cdot 4SiO_2$ + excess H_2O in part represent metastable equilibrium. In particular, univariant curve (d), for the reaction lawsonite + 2 quartz + $2H_2O$ = wairakite, apparently lies within the field of stability for the more hydrous condensed assemblage Ca-montmorillonite + prehnite + quartz. This modification of Figure 6.16 is presented as Figure 6.19. The Ca-montmorillonite + prehnite + quartz P–T field has not been explored in detail, but to some extent it may be used to model yet another low-grade metamorphic facies, the prehnite–pumpellyite facies.

Before leaving the lowest grades of metamorphism, we will look briefly at two additional equilibria appropriate for the zeolite metamorphic facies, as summarized in Figure 6.20. The thermal stability range of analcime + quartz (see also Figure 3.8) is clearly a function of the degree of order-disorder in the participating phases, because the cation distribution affects the molar Gibbs free energy of a phase. Both this diagram and the one for stilbite show an interesting feature not illustrated previously: these dehydration curves possess negative P–T slopes at elevated fluid pressures, because the aqueous fluid phase (essentially liquid water) apparently is more compressible than the relatively rigid open-framework structures which constitute the low-temperature assemblages. Accordingly, although ΔV may be slightly positive at low temperatures, it becomes negative at values of P_{H_2O} above about 3 or 4 kilobars. Thus we expect the development of zeolites such as these to be confined to the upper levels of the Earth's crust.

We turn now to the greenschist facies, in which increase in grade is signaled by the appearance of biotite in quartzose rocks of pelitic composition which, at lower grades, are characterized by white mica and chlorite. The lower thermal-stability limit of biotite has not been encountered experimentally, but apparently it, and the other layer silicates, are stable on their own bulk compositions down to very low temperatures. Thus the biotite isograd seemingly represents a tie-line shift which involves the bulk compositions of common

---

FIGURE 6.19. *(facing page)*
Revised stability relations for the bulk composition $CaO \cdot Al_2O_3 \cdot 4SiO_2$ + excess H_2O, taken from Liou (1971b, Figures 1 and 2). (a) shows compositions of condensed phases in the system $CaO \cdot SiO_2$–Al_2O_3–SiO_2–H_2O projected onto the anhydrous base. Open circles represent hydrous phases, solid circles anhydrous phases. Binary and pseudobinary joins are shown by dashed lines. (b) presents modified P_{fluid}–T stability fields for the silica-saturated Ca-Al H_2O-rich silicates. Univariant P–T curves for metastable equilibria are indicated by dotted lines.

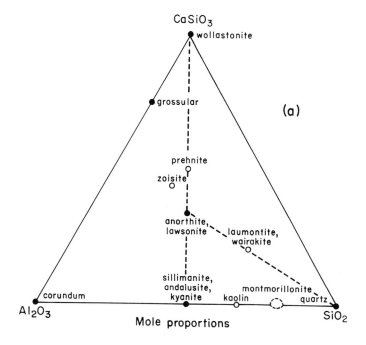

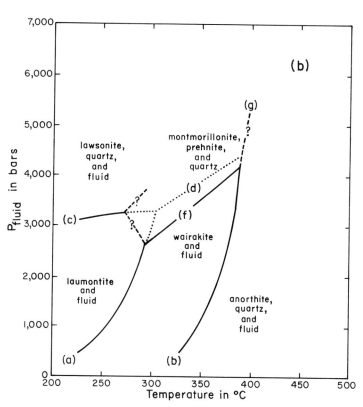

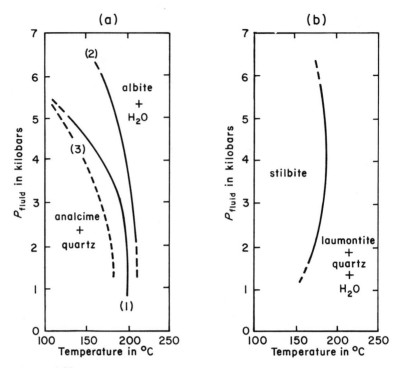

FIGURE 6.20.

Experimental P_{fluid}–T diagrams for some zeolite metamorphic grade equilibria. Compositions of the newly considered phases are: analcime = $NaAlSi_2O_6 \cdot H_2O$; stilbite = $CaAl_2Si_7O_{18} \cdot 7H_2O$. (a) presents phase equilibria for the bulk compositions $Na_2O \cdot Al_2O_3 \cdot 6SiO_2$ + excess H_2O. Curve (1) was investigated by Thompson (1971) using a weight-change method and structurally ordered natural minerals. Curve (2) was determined by Liou (1971c) employing natural analcime but synthetic quartz and somewhat disordered synthetic albite. Curve (3) was calculated by Liou using calorimetric data for low albite. (b) presents phase equilibria for the bulk composition $CaO \cdot Al_2O_3 \cdot 7SiO_2$ + excess H_2O, after Liou (1971d).

rocks. Phase relations are illustrated diagrammatically in Figure 6.21. In this figure, ferrous iron is combined with magnesium (= R^{2+}), and Fe^{3+} with Al (= R^{3+}); all assemblages are postulated to be saturated with an H_2O-rich fluid phase, quartz, and albite. At low temperatures, white mica exhibits extensive solid solution from muscovite, $KAl_2Si_3AlO_{10}(OH)_2$, toward celadonite, $K(Fe^{2+},Mg)(Fe^{3+},Al)Si_4O_{10}(OH)_2$. These solid solutions are termed phengite. Apparently it is the contraction of the phengite$_{ss}$ field and therefore the gradual restriction of white mica compositions to more nearly pure muscovite with increased grade which causes the tie-line shifts to involve pelitic bulk compositions. For instance, at low grade (Figure 6.21a) bulk composition x consists of phengite$_{ss}$ + chlorite (+ albite + quartz); whereas at higher grade (Figure 6.21b) the condensed assemblage also contains biotite. The reaction of concern

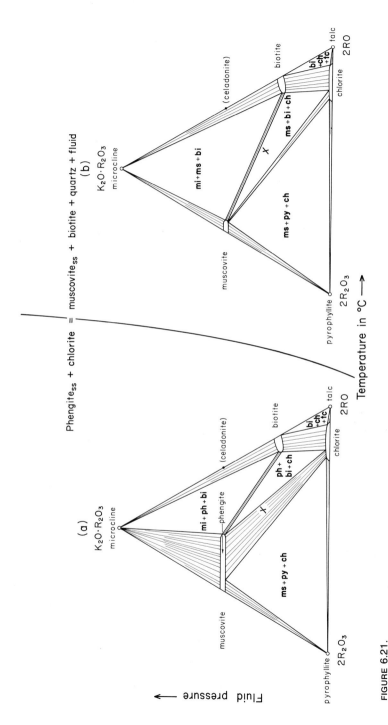

Phengite$_{ss}$ + chlorite = muscovite$_{ss}$ + biotite + quartz + fluid

FIGURE 6.21.

Schematic P_{fluid}–T diagram and chemographic relations in a portion of the system K_2O-R_2O_3–$2R_2O_3$–$2RO$, after Ernst (1963b); here $R_2O_3 = Fe_2O_3 + Al_2O_3$, and $RO = FeO + MgO$. Excess phases are albite, quartz, and an aqueous fluid phase. Abbreviations are: **bi** = biotite; **ch** = chlorite; **mi** = microcline; **ms** = muscovite; **ph** = phengite; **py** = pyrophyllite; **tc** = talc. Solid solution from muscovite toward celadonite decreases at higher temperatures. The pelitic bulk composition x is devoid of biotite at low grade, but carries this phase in addition at temperatures exceeding the biotite isograd (heavy line). The thermal stability range of synthetic phengite$_{ss}$ was investigated by Velde (1965). Velde also suggested that the join **mi-chl** is stable under the conditions shown in this figure, rather than **ms-bi**. Both constructions may be correct for different bulk compositions, but in any case the phenomenon of the incoming of biotite is modeled by either topology.

here, ignoring Mg/Fe^{2+} and Al/Fe^{3+} disproportionation is:

(phengite)   (chlorite)
$$8KR^{3+}_{1.5}R^{2+}_{0.5}Si_{3.5}R^{3+}_{0.5}O_{10}(OH)_2 \; + \; R^{2+}_5 R^{3+}Si_3 R^{3+}O_{10}(OH)_8$$

(muscovite)   (biotite)   (quartz)   (fluid)
$$= 5KR^{3+}_2 Si_3 R^{3+}O_{10}(OH)_2 \; + \; 3KR^{2+}_3 Si_3 R^{3+}O_{10}(OH)_2 \; + \; 7SiO_2 \; + \; 4H_2O.$$

All the layer silicates, of course, are solid solutions. As you can see from Figure 6.21, the particular physical conditions under which biotite first appears with increasing temperature and pressure (i.e., the biotite isograd) is a sensitive function of the rock bulk composition. For some rather potassic compositions, even microcline may be generated in the higher-grade rocks.

The entrance of garnet is modeled by relations presented in Figure 6.22 for

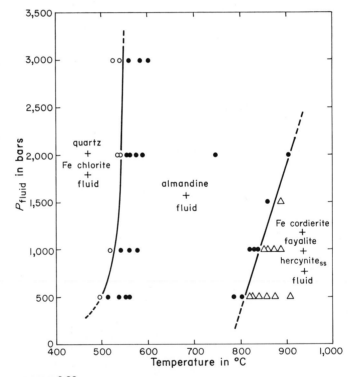

FIGURE 6.22.

Conventional P_{fluid}–T diagram for the bulk composition $3FeO \cdot Al_2O_3 \cdot 3SiO_2$ + excess H_2O at the relatively low oxygen fugacities defined by the iron + quartz–fayalite buffer, after Hsu (1968, Figure 5). Four components are required to specify the compositions of the phases in this system under these f_{O_2} conditions.

the bulk composition $3FeO \cdot Al_2O_3 \cdot 3SiO_2$ + excess H$_2$O. Because iron, an element of variable valency, is involved in the equilibria, the oxygen fugacity is an important parameter and must be specified. The phase diagram illustrated is for the relatively reducing conditions in which values of f_{O_2} are defined by the iron + quartz-fayalite buffer (see Figure 2.5, curve 8). From Figure 6.22 it is seen that the assemblage quartz + iron chlorite reacts at higher grade to produce almandine + H$_2$O, the common garnet of regional metamorphic terranes. At even higher grade, this phase in turn gives way to an iron-cordierite-bearing assemblage, more or less characteristic of the pyroxene hornfels facies. Of course, solid solution with other garnet end-members will modify appreciably the relationships presented in the diagram. Many garnets do show evidence of retrograde reaction to chlorite and, on the other hand, chlorite generally decreases substantially as garnet joins the metamorphic assemblage—hence the conversion of quartz + iron chlorite to almandine + fluid presented in the figure is a reasonable model for the garnet isograd. In the next section, dealing with oxidation-reduction reactions, it will be shown that the garnet isograd, like the phase relations for all iron-bearing minerals, is a function of f_{O_2} as well.

The stability relations of the iron staurolite end-member in the presence of excess SiO$_2$ and H$_2$O are presented in Figure 6.23. As with phase equilibria for almandine, oxygen fugacity must be buffered in the experiments to provide control of the oxidation state. In this case, the fayalite–magnetite + quartz buffer (see Figure 2.5, curve 4) has been selected; this f_{O_2} range is sufficient to maintain all the iron in the divalent state. Behavior depicted in Figure 6.23 therefore may be described in terms of the quaternary system FeO–Al$_2$O$_3$–SiO$_2$–H$_2$O. Also somewhat similar to almandine in its stability relations, staurolite exhibits low-temperature and high-temperature stability limits and, in addition, a low-pressure stability limit for equilibria involving the bulk composition staurolite + excess quartz + aqueous fluid. The specific reactions illustrated in Figure 6.23 and shown in heavy solid lines are:

(I) 6 staurolite + 25 quartz = 8 almandine + 46 aluminosilicate + 12 H$_2$O;
(II) 2 staurolite + 15 quartz = 4 iron cordierite
$\qquad\qquad\qquad$ + 10 aluminosilicate + 4 H$_2$O;
(III) 8 chloritoid + 10 aluminosilicate = 2 staurolite + 3 quartz
$\qquad\qquad\qquad$ + 4 H$_2$O; and
(IV) 3 iron cordierite = 2 almandine + 4 aluminosilicate
$\qquad\qquad$ + 5 quartz (curve g of Figure 6.9).

Whether kyanite, sillimanite, or andalusite is the stable aluminosilicate participating in the reaction is a function of the physical conditions, as is clear from Figure 2.7 (see also Figure 6.5). If we consider bulk compositions having higher FeO/Al$_2$O$_3$ ratios than staurolite, the following reactions are also

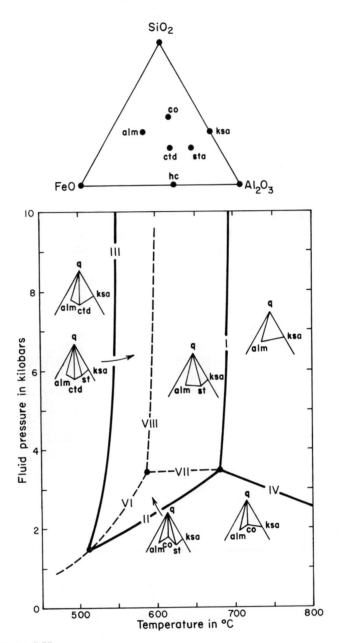

FIGURE 6.23.

Phase relations for staurolite in the silica-rich portion of the system
$FeO-Al_2O_3-SiO_2-H_2O$ with oxygen fugacities defined by the fayalite–
magnetite + quartz buffer, after Richardson (1968, Figure 5; see also
Hoschek, 1969; and Ganguly, 1972). The ternary diagram shows compositions
of phases in the quaternary system projected onto the anhydrous base
where necessary. Abbreviations are: **alm** = almandine; **co** = iron
cordierite; **ctd** = chloritoid; **hc** = hercynite$_{ss}$; **ksa** = Al_2SiO_5 polymorph
(see Figures 2.7 and 6.5); **q** = quartz; and **sta** = staurolite.

encountered:

(VI) 2 chloritoid + 3 quartz = iron cordierite + 2 H$_2$O;

(VII) 4 staurolite + 10 almandine + 55 quartz = 23 iron cordierite
$$+ \: 8 \: H_2O; \: and$$

(VIII) 23 chloritoid + 7 quartz = 2 staurolite + 5 almandine + 19 H$_2$O.

The P–T locations of these univariant equilibria are indicated by dashed light lines in Figure 6.23. It is obvious from the phase diagram that staurolite is restricted to a specific range of physical conditions, bounded by rather well-defined reactions. Moreover, its appearance even given the appropriate P–T values is strongly a function of bulk composition; hence staurolite occurrences are restricted even more as the chemistry of the enclosing rock departs from that of the mineral itself.

Aluminosilicate polymorphism was discussed earlier in this chapter in the section dealing with solid-solid reactions. An Al$_2$SiO$_5$ phase, of course, occurs only in rather aluminous rock bulk compositions such as typify certain alkali-poor pelitic schists. For instance, the breakdown of pyrophyllite, Al$_2$Si$_4$O$_{10}$(OH)$_2$, on heating would produce kyanite or andalusite as well as H$_2$O and quartz, as indicated in Figure 6.24. However, more potassic Al-rich bulk compositions would be characterized by the occurrence of muscovite rather than pyrophyllite. At somewhat higher grade than shown in Figure 6.24a, the thermal decomposition of muscovite in the presence of quartz generates additional Al$_2$SiO$_5$, the specific polymorph depending on the physical conditions, as shown in Figure 6.24b (see also Figure 3.7). This reaction models the so-called "second sillimanite isograd" where, in Al$_2$SiO$_5$-bearing quartzose metamorphic rocks, additional aluminosilicate growth is occasioned by the breakdown of muscovite. If the protolith contains Na-rich feldspar$_{ss}$ accompanying the K-rich feldspar$_{ss}$, a granitic minimum melt will form at elevated values of P_{H_2O} (see Figure 4.41b); depending on the rock chemistry, one or two of the solid phases shown in Figure 6.24b would dissolve completely in the melt. Of course, we have assumed H$_2$O saturation of the condensed assemblage, in harmony with the most commonly employed experimental techniques—but this is no assurance that such conditions prevail in nature. If the H$_2$O activity were reduced at constant total pressure, this phase diagram would be modified considerably. For instance, the granite solidus curve would be displaced to higher temperatures, whereas the dehydration reactions involving pyrophyllite and muscovite + quartz would decline to lower temperatures; the result would be a broader P–T exposure for the anhydrous aluminosilicate polymorphs than illustrated in Figure 6.24.

Certain other, high-grade reactions which model eclogite, garnet granulite, and pyroxene hornfels facies have been discussed previously and will not be elaborated on here; for examples, see Figures 5.2, 5.3, 5.4, 5.13, 6.9 (especially curves f and g), 6.15, 6.22, and 6.23. As is evident from our brief mention of the

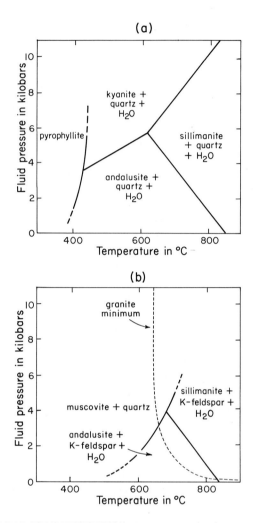

FIGURE 6.24.
Conventional P_{fluid}–T diagrams for portions of the system K_2O–Al_2O_3–SiO_2–H_2O, with aluminosilicate polymorphism taken from Richardson, Gilbert and Bell (1969, Figure 4). (a) Stability relations for pyrophyllite, and the reaction $Al_2Si_4O_{10}(OH)_2$ $= Al_2SiO_5 + 3SiO_2 + H_2O$, after Kerrick (1968, Figure 4). (b) The dehydration of muscovite in the presence of excess quartz by the reation $KAl_2Si_3AlO_{10}(OH)_2 + SiO_2$ $= KAlSi_3O_8 + Al_2SiO_5 + H_2O$, according to Althaus *et al.* (1970, p. 331). Rather comparable results were obtained by Evans (1965). For rocks containing sodic feldspar$_{ss}$ in addition to potassic feldspar$_{ss}$, the beginning of melting assuming H_2O saturation is also shown, after Luth, Jahns, and Tuttle (1964).

granite minimum-melting relations just above (Figure 6.24b), where quartzose, feldspathic crustal rocks are subjected to high aqueous-fluid pressures in deeper levels of the continental crust, anatexis is to be expected; such phenomena are widespread in tracts of migmatite, which characterize portions of many shield areas. Because, in contrast, many granulite terranes bear evidence of rather limited degrees of partial fusion, it is presumed that large sections of the deeper crust have experienced conditions involving low values of a_{H_2O} at high confining pressures and temperatures.

REACTIONS INVOLVING O$_2$

Most metamorphic reactions involve the participation of minerals representing complex iron-bearing solid solutions. An attempt has been made thus far to avoid such chemical complexities wherever possible by considering equilibria among magnesian end-members or, where Fe is involved, by defining the attending conditions in such a way that the oxidation state of the system does not change during the course of the reaction. A large number of metamorphic reactions do produce or consume O$_2$, however. This is why the oxygen fugacity is considered to be an important factor influencing the phase equilibria, as are temperature, confining pressure and the partial pressures, chemical potentials, activities, or fugacities of the other volatile species—the two principal species being CO$_2$ and H$_2$O, as already discussed. Most experimental investigators so far have used laboratory control of three variables: temperature, total (= aqueous fluid) pressure, and oxygen fugacity. Their results are presented in two-dimensional diagrams which, in fact, are sections through a volume defined by these three variables, and are given either as conventional P_{fluid}–T plots with oxygen fugacity specified by a particular buffer (see Chapter 2, pp. 32–33, and Figure 2.5), or as log f_{O_2}–T sections at some conveniently chosen total pressure.

First let us examine phase relations for ferrotremolite in a system closely approximating its own bulk composition, namely, 2CaO·5FeO·8SiO$_2$ + excess H$_2$O and with variable amounts of oxygen. An isobaric log f_{O_2}–T diagram at three kilobars fluid ($\approx$ H$_2$O) pressure is given in Figure 6.25. Also shown are oxygen-buffer curves where they coincide with field boundaries in the investigated system. As defined by the condensed oxygen-buffer assemblages, and with numbering the same as in Figure 2.5, these are: (1) magnetite–hematite; (4) fayalite–magnetite + quartz; and (8) iron + quartz–fayalite.

It is evident that the phase Ca$_2$Fe$_5^{2+}$Si$_8$O$_{22}$(OH)$_2$ is confined to relatively low temperatures, even at high aqueous-fluid pressures in a "rock" of essentially its own bulk composition. Under extremely reducing conditions, the amphibole decomposes, evolving oxygen and a more reduced iron-bearing condensed assemblage, iron + quartz + hedenbergitic pyroxene$_{ss}$ + oxygen. The pyroxene shows extensive solid solution from CaFe^{2+}Si$_2$O$_6$ toward Fe$^{2+}_2$Si$_2$O$_6$. The system may be considered as open to the aqueous fluid phase at 3 kilobars P_{total} (= P_{fluid}); hence the volume of H$_2$O need not be considered in the reaction (see discussion in Chapter 3, pp. 51–53). The thermodynamic relationship for the redox reaction, written to evolve or consume one mole of oxygen, follows from equation (3.15):

$$d\,\Delta G = 0 = (\Delta V_{condensed})\,dP_{total} + (\Delta V_{O_2})\,dP_{O_2} - (\Delta S_{total})\,dT.$$

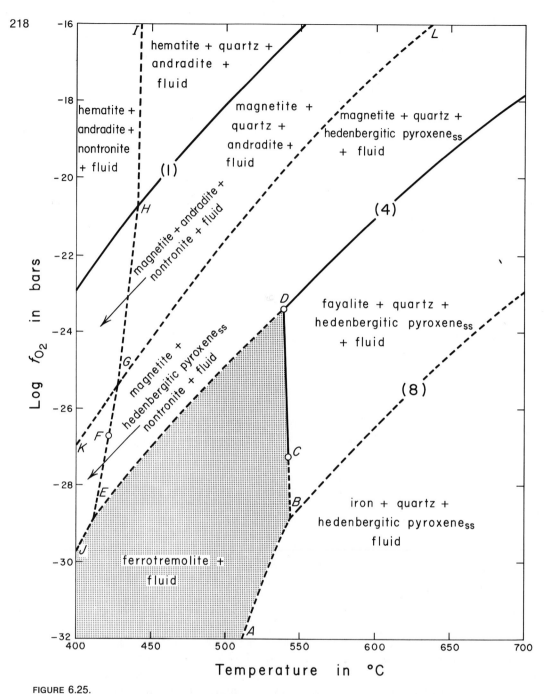

FIGURE 6.25.

$\text{Log} f_{O_2}$–T diagram at 3,000 bars fluid pressure for the bulk composition $2\text{CaO}\cdot5\text{FeO}\cdot8\text{SiO}_2$ + excess H_2O and variable proportions of ferrous and ferric iron, after Ernst (1966, Figure 11). The stability field of ferrotremolite, $\text{Ca}_2\text{Fe}^{2+}_5\text{Si}_8\text{O}_{22}(\text{OH})_2$, is stippled. Buffer curves (1), (4), and (8) are from Figure 2.5. It should be noted that, for intermediate iron–magnesium compositions in the system $\text{Ca}_7\text{Si}_8\text{O}_{23}$–$\text{Mg}_7\text{Si}_8\text{O}_{23}$–$\text{Fe}_7\text{Si}_8\text{O}_{23}$–$\text{H}_2\text{O}$, experiments conducted by Cameron (1975, Figure 5) under appropriate oxidation states produced clinopyroxene$_{ss}$ + cummingtonite$_{ss}$ + quartz + fluid on the thermal decomposition of actinolite$_{ss}$ rather than the fayalite + hedenbergitic clinopyroxene$_{ss}$ + quartz + fluid assemblage shown above.

At constant total pressure and equilibrium conditions, this reduces to

$$(\Delta V_{O_2}) \, dP_{O_2} = (\Delta S_{total}) \, dT. \tag{6.3}$$

Rearranging and substituting fugacity for partial pressure yields the expression:

$$\left(\frac{\partial f_{O_2}}{\partial T}\right)_{P_{total}} = \frac{\Delta S_{total}}{\Delta V_{O_2}}. \tag{6.4}$$

If we substitute RT/f_{O_2} for the volume of oxygen involved in the reaction, we arrive at the expression

$$\left(\frac{\partial \ln f_{O_2}}{\partial T}\right)_{P_{total}} = \frac{\Delta S_{total}}{RT}; \tag{6.5}$$

however, care must be exercised in this substitution because the sign of the righthand term depends on whether one mole of oxygen is evolved $(= +)$ or consumed $(= -)$ when reactants are converted to products. For our purposes, equation (6.4) is clear enough to account for the observed slopes of the $\log f_{O_2}$-T curve.

Because the ferrotremolite assemblage is both the low-entropy and the small-volume configuration (more oxidized condensed assemblage) relative to the high-temperature assemblage of equivalent bulk composition (including both the more reduced solid phases plus the evolved O_2), the slope of the $\log f_{O_2}/T$ curve for field boundary AB in Figure 6.25 is positive: ΔS_{total} is positive, as is ΔV_{O_2}. Similarly, curves ED and JE have positive slopes in $\log f_{O_2}$-T coordinates because the low-temperature assemblages, which contain magnetite, are more condensed (i.e., is more oxidized, hence has a smaller volume) than the higher-entropy, large-volume ferrotremolite $+ O_2$ assemblage of equivalent composition: again, $\Delta S_{total}/\Delta V_{O_2}$ has a positive sign. Between B and D, the dehydration of ferrotremolite produces a higher-grade assemblage of virtually identical oxidation state. Hence, because O_2 is neither generated nor consumed in this reaction, the value of the expression for the curve slope,

$$\left(\frac{\partial f_{O_2}}{\partial T}\right)_{P_{total}} = \frac{\Delta S_{total}}{\Delta V_{O_2}},$$

might be expected to be infinite. In a general way this is true; however, because at very low oxygen fugacities, such as defined by the iron $+$ quartz-fayalite buffer, there is extremely little O_2 in the fluid phase, the equilibrium molecular dissociation of H_2O to hydrogen and oxygen requires that f_{H_2} be substantial (see the discussion of the law of mass action, Chapter 1, pp. 18–21, especially equation 1.73). This in turn reduces the value of f_{H_2O} by dilution, so that at constant total ($=$ fluid) pressures the thermal-stability limits of hydrous phases tend to be reduced at very low oxygen fugacities.

Other items of interest in Figure 6.25 include the presence of a low-temperature stability field for clay minerals throughout a wide range of oxygen fugacities, and the mutual restriction of andradite garnet, $Ca_3Fe^{3+}_2Si_3O_{12}$ and hedenbergitic pyroxene$_{ss}$ to opposite sides of the field boundary curve KGL. In the system $2CaO \cdot 5FeO \cdot 8SiO_2(+O_2)$, hedenbergitic pyroxene$_{ss}$ characterizes the more reducing conditions, whereas ferric iron-rich garnet is favored by oxidation. In this isobaric section, ferrotremolite is seen to give way, both at higher temperatures and at higher f_{O_2} values, to a pyroxene solid solution-bearing condensed assemblage; the latter in turn yields to garnetiferous phase compatibilities at even higher values of f_{O_2}. Thus these critical metamorphic minerals evidently may be produced by changes in oxidation state as well as by changes in temperature and pressure.

Even under the most favorable circumstances, with oxygen fugacities at relatively low values near the f_{O_2} base of the magnetite field, ferrotremolite possesses a rather restricted thermal-stability range compared to its magnesian analogue, tremolite. The dehydration curves for these two end-members are presented as a conventional P_{fluid}–T diagram in Figure 6.26. The complete replacement of Mg by Fe^{2+} results in a diminution of the high-temperature stability limit of amphibole by about 350–375C°.

Figure 6.27 illustrates the sympathetic change in composition for magnesium-free sodic amphiboles of the riebeckite–arfvedsonite solid solution series as a function of oxygen fugacity. It is a $\log f_{O_2}$–T plot of phase relations for the bulk composition $Na_2O \cdot 3FeO \cdot Fe_2O_3 \cdot 8SiO_2$ + excess H_2O and variable ferrous-ferric ratio at an aqueous-fluid pressure of 2,000 bars. Again, as in Figure 6.25, oxygen buffer curves (1), (4), and (8) of Figure 2.5 coincide with the field boundaries that delineate the high-temperature assemblages of equivalent bulk composition along which the phases magnetite + hematite, fayalite + magnetite + quartz, and iron + quartz + fayalite are stable with an oxygen-bearing fluid. This phenomenon, which occurs because the investigated bulk composition is appropriate for the production of the critical oxide and iron silicate phases in the presence of free silica, indicates that these high-temperature assemblages themselves define the oxygen fugacity at any given P and T. The phase rule can be used to demonstrate that this is so. For instance, for this five-component system (Na_2O, FeO, Fe_2O_3, SiO_2, H_2O), five phases are present under subsolidus conditions defined by oxygen buffer curve (1)—hence the variables T and P_{fluid} may be chosen arbitrarily without changing the phase assemblage; oxygen fugacity is then fixed at any selected values of T and P_{fluid}. However, in the region $FHCA$, for instance, only four phases are stable, and f_{O_2} must be specified as well as T and P_{fluid} before the system is completely defined.

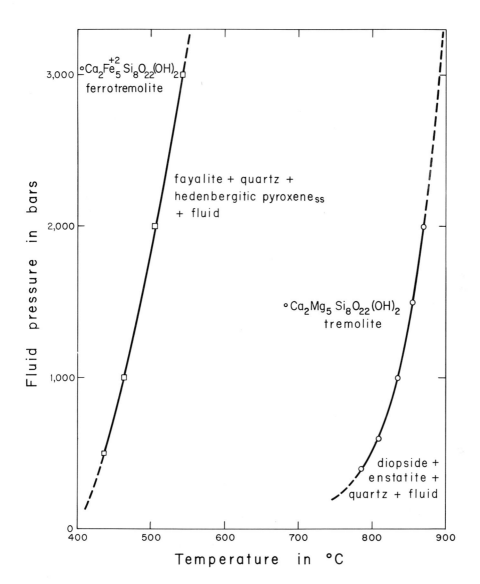

FIGURE 6.26.

Conventional P_{fluid}–T diagram for the bulk compositions $2CaO \cdot 5MgO \cdot 8SiO_2$ + excess H_2O (Boyd, 1959, Figure 3) and $2CaO \cdot 5FeO \cdot 8SiO_2$ + excess H_2O, with f_{O_2} defined by the condensed buffer assemblages wüstite–magnetite and iron–magnetite (curves 5 and 6, respectively, of Figure 2.5), after Ernst (1966, Figure 8).

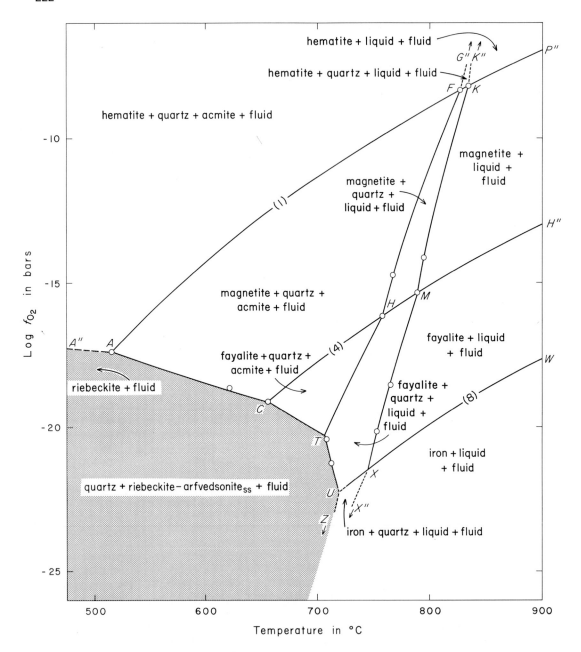

FIGURE 6.27.

Log f_{O_2}-T diagram at 2,000 bars fluid pressure for the bulk composition $Na_2O \cdot 3FeO \cdot Fe_2O_3 \cdot 8SiO_2$ + excess H_2O and variable proportions of ferrous and ferric iron, after Ernst (1962, Figure 11a). The stability field of riebeckite and of riebeckite-arfredsonite solid solutions is stippled. Buffer curves (1), (4), and (8) are from Figure 2.5.

Under relatively oxidizing conditions where values of f_{O_2} are specified by the magnetite–hematite buffer, the amphibole is riebeckite, $Na_2Fe^{2+}_3Fe^{3+}_2Si_8O_{22}(OH)_2$. With more reducing conditions below curve $ACTU(Z)$ it co-exists with progressively greater amounts of quartz and becomes enriched in ferrous iron; where f_{O_2} is defined by the assemblage iron + either magnetite or wüstite, the riebeckite–arfvedsonite$_{ss}$ achieves the approximate composition $Na_{2.4}Fe^{2+}_{4.9}Fe^{3+}_{0.7}Si_{7.7}Fe^{3+}_{0.3}O_{22}(OH)_2$, assuming only two hydroxyl ions per formula unit. It may be observed from Figure 6.27 that the sodic amphibole$_{ss}$ thermal-stability limit rises to higher temperatures at constant fluid pressure as f_{O_2} values decline. It does so because the high-temperature product assemblage of equivalent bulk composition is more oxidized (hence more condensed) than the ferrous iron-rich, low-temperature, low-entropy amphibole + O_2 reactant assemblage; therefore, the value of the expression

$$\left(\frac{\partial f_{O_2}}{\partial T}\right)_{P_{total}} = \frac{\Delta S_{total}}{\Delta V_{O_2}},$$

is negative, because for the reaction, although ΔS is positive, ΔV_{O_2} is negative.

A three-dimensional diagram of the variables $\log f_{O_2}$, P_{fluid}, and T is presented in Figure 6.28 with buffer curves (1), (4), and (8) shaded. The top face is the 2,000 bar P_{fluid} section illustrated in Figure 6.27, with the riebeckite–arfvedsonite$_{ss}$ thermal-stability limit indicated by curve $A''ACTUZ$. The surfaces representing initiation of fusion, where acmite and, at low oxygen fugacities, aenigmatite and arfvedsonitic amphibole disappear, are also shown (at 2,000 bars fluid pressure for Figures 6.27 and 6.28, this is curve $G''FHTU$). For simplicity, however, the "quartz out" surface is omitted; at 2,000 bars P_{fluid} in Figure 6.27, this is represented by curve $K''KMXX''$. The influence of the several variables should be clear from this diagram.

To summarize, study of the system $Na_2O \cdot 3FeO \cdot Fe_2O_3 \cdot 8SiO_2$ under conditions of H_2O saturation and variable oxidation states demonstrates that, even in the absence of magnesium, change in f_{O_2} exerts a powerful control on the phase equilibria and on the compositions of complex solid solutions. Naturally, lowered oxygen fugacity tends to favor the expansion of the thermal-stability ranges for the more reduced condensed assemblages.

Before leaving the double-chain silicates, let us briefly investigate the phase relations of the iron-bearing calcic amphibole ferropargasite, $NaCa_2Fe^{2+}_4AlSi_6Al_2O_{22}(OH)_2$. A $\log f_{O_2}$–T diagram at 2,000 bars P_{fluid} is presented in Figure 6.29. As in the studies of ferrotremolite and riebeckite–arfvedsonite solid solutions, experiments were carried out under H_2O-saturated conditions and with oxygen fugacities controlled by condensed buffer assemblages as illustrated in Figure 2.5. In this diagram, all the experimentally employed buffer curves are indicated by light lines; only where field boundaries coincide with the buffer curves are the latter shown in heavy lines. Ferropargasite has

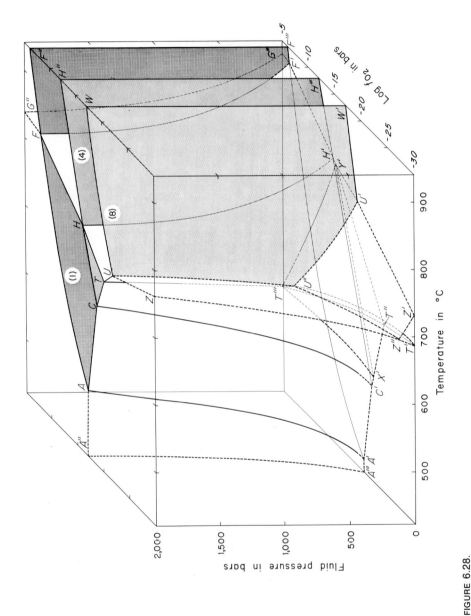

FIGURE 6.28.

Stability volumes defined by $\log f_{O_2}$, P_{fluid}, and T for the bulk composition $Na_2O \cdot 3FeO \cdot Fe_2O_3 \cdot 8SiO_2$ + excess H_2O and variable proportions of ferrous and ferric iron, after Ernst (1962, Figure 12c). Dashed curves indicate inferred or computed field boundaries. Field boundary surfaces for (1) magnetite–hematite, (4) fayalite–magnetite + quartz, and (8) iron + quartz–fayalite are stippled. The top face of the diagram is a perspective view of Figure 6.7. The bottom face of the diagram is an isobaric section at a P_{fluid} value only slightly greater than zero; technically, the dehydration surfaces illustrated, such as $ACC'A'$, should curve in to this asymptotic plane at zero fluid pressure, becoming tangent to it at $0°$Kelvin.

225

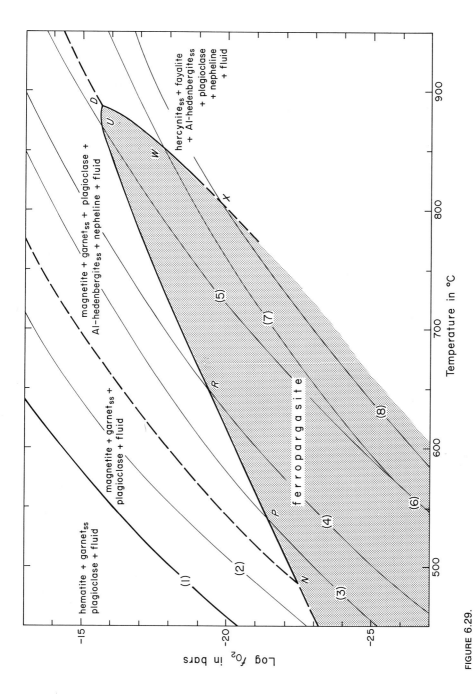

FIGURE 6.29.

Log f_{O_2}–T diagram at 2,000 bars fluid pressure for the bulk composition $Na_2O\cdot4CaO\cdot8FeO\cdot3Al_2O_3\cdot12SiO_2$ + excess H_2O and variable proportions of ferrous and ferric iron, after Gilbert (1966, Figure 9). The stability field of ferropargasite, $NaCa_2Fe^{2+}_4AlSi_6Al_2O_{22}(OH)_2$ is stippled. Buffer curves are from Figure 2.5.

its maximum thermal-stability range at relatively low oxidation states, where, in the binary Fe–O system, wüstite would be stable (i.e., in the f_{O_2}–T region between buffer curves 5 and 7). With more reducing conditions, ferropargasite decomposes to a high-temperature assemblage of nearly the same oxidation state: hercynite$_{ss}$ + fayalite + aluminous hedenbergite$_{ss}$ + plagioclase + nepheline + aqueous fluid. The P_{fluid} isobaric decrease in thermal-stability range of the calcic amphibole with diminished f_{O_2} along curve DWX chiefly reflects the decline in f_{H_2O} produced by dilution of the fluid phase by hydrogen at constant total (fluid) pressure. Under progressively more oxidizing conditions, ferropargasite (+ oxygen) dehydrates to form the successively more oxidized condensed assemblages magnetite$_{ss}$ + garnet$_{ss}$ + plagioclase + aluminous hedenbergite$_{ss}$ + nepheline + fluid, and magnetite + garnet$_{ss}$ + plagioclase + fluid. Because the SiO_2-undersaturated phases nepheline and hercynite occur in the high-temperature assemblage equivalent in composition to ferroparagsite, the presence of excess silica would diminish the P_{fluid} isobaric $\log f_{O_2}$–T stability field of ferropargasite (e.g., see Chapter 4, p. 78, and Figures 6.9 and 6.11). Basically this reflects a lower total G for the new high-temperature assemblage relative to ferroamphibole + quartz.

Interestingly enough, Figure 6.29 shows that, at constant P, T, and bulk composition (except for hydrogen and oxygen contents), the amphibole- and garnet-bearing assemblages may be accounted for by systematic differences in the attending f_{O_2}. Thus the amphibolite and garnet granulite mineral facies under some conditions may be considered as functions of oxygen fugacity as well as of the fugacity of H_2O, confining pressure, temperature, and rock bulk composition. It is, of course, true that if limited amounts of fluid have access to a rock, the particular mineral assemblages present tend to buffer the fluid phase, instead of vice versa (Greenwood, 1975).

Phase equilibria for annite, $KFe^{2+}_3Si_3AlO_{10}(OH)_2$, the ferrous iron end-member of the biotites, is presented in Figure 6.30. Relationships are illustrated in $\log f_{O_2}$–T diagrams at about 2,000 bars fluid pressure for the bulk compositions (a) $K_2O \cdot 6FeO \cdot Al_2O_3 \cdot 6SiO_2$ and (b) $K_2O \cdot 6FeO \cdot Al_2O_3 \cdot 12SiO_2$, both with excess H_2O and variable ferrous–ferric ratios. Where the bulk composition is that of annite itself (Figure 6.30a), the 2,000-bar isobaric thermal-stability range of annite extends to temperatures exceeding 800°C (point C_1) at relatively low oxidation states. Depending on the f_{O_2} range involved, the dehydration of this trioctahedral mica produces magnetite and/or leucite and/or kalsilite. However, magnetite and quartz cannot coexist stably at values of f_{O_2} less than those defined by the assemblage fayalite-magnetite + quartz (curve 4 of Figure 2.5), and both leucite and kalsilite are incompatible with excess SiO_2 no matter what the oxygen fugacity. Therefore, the presence of additional silica (Figure 6.30b) promotes the formation of fayalite at much higher oxidation states, and sanidine over a very broad f_{O_2}–T region compared to the situation for the pure annite bulk composition. This in turn means a reduction in the total Gibbs free energy of the high-temperature fayalite- and sanidine-

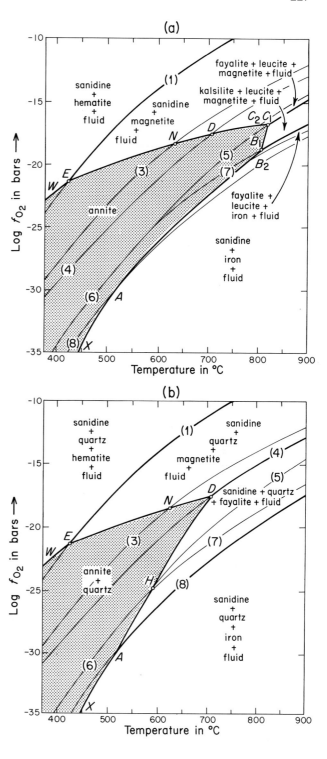

FIGURE 6.30.

Log f_{O_2}-T diagrams at 2,070 bars fluid pressure with excess H_2O and variable proportions of ferrous and ferric iron for the bulk compositions (a) $K_2O·6FeO·Al_2O_3·6SiO_2$ and (b) $K_2O·6FeO·Al_2O_3·12SiO_2$ after Eugster and Wones (1962, Figures 4 and 6, respectively). The stability fields of annite and annite + quartz are stippled. Buffer-curve numbering corresponds to Figure 2.5.

bearing assemblages in the system $K_2O \cdot 6FeO \cdot Al_2O_3 \cdot 12SiO_2 + H_2O + O_2$ relative to that of annite + quartz—and a consequent diminution in the thermal-stability limit of the latter at about 2,000 bars P_{fluid} to near 700°C (point D in Figure 6.30b). The annite ± quartz field boundaries $WEND$ and XA are unaffected by departure of the system from the composition of annite itself to a quartz-normative value because both high- and low-temperature assemblages are already silica-saturated.

The stability relations of the biotites, $K(Fe^{2+},Mg)_3Si_3AlO_{10}(OH)_2$, intermediate members of the phlogopite–annite solid solution series, are shown in Figure 6.31 in terms of log f_{O_2} and T at about 2,000 bars P_{fluid}. The curve labeled 100 represents the thermal-stability limit of pure annite, discussed as curve $WENDC_2C_1B_1B_2AX$ of Figure 6.30a. The field boundary designated O represents the thermal-stability limit of iron-free phlogopite. As you would expect, this latter curve is independent of oxygen fugacity except at very low oxidation states near and below the f_{O_2} base of the magnetite stability field; here, because of the relative abundance of hydrogen in the fluid phase at constant total pressure, f_{H_2O} declines to a small value, causing the pure phlogopite stability range to decrease proportionately. Intermediate iron-magnesium

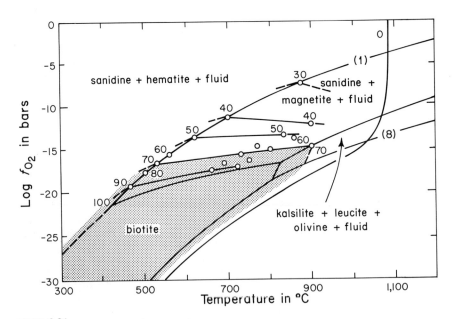

FIGURE 6.31.

Log f_{O_2}–T diagram at 2,070 bars fluid pressure for the intermediate biotities in the system $K_2O \cdot 6MgO \cdot Al_2O_3 \cdot 6SiO_2$–$K_2O \cdot 6FeO \cdot Al_2O_3 \cdot 6SiO_2$ with excess H_2O and variable proportions of ferrous and ferric iron, after Wones and Eugster (1965, Figure 4). Isopleths for the beginning of dehydration are shown principally in 10 mole per cent intervals; for instance, the stippled region shows the log f_{O_2}–T stability field for $Phlog_{30}Ann_{70}$. Buffer-curve numbering corresponds to Figure 2.5.

biotites dehydrate over a finite $\log f_{O_2}$–T interval, and isopleths signaling the onset of breakdown are shown; for instance, the stippled area in Figure 6.31 indicates the $\log f_{O_2}$–T region where, in a system of its own chemical composition, $Phlog_{30}Ann_{70}$ would constitute the total condensed assemblage.

As this section has abundantly demonstrated, where an element of variable valency is involved—such as iron (and, of much lesser importance, manganese, copper, nickel, carbon, or sulfur)—metamorphic equilibria are markedly influenced by the oxygen fugacity, because the stability relations of the participating Fe-bearing minerals are strong functions of f_{O_2}. And, as we have also seen, where f_{O_2} is very low in a fluid phase composed dominantly of H_2O, H_2, and O_2, the value of f_{H_2} becomes appreciable; hence, because of the diminished fugacity or activity of H_2O, the stabilities of hydrous phases are correspondingly reduced. This subject will be explored more fully in the following section, which deals with the general phenomenon of the influence of a multicomponent fluid on metamorphic equilibria.

REACTIONS INVOLVING A MULTICOMPONENT FLUID PHASE

In our discussion of the role of oxidation state and its effect on the stability relations and compositions of iron-bearing phases, we have considered volatile mixtures in the binary system O–H. Here the activities, chemical potentials, partial pressures, or fugacities of the species O_2, H_2, and H_2O are related by the equilibrium constant for the molecular dissociation of H_2O. Now we turn our attention to a multicomponent fluid phase characterized by mixtures of chemically independent (i.e., nonreacting) volatile species, such as the simple binary system CO_2–H_2O. Other simple gaseous systems, such as Ar–H_2O, or more complex multicomponent interreacting volatile systems, for instance, gaseous mixtures consisting of the components C, O, and H (see French, 1966; and Eugster and Skippen, 1967) are of theoretical and practical significance, but obey the same principles and will not be considered here. In the simple binary situation, assuming ideal gaseous solutions, the partial pressure, fugacity, activity, or chemical potential of the volatile component i is reduced from the value appropriate to a pure, one-component gas phase strictly because of dilution by the contaminant component, j. The interaction (i.e., nonideality of behavior) of volatile species for a mixed-component homogeneous gas phase results in changes in the f_i/P_i ($= \gamma_i$) ratio under constant physical conditions, but for simplicity this phenomenon will be ignored. Although we have assumed no reaction takes place between components in, say, a binary gas phase itself, this additional species, like the one-component gas, may interact with the condensed assemblage, or it may not.

An example of contamination of the fluid phase by NaOH, and its effect on the equilibrium $Mg(OH)_2 = MgO + H_2O$, was illustrated in Figure 2.6. In

Figure 3.5, the influence of values of confining pressure exceeding that of CO_2 was computed for the reaction $CaCO_3 + SiO_2 = CaSiO_3 + CO_2$; here the difference between P_{CO_2} and P_{total} may be due to equilibrium between a low-pressure fracture system containing gaseous carbon dioxide and a rock at lithostatic pressure, or it could reflect the presence of a multicomponent fluid phase. Figure 5.13b shows the melting interval of moist basalt in which the activity of H_2O is at a very low value—in other words, the melt is not saturated with respect to H_2O.

Reactions which consume or evolve one or two gaseous constituents have been treated rigorously by Greenwood (1967c), and the principles of metamorphic mixed volatile equilibria (CO_2–H_2O) have been reviewed by Kerrick (1974). Consider a heterogeneous reaction of the sort

$$\text{condensed reactants} = \text{condensed products} + n_i + n_j,$$

where n_i and n_j are the numbers of moles of volatile components i and j, respectively. The following relationships were derived by Greenwood for a system characterized by ideal mixtures of these volatile components:

$$\left(\frac{\partial T}{\partial X_j}\right)_P = \frac{RT}{\Delta S}\left(\frac{n_j}{X_j} - \frac{n_i}{X_i}\right) = \frac{RT^2}{\Delta H}\left(\frac{n_j}{X_j} - \frac{n_i}{X_i}\right); \tag{6.6}$$

and

$$\left(\frac{\partial P}{\partial X_j}\right)_T = -\frac{RT}{\Delta V}\left(\frac{n_j}{X_j} - \frac{n_i}{X_i}\right). \tag{6.7}$$

In these equations, X_j is the mole fraction of component j in the fluid phase (of course, $X_i + X_j = 1$). The topologies of phase relations, plotted on a constant total pressure T–X_j diagram, are presented in Figure 6.32; to study these equilibria, refer to equation (6.6).

Three main types of equilibrium are distinguishable, depending on whether one or both volatile components participate in the reaction, and whether they appear in the equations as reactants or products. They may be described briefly as follows:

(1a) condensed reactants = condensed products + n_j ($n_i = 0$). Because ΔS is positive for the reaction as written, the curve slope, $\left(\frac{\partial T}{\partial X_j}\right)_P$, is also positive, as shown by univariant curve (1a) in Figure 6.32.

(1b) condensed reactants = condensed products + n_i ($n_j = 0$). Because the entropy change is positive, the curve slope, which equals $-\frac{RT}{\Delta S}\left(\frac{n_i}{X_i}\right)$, is negative, as seen from curve (1b) in Figure 6.32.

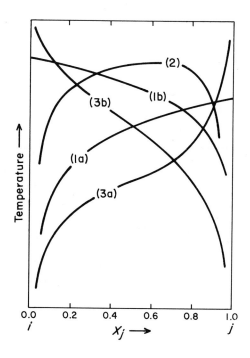

FIGURE 6.32.
Schematic $T - x$ diagram at constant fluid pressure in which the homogeneous volatile phase consists of a binary mixture of components i and j, after Greenwood (1967c, Figure 1); X_j is the mole fraction of component j in the fluid phase. Curves 1a, 1b, 2, 3a, and 3b exemplify phase equilibria as described by the correspondingly numbered reactions in the text.

(2) condensed reactants = condensed products $+ n_i + n_j \, (n_i > 0; \, n_j > 0)$. According to equation (6.6), this univariant equilibrium will possess a maximum in $T\text{-}X_j$ coordinates—that is,

$$\left(\frac{\partial T}{\partial X_j}\right)_P = 0, \text{ where } X_j = \frac{n_j}{n_i + n_j}.$$

In other words, the thermal crest on this curve will be located on the abscissa where the composition of the fluid phase is precisely that produced by the stoichiometric reaction. Demonstration of the validity of this relationship may be seen by setting the expression $\left(\dfrac{n_j}{X_j} - \dfrac{n_i}{X_i}\right)$ equal to zero. Curve (2) in Figure 6.32 exhibits this reaction topology.

(3a) condensed reactants $+ n_i$ = condensed products $+ n_j \, (n_j > 0 > n_i)$. The $T\text{-}X_j$ curve for this equilibrium has a positive $\left(\dfrac{\partial T}{\partial X_j}\right)_P$ slope as illustrated in Figure 6.32, because, from equation (6.6), it is clear that the expression $\dfrac{n_j}{X_j} - \dfrac{n_i}{X_i}$ exceeds zero.

(3b) condensed reactants $+ n_j =$ condensed products $+ n_i \, (n_i > 0 > n_j)$.
This is the converse of (3a); hence (3b) possesses a negative $T\text{–}X_j$ curve slope.

Both (3a) and (3b) possess intermediate inflection points in $T\text{–}X_j$ coordinates, and are asymptotic to the ordinate axes. By employing equation (6.7), one can predict the influence of changing composition of the fluid phase on the pressure of equilibrium between reactant and product assemblages for the various reactions listed above.

As an example of the first type of reaction, in which the product assemblage includes an evolved volatile species, a 1,000-bar P_{fluid} isobaric section for the system $CaO\text{–}SiO_2\text{–}CO_2\text{–}H_2O$ is presented as a $T\text{–}X_{CO_2}$ diagram in Figure 6.33. The reaction of concern, $CaCO_3 + SiO_2 = CaSiO_3 + CO_2$, was previously illustrated in Figure 3.5, where the partial pressure of carbon dioxide was calculated as less than the confining pressure, and in Figures 6.14 and 6.15, where

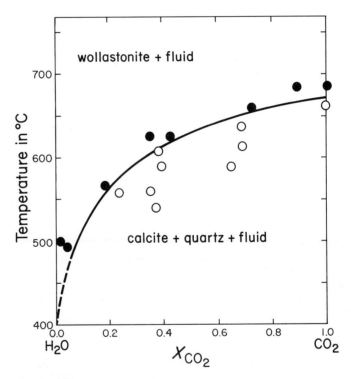

FIGURE 6.33.

Experimentally established $T\text{–}x$ diagram for the condensed bulk composition $CaO\cdot SiO_2$ in the quaternary system $CaO\text{–}SiO_2\text{–}CO_2\text{–}H_2O$ at 1,000 bars fluid pressure, after Greenwood (1967a, Figure 1); X_{CO_2} is the mole fraction of CO_2 in the fluid phase.

$P_{CO_2} = P_{total}$. Figure 6.33 shows the effect of diluting the volatile phase with H_2O; this latter component does not react with the condensed assemblage throughout the range of conditions considered. Quite clearly, the presence of H_2O in the fluid markedly influences the temperature at which calcite + quartz become incompatible. The phases calcite + quartz + wollastonite + fluid coexist along the curve as shown. In the isobaric four-component system, this means that there is a single degree of freedom; as discussed in Chapter 1, the phase rule here may be given as $F = C - P + 1$. Either the temperature or the composition of the volatile phase may be regarded as the independent variable, but once one is selected, the other is specified, provided chemical equilibrium is attained.

The second kind of reaction, in which the product assemblage includes two evolved volatile components, is presented in Figure 6.34. Here a 4,000-bar P_{fluid} diagram for the equilibrium $5KMg_3Si_3AlO_{10}(OH)_2 + 6CaCO_3 + 24SiO_2$

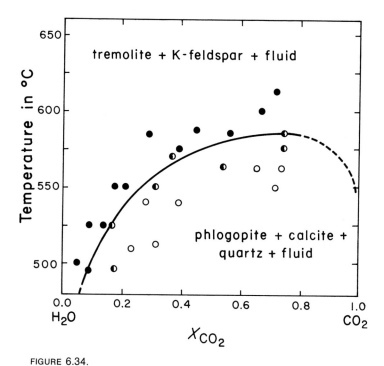

FIGURE 6.34.

Experimentally established T–x diagram for a portion of the system $K_2O\cdot Al_2O_3$–MgO–CaO–SiO_2–H_2O–CO_2 at 4,000 bars fluid pressure, after Hoschek (1973, Figure 2); X_{CO_2} is the mole fraction of CO_2 in the fluid phase. Divided-run symbols indicate lack of clear growth of either high-temperature or low-temperature assemblage.

$= 3Ca_2Mg_5Si_8O_{22}(OH)_2 + 5KAlSi_3O_8 + 2H_2O + 6CO_2$ is shown as a T-X_{CO_2} section. Because the stoichiometric ratio of evolved CO_2 to H_2O is $3:1$, the isobaric thermal maximum for this equilibrium should occur when the mole fraction of carbon dioxide in the fluid phase is 0.75. Evidently this relationship is compatible with the experimental data.

An example of the third type of reaction, in which the product assemblage involves both the consumption of a volatile species and the concomitant generation of another, is presented as Figure 6.35. This is a 5,000-bar P_{fluid} isobaric T-X section for the reaction $2Ca_2Al_3Si_3O_{12}(OH) + CO_2 = 3CaAl_2Si_2O_8 + CaCO_3 + H_2O$. Because the number of moles of gas evolved and used up by the reaction are the same, entropy changes for the equilibrium tend to cancel one another, and the over-all ΔS value for the reaction is an extremely small number; accordingly, from equation (6.6), $(\partial T/\partial X_j)_P \approx \infty$.

More complex isobaric P_{total} T-X_{CO_2} diagrams encompasing all the univariant reactions throughout the range of conditions considered are presented in Figures 6.36 and 6.37 for portions of the systems CaO-Al_2O_3-SiO_2-H_2O-CO_2 and MgO-SiO_2-H_2O-CO_2, respectively. Figure 6.36 is schematic, and equilibria involving quartz and wollastonite are not considered; at 2,000 bars fluid pressure, the univariant point is located at about 540°C, with the volatile phase consisting of roughly 20 mole per cent CO_2 (see Storre, 1970, Figure 5; Gordon, 1973). In this five-component system, at such an isobaric invariant point, six

FIGURE 6.35.

Experimentally established T-x diagram for a portion of the system $CaO \cdot Al_2O_3 \cdot 2SiO_2$-$CaO$-$CO_2$-$H_2O$ at 5,000 bars fluid pressure, after Storre and Nitsch (1972, Figure 2; see also Gordon and Greenwood, 1971, and Gordon, 1973); X_{CO_2} is the mole fraction of CO_2 in the fluid phase.

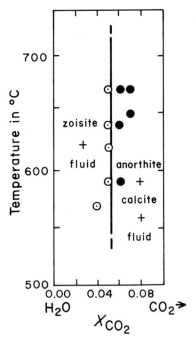

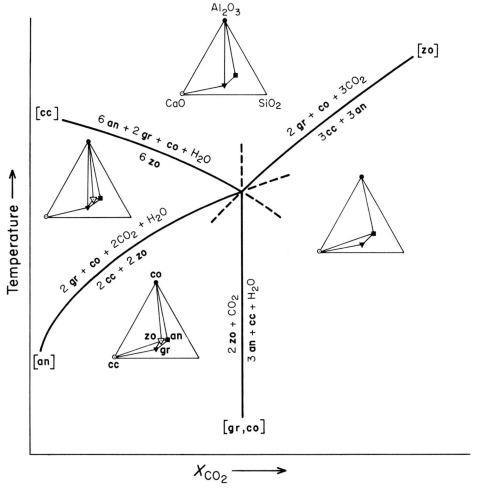

FIGURE 6.36.
Diagrammatic T–x section of the silica-poor portion of the system CaO–Al$_2$O$_3$–SiO$_2$–H$_2$O–CO$_2$ at constant total pressure, after Storre and Nitsch (1972, Figure 1). Compositions in this quinary system are projected onto the nonvolatile base, CaO–Al$_2$O$_3$–SiO$_2$. Closed symbols indicate phases which lie on this base; open symbols signify volatile-bearing phases. Brackets enclosing the nonparticipating phase define the univariant curves; metastable extensions of these curves are dotted. Abbreviations are: **an** = anorthite; **cc** – calcite; **co** – corundum; **gr** – grossular; **zo** = zoisite; and **f** = fluid.

phases are stable, namely anorthite + calcite + corundum + grossular + zoisite + fluid. Schreinemakers' treatment indicates that, in the general case, six univariant curves intersect at this point. However, one of the reactions, 2 zoisite + CO$_2$ = 3 anorthite + calcite + H$_2$O, is degenerate, hence two curves [**gr**] and [**co**] coincide; moreover, excess volatile constituents are stipulated as everywhere present (i.e., $P_{CO_2} + P_{H_2O} = P_{total}$), and a fluid-absent

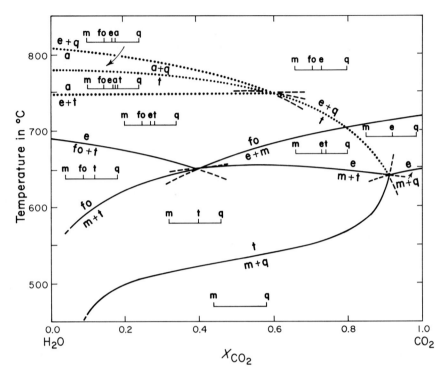

FIGURE 6.37.

A T–x diagram, experimentally established in part, for the system MgO–SiO_2–H_2O–CO_2 at 7,000 bars fluid pressure, after Evans and Trommsdorff (1974, Figure 2). Compositions in this quaternary system are projected onto the nonvolatile base, MgO–SiO_2. The condensed phases, but not necessarily their stoichiometric proportions, are indicated for each reaction; metastable extensions of univariant curves are dashed. Location of the equilibrium $Mg_3Si_4O_{10}(OH)_2 + 4MgSiO_3 = Mg_7Si_8O_{22}(OH)_2$ is very uncertain (e.g., see Figure 6.7); hence relationships illustrated are tentative and are shown by dotted lines. Abbreviations are: **a** = authophyllite; **e** = enstatite; **fo** = forsterite; **m** = magnesite; **q** = quartz; and **t** = talc.

reaction for this part of the system is chemographically impossible unless other volatile-bearing condensed phases participate in the reactions. For these reasons, only four curves appear in Figure 6.36.

A rather complicated web of univariant T–X curves for a portion of the system MgO–SiO_2–H_2O–CO_2 is presented in Figure 6.37, reflecting the large number of solid phases of concern. Accordingly, several isobaric invariant points are present in this quaternary system: each invariant point is characterized by the compatibility of five phases. (In the upper righthand portion of the diagram, the two curves defined by the reactions talc = 3 enstatite + quartz + H_2O, and enstatite + magnesite = forsterite + CO_2, cross in projection only, hence this is not a true isobaric invariant point, but rather an indifferent intersection.) Both this figure and the previous one illustrate the great importance of the fluid phase composition in determining the temperature of a particular metamorphic reaction at some given pressure. For this reason, isogradic surfaces and zone boundaries of mineral facies in rocks must be viewed in terms of the chemistry of the attending fluid. Analysis of fluid inclusions (e.g., see Touray, 1968; Poty, Stalder, and Weisbrod, 1974; Weisbrod and Poty, 1975) may provide the necessary constraints, as can the observed sequence and nature of the progressive metamorphic reactions themselves.

In general we can distinguish two principal types of behavior of a rock column during progressive metamorphism. (1) Where permeability is sufficiently large, the composition of the metamorphic solution passing through the system may be determined by reactions occurring outside the particular lithologic section being considered; by analogy with the diagrams just presented (e.g., Figures 6.33 through 6.37), X_{CO_2} may be maintained, or even buffered, at some arbitrary value. With rising temperature—assuming fixed P_{total} or a constant range of confining pressures—the mineral paragenetic sequence for such an open system will then reflect the individual condensed-rock bulk compositions and the particular value of X_{CO_2} imposed on the system. This behavior would be shown by a vertical line reflecting increasing temperature at constant pressure and fluid chemistry in an isobaric $T–X_{CO_2}$ diagram. (2) On the other hand, where the permeability of a lithologic unit is low enough to inhibit the circulation of a fluid phase, the system may be considered as closed to the import or export of matter. In this case the total composition of each rock determines the value of X_{CO_2} for changing physical conditions as well as the mineral paragenesis (e.g., see Greenwood, 1975). In general, with rising temperature (at constant P_{total}), no change in assemblage occurs for a given fluid composition until a reaction curve is encountered. At this stage, incipient formation of the product assemblage defines the composition of the fluid as a univariant function of temperature. With continued heating, the proportions of the condensed phases vary in response to the changing composition of the volatile phase, the chemistry of which is controlled by the reaction indicated by the univariant $T–X_{CO_2}$ curve. Eventually the assemblage will achieve an isobaric invariant point or a maximum on an isobaric univariant curve; at this stage, further heating will result in the replacement of one or more of the condensed phases, and in a discontinuity in the $T–X_{CO_2}$ path followed by the fluid.

We may conclude this section by reemphasizing the observation that the nature and sequence of mineral assemblages formed under a particular range of physical conditions are critically related to the fluid phase composition. In some cases the solid assemblage may buffer the chemistry of the fluid phase; whereas in others the chemical potentials of mobile components may control the nature of the mineral parageneses. In yet other circumstances (and this is perhaps the most general case), both condensed assemblage and fluid undergo chemical changes sympathetically during recrystallization. Fluid chemistry in turn is related to the composition and state of aggregation of the protoliths, and especially to the degree of "openness" displayed by the recrystallizing units.

ELEMENT FRACTIONATION BETWEEN COEXISTING PHASES

In previous chapters we have dealt qualitatively with the concept that in general, where solid, liquid, or gaseous solution is possible, elements are systematically but unequally partitioned between coexisting phases. For instance, study of the system $2MgO \cdot SiO_2$–$2FeO \cdot SiO_2(-SiO_2)$ demonstrates that, under equilibrium conditions, ferrous iron is strongly concentrated in the melt relative to olivine (see Figures 4.13, 4.32, 4.33, 4.44, and 5.9), at least at low oxidation states. Moreover, for the crystalline assemblage, orthopyroxene typically is slightly more magnesian than the olivine with which it is stable. The binary and ternary feldspars exhibit a pronounced enrichment in refractory constituents relative to the liquid in which the more fusible constituents are concentrated (see Figures 4.15, 4.16, 4.20, 4.47, and 4.48).

Although we have referred to solid-solid and solid-liquid element fractionation, this relationship holds for equilibria involving a volatile phase as well. For an example, consider the partitioning of alkalis between a supercritical fluid and alkali feldspar solid solution(s) at 2,000 bars total pressure as illustrated in Figure 6.38. The volatile phase in this particular case is a homogeneous, 2 molar aqueous solution of $(Na,K)Cl$. Moderate changes in the strength of the aqueous solution do not significantly affect the solid assemblage. The situation is complicated by the fact that the fluid phase is in equilibrium with two alkali feldspar solid solutions below about 680°C at this pressure. It is clear, however, that, for sodium-rich fluid compositions (i.e., less than about 20 to 25 mole per cent K), potassium is enriched in the volatile phase with respect to sodium; whereas for intermediate and potassic fluid compositions, K is enriched in the alkali feldspar relative to Na. As is clear from Figure 6.38, the distribution of alkalis between the aqueous solution and the solid assemblage is quite dependent on temperature, except for very sodic compositions. Therefore, in the presence of a thermal gradient, such as is imposed by the emplacement of a granitic pluton, or attending high-temperature regional metamorphism, the free circulation of an alkaline aqueous fluid of nearly constant intermediate Na–K composition would tend to enrich the higher grade sections in more sodic feldspar$_{ss}$; in contrast, alkali feldspars occurring in the cooler rocks would tend to be converted to more potassic compositions.

The general phenomenon of element fractionation between coexisting minerals was systematically treated first by Ramberg and de Vore (1951); the subject has since been investigated in considerable detail by numerous workers (e.g., see the synthesis by Saxena, 1973). In principle, one attempts to examine the disproportionation of elements between specific phases as a function of their compositional ranges, it being assumed that all assemblages formed in equilibrium under a common set of physical conditions. The fundamental relationship for the over-all partitioning is that given by the Van't Hoff reac-

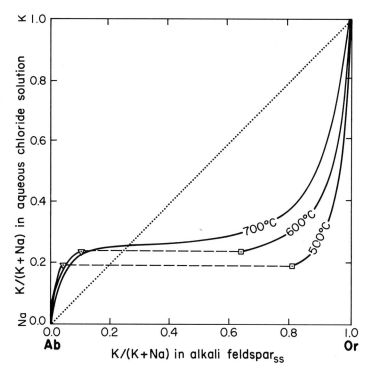

FIGURE 6.38.

Experimentally established mole fractions of potassium (K/(K + Na)) in coexisting 2 molar aqueous chloride and alkali-feldspar solutions at 2,000 bars fluid pressure, after Orville (1963, Figure 11). Pairs of alkali-feldspar solid solutions, linked by tie lines (long dashes), outline the shape of the $NaAlSi_3O_8$–$KAlSi_3O_8$ solvus (see also Figures 4.16 and 4.20); the dotted curve is provided for reference only, and would signify equal distribution of sodium and potassium between the alkali-feldspar solid solution(s) and the fluid phase.

tion isotherm (equation 1.71) as discussed on pp. 19–21:

$$\Delta G = \Delta G^0 + RT \ln \frac{a_L{}^l a_M{}^m}{a_A{}^a a_B{}^b},$$

for a heterogeneous equilibrium of the sort, $aA + bB = lL + mM$. Here A, B, L, and M may represent the individual participating phases, or they may stand for species involved in one or more solid, liquid, or gaseous solutions; a, b, l, and m are the stoichiometric coefficients of the reaction. A typical partition reaction involves the exchange of two components between a pair of phases, and may be indicated more simply as $lA_I + mB_{II} = lB_I + mA_{II}$, where A and

B refer to the discrete chemical species involved (for example, ferrous iron and magnesium ions), l and m indicate the moles of ions being exchanged, and I and II represent the two coexisting phases. For simplicity we have ignored the possibility that either or both of the participating phases contain structurally distinct sites which can accommodate the exchangable ions (see pp. 245–248). Let us now divide by l, letting $m/l = n$. Our exchange equation becomes $A_I + nB_{II} = B_I + nA_{II}$. This reaction is related to the Gibbs-free-energy change by means of the Van't Hoff reaction isotherm as just presented:

$$\Delta G = \Delta G^0 + RT \ln \frac{(a_B)_I(a_A)^n_{II}}{(a_A)_I(a_B)^n_{II}}, \tag{6.8}$$

or, on rearranging,

$$\Delta G = \Delta G^0 + RT \ln (a_B/a_A)_I(a_A/a_B)^n_{II}, \tag{6.9}$$

where the activity of A in phase I is designated $(a_A)_I$. At equilibrium, the Gibbs-free-energy change of the reaction is zero, and expression (6.9) reduces to

$$-\Delta G^0 = RT \ln (a_B/a_A)_I(a_A/a_B)^n_{II}. \tag{6.10}$$

Assuming that the activity coefficients are of unit value (i.e., ideal solution behavior), we can replace the activities by mole fractions or concentrations of the ions, X_A and X_B. The advantage here is that mole fractions and concentrations are readily measurable properties of the system in question. The distribution constant, K_D, may be defined in terms of the standard Gibbs-free-energy change of the partition reaction (see also equation 1.72):

$$e^{-\Delta G^0/RT} = K_D = (X_B/X_A)_I(X_A/X_B)^n_{II}. \tag{6.11}$$

Where component B approaches infinite dilution, $(X_A)_I \approx (X_A)_{II} \approx 1.0$ and expression (6.11) reduces to

$$K_{ND} = (X_B)_I/(X_B)_{II}, \tag{6.12}$$

a statement of the Nernst distribution law. The exponent n has unit value in dilute liquid solutions provided the exchange is ion-for-ion. As an example, the fractionations of TiO_2 and of MnO between competing sodic amphiboles and sodic pyroxenes from blueschist facies metamorphic rocks are illustrated in Figures 6.39a and 6.39b, respectively. The concentrations of TiO_2 and MnO clearly exceed trace amounts, but nonetheless, a fairly systematic straight-line infinite-dilution type of fractionation may be discerned. Titanium is slightly concentrated in the sodic amphiboles relative to sodic pyroxenes. In contrast, manganese disproportionates slightly toward enrichment in the single-chain silicates relative to the amphiboles.

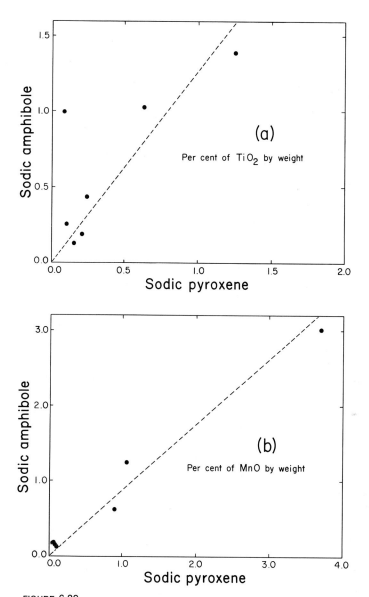

FIGURE 6.39.

Fractionation of TiO_2 and MnO between coexisting sodic amphiboles and sodic pyroxenes from blueschist facies rocks of California and Shikoku, after Onuki and Ernst (1969, Figure 5). The partitioning crudely obeys a Nernst infinite-dilution type of behavior, as described by equation (6.12).

Differentiating equation (1.72) or (6.11) with respect to temperature and pressure yields the previously derived relationships,

$$(\partial \ln K_D / \partial T)_P = \Delta H^0 / RT^2$$

(equation 1.76), where ΔH^0 is the enthalpy change of the reaction (see also equation 1.77), and

$$(\partial \ln K_D / \partial P)_T = -\Delta V^0 / RT$$

(equation 1.79), where ΔV^0 is the volume change of the reaction. It is apparent from equation (1.76) that, provided the enthalpy change is not strongly dependent on the temperature, variation of the distribution constant with temperature should be less pronounced at elevated temperatures than at lower temperatures. More importantly, as is seen from equation (6.11) K_D will more closely approach unit value at high temperatures unless the absolute change of $-\Delta G^0$ is proportional to the increment in temperature. Change in total pressure usually does not affect K_D significantly, because volume changes in exchange reactions are generally negligible (see equation 1.79). By rearranging the terms of equation (6.11), we obtain the expression

$$K_D (X_B / X_A)_{\text{II}}{}^n = (X_B / X_A)_{\text{I}}.$$

This equation defines a straight line on a log-log graphical plot of $(X_B / X_A)_{\text{I}}$ on the ordinate versus $(X_B / X_A)_{\text{II}}$ on the abscissa, provided that the values of K_D and n are not a function of composition of either phase I or phase II. Where $(X_B / X_A)_{\text{II}}$ is unity, the horizontal intercept on the ordinate indicates the numerical value of K_D; the slope of the line gives the numerical value of n.

Examples of ferric iron (+ titantium) versus total aluminum, and of ferrous iron (+ manganese) versus magnesium, between coexisting sodic amphiboles and sodic pyroxenes are presented in Figures 6.40a and 6.40b, respectively. It appears that the Fe^{3+} (+ minor Ti) $\leftrightarrow$ Al exchange is essentially ion-for-ion ($n = 0.9$), as if these species were being fractionated between pairs of virtually structureless, homogeneous solutions; Fe^{3+} is weakly concentrated in the Na-pyroxene relative to the Na-amphibole ($K_D = 1.2$) only for very ferric iron-rich compositions. In contrast, the Fe^{2+} (+ minor Mn) $\leftrightarrow$ Mg fractionation is more complex ($n = 1.5$); Fe^{2+} is markedly concentrated in the amphibole for intermediate and iron-rich bulk compositions, but is disproportionated into the pyroxene at magnesian bulk compositions ($K_D = 2.3$).

Systemmatic Mg/Fe^{2+} and $Mn/(Mg + Fe^{2+})$ ratios for coexisting chlorites and various other mafic phases from kyanite-zone peltic schists of central Vermont are illustrated on log-log cation ratio plots in Figure 6.41. For the rocks selected from a small area (P and T essentially constant), the number of phases is equal to the number of components required to describe them; hence the assemblages appear to be isothermal, isobaric invariant. Small variations

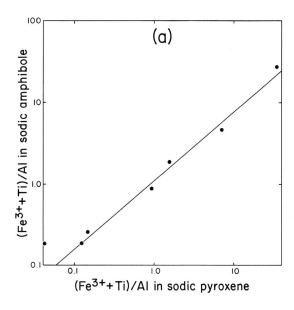

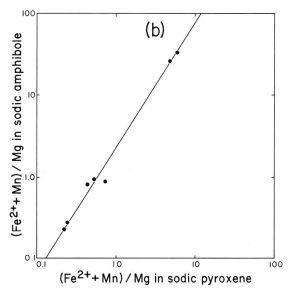

FIGURE 6.40.

Partitioning in coexisting sodic amphiboles and sodic pyroxenes from
blueschist facies rocks of California and Shikoku of (a) ferric iron
+ minor titanium versus aluminum, and (b) ferrous iron + minor
manganese versus magnesium, after Onuki and Ernst (1969, Figures 4
and 3, respectively). Ratios are in atomic proportions.

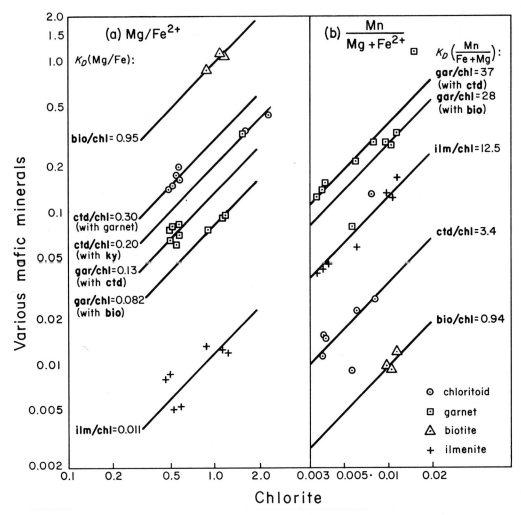

FIGURE 6.41.

Fractionation of (a) ferrous iron versus magnesium and (b) manganese versus ferrous iron and magnesium, between chlorite (abscissa) and various other mafic minerals (ordinate) from the Lincoln Quadrangle, Vermont, after Albee (1965, Figure 12). Ratios are presented on an atomic basis. Abbreviations are: **bio** = biotite; **chl** = chlorite; **ctd** = chloritoid; **gar** = garnet; **ilm** = ilmenite.

in Mg/Fe^{2+} and $Mn/(Mg + Fe^{2+})$ ratios for a phase such as chlorite from a particular assemblage demonstrate that, although this statement is essentially correct, it cannot be applied in detail. Although ferrous iron, magnesium, and manganese are strongly disproportionated among the several minerals, the 45° curve slopes does reflect an ion-for-ion type of cation fractionation behavior for the complex and diverse crystalline solutions.

An alternative method of graphical portrayal commonly employed in the literature (e.g., see Kretz, 1959; Mueller, 1960; Orville, 1963; Ernst, 1964) makes use of the fact that there is only one independent chemical parameter under consideration—that is, $(X_B)_I = (1 - X_A)_I$. Here one plots the mole fraction of A in phase I along the ordinate, $(X_A)_{II}$ along the abscissa, as was done in Figure 6.38. This type of diagram has the advantage of minimizing the large percentage errors which are present in mole fraction quotients at low concentrations; constant K_D values fall on regular curves rather than along straight lines, however, as in the diagrams presented as Figures 6.40 and 6.41. The distributions of Fe^{2+} and Mg between coexisting hypersthene and augite for igneous and high-grade metamorphic environments are presented in Figure 6.42, cast in terms of mole fractions, $X_{Mg}[= Mg/(Mg + Fe^{2+})]$. As is predictable from equation (6.11) and the discussion on p. 242, the partitioning of cations between the competing phases is less pronounced for the higher-temperature magmatic pairs ($K_D \approx 0.73$) than for the lower-temperature metamorphic pairs ($K_D \approx 0.54$). The fractionation is essentially ideal ($n = 1$), however, instead of regular but nonideal ($n \neq 1$), as in the situation illustrated in Figure 6.38.

Inasmuch as we have used both mole fraction and log-log ratio plots, these are compared for two values of K_D and n in Figure 6.43.

Thus far we have regarded the different phases competing for specific ions as individually characterizable, but as essentially homogeneous and without variability in terms of the local atomic environment. Where the configurational entropy of mixing of A and B in each phase is maximized (i.e., the exchangeable ions are structurally disordered), the exchange reactions may be considered as an ion-for-ion replacement; accordingly l and m (hence n) are unity. On the other hand, if the ions are partially or completely ordered in the various crystallographic sites (and this is the more common situation), then exchange

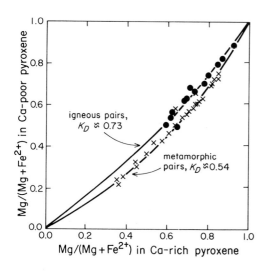

FIGURE 6.42.

Fractionation of magnesium and ferrous iron between hypersthene and augite from mafic igneous rocks (filled circles) and high-grade metamorphic rocks including granulites, gneisses, chanockites, and eclogites (x symbols), after Kretz (1963, Figure 1). Constant values of K_D plot as smooth curves on such a mole fraction diagram.

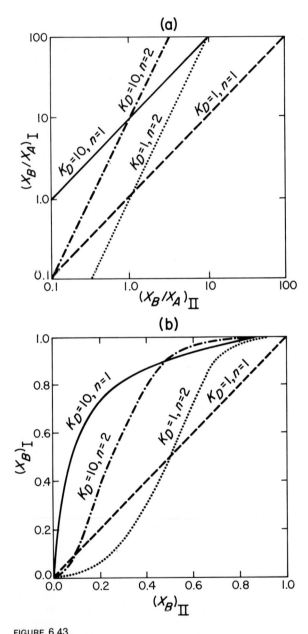

FIGURE 6.43.

Comparison of element-partitioning diagrams for components
A and B between phases I and II based on (a) log-log ratios,
and (b) mole fractions. The exchange reaction considered is
$A_I + nB_{II} = B_I + nA_{II}$.

equilibrium becomes more complicated, and may involve two or more partition coefficients for ions ordered in specific, contrasting structural sites of one phase being exchanged for an equivalent number of ions in one or more sites in the other phase.

Equilibrium partitioning between olivine and calcium-poor pyroxene was regarded by Ramberg and de Vore (1951) as of the ion-for-ion type (i.e., assuming that $n = 1$):

$$FeSi_{0.5}O_2 + MgSiO_3 = MgSi_{0.5}O_2 + FeSiO_3.$$

Bartholomé (1962) wrote the same over-all reaction as

$$Fe_2SiO_4 + 2MgSiO_3 = Mg_2SiO_4 + 2FeSiO_3.$$

In the first case, ΔG^0 for the reaction involves $\frac{1}{2}$ mole of olivine and one mole of hypersthene, whereas in the second it involves twice as much. As is clear from equation (6.11), this is reflected in different values of K_D, but not in the observed partitioning, because the mole fraction quotients involve a square term, i.e., $(X_{Fs}^2/X_{En}^2)_{hypersthene}$, in the second case. Analyzed natural assemblages yield values of n intermediate between one and two, probably because calcium-poor pyroxenes exhibit significant cation ordering, although seemingly olivines do not. As pointed out by Mueller (1962), because layer-lattice and single- and double-chain silicates, for instance, possess a variety of sites which accommodate the cations in question, the values of K_D are a composite function of the sublattice distributions, which in turn depend on order-disorder relations. This subject has been explored in detail by Matsui and Bano (1965), Mueller (1969), Thompson (1969), and Grover and Orville (1969).

To elucidate the principle, let us consider the simplest possible case in which ordering exerts a significant effect on fractionation, that of the distribution of ions between a phase I characterized by a single structural site, such as fluid or olivine, and a double-site phase II, such as hypersthene or calcic pyroxene, following the treatment by Grover and Orville (1969). Standard Gibbs free energies of exchange are shown diagrammatically in Figure 6.44. In general,

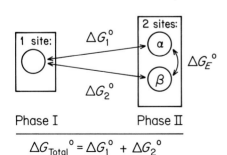

FIGURE 6.44.
Schematic exchange energies for a single-site phase, I, and a double-site phase, II, after Grover and Orville (1969, Figure 1). ΔG_{total}^0 may be obtained by using equation (6.11) and the measured intercrystalline partitioning.

where fractionation takes place between a phase of q energetically distinct sites and another phase of r sites, the total number of partition coefficients is qr, of which only $q + r - 1$ are independent. For iron-magnesium distribution between olivine and hypersthene, three different partition reactions may be written, only two of which are independent:

(a) $\Delta G_1{}^0$ for the equilibrium $A_I + B_{II\alpha} = B_I + A_{II\alpha}$;

(b) $\Delta G_2{}^0$ for the equilibrium $A_I + B_{II\beta} = B_I + A_{II\beta}$; and

(c) $\Delta G_E{}^0$ for the equilibrium $A_{II\alpha} + B_{II\beta} = B_{II\alpha} + A_{II\beta}$.

Reactions (a) and (b) represent the fractionation of components A and B between the single-site phase and site α of the double-site phase, and between the single-site phase and site β of the double-site phase, respectively. The sum of $\Delta G_1{}^0$ and $\Delta G_2{}^0$ is $\Delta G_{total}{}^0$, the observed over-all fractionation between the coexisting phases. Reaction (c) indicates the standard free energy change of the intracrystalline exchange for the double-site phase. Obviously, where $\Delta G_E{}^0$ is zero, ion-for-ion partitioning behavior results.

Isothermal element fractionation between a single-site phase I and a double-site phase II, assuming ideal solution on each site, is presented in Figure 6.45. Here the control of over-all fractionation is shown by curves a–g for different numerical values of $\Delta G_{total}{}^0$. The shape of each partition curve is a function of the relative intensity of competition between sites α and β in the double-site phase for the species A and B, as reflected by the numerical value of the intracrystalline exchange energy, $\Delta G_E{}^0$. Clearly, the behavior of solid solutions is a strong function of contrasting structural-site energies.

The departure of accurately analyzed mineral pairs from a log-log straight-line plot or a mole-fraction ideal distribution curve may result from: (1) failure of the coexisting phases to attain chemical equilibrium; (2) marked deviation of activity coefficients from unit values; (3) the presence of differing amounts of an additional component which influences K_D; or (4) disparate physical conditions for the different mineral pairs. In general, the phases of natural lithologic environments are commonly subject to all these phenomena; hence only under favorable circumstances are the element fractionation results of unequivocal acceptability. In any event, the principle of disproportionation of elements between competing phases is understandable and the partitioning can be treated quantitatively.

Although the discussion presented above has been concerned solely with the partitioning of cations between coexisting phases, obviously the derived relationships are equally applicable to a study of the fractionation of anions. Moreover, a profoundly important process for petrology, that of stable isotope partitioning, likewise obeys the laws just discussed. Because mass differences of the isotopes of an element do not affect its cross section or size (i.e., its partial molar volume), the disproportionation of isotopes of an element under equi-

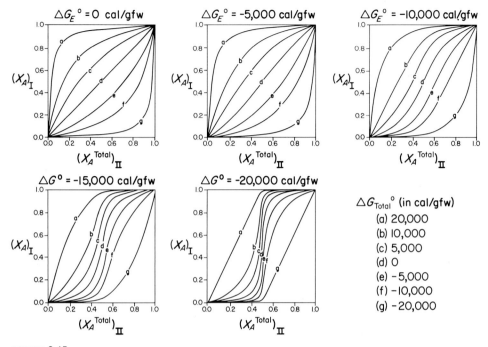

FIGURE 6.45.

Diagrammatic isothermal partitioning (at 1,000°C) of components A and B between an ideal single-site phase I and an ideal double-site phase II, after Grover and Orville (1969, Figure 2). Here $(X_A^{total})_{II}$ represents the mole fraction of A [i.e., $A/(A + B)$] in phase II, irrespective of site occupancy.

librium conditions between competing phases is almost exclusively a function of temperature and the contrasting crystallochemical environments. Therefore, natural partition data are extremely useful as petrologic geothermometers once the fractionation has been established experimentally or calculated theoretically as a function of temperature. A good summary of the phenomenon and the conventions employed has been given by Epstein and Taylor (1967) for 0^{18}–0^{16} fractionations in rocks.

PETROGENESIS OF METAMORPHIC ROCKS AND PLATE TECTONICS

We have seen that the phase assemblages developed in metamorphic rocks, and the compositions of the coexisting minerals, are sensitive functions of the chemical potentials of all the components in the system (i.e., the aggregate bulk composition of the condensed assemblage + the fluid phase) and of the state variables, temperature and pressure. Among the most important chemical variables are the μ values for CO_2, O_2, H_2O, and H_2, because these are the species most likely to undergo strong fluctuations or systemmatic changes with time and therefore may behave as mobile components. The condition that equilibrium be attained locally has been implicit in this discussion, but obviously, where reactions proceed slowly, as in most geological systems, meta-stability and marked departures from the lowest Gibbs-free-energy configurations are not only possible but probable. The presence of relict minerals, incompatible phases, and mineralogic zoning all testify to the failure of the system to reach the equilibrium state. Within a rock, such variations or gradients in chemical potentials reflect changes in μ values due to changing reacting compositions, temperatures, and/or pressures, which exceed the rates of reaction tending to obliterate these disequilibrium arrangements.

It is clear from Chapter 1 that all such reactions drive toward the lowest Gibbs-free-energy configuration. Accordingly, under favorable circumstances, an examination of metamorphic mineral parageneses may reveal the series of stable assemblages which characterize individual rock bulk compositions within a discrete range of P–T conditions. Comparison of the observed phase compatibilities and the compositions of the minerals with experimentally established phase-equilibrium diagrams, with stable isotope geothermometry (especially hydrogen and oxygen disproportionations), and with computations based on calorimetric data, allow the construction of a petrogenetic grid such as is illustrated in Figure 6.46. In this figure, the equilibrium mineral assemblages characteristic of the several metamorphic facies have been placed in a conventional P–T diagram. The presence of an aqueous fluid phase, including only negligible CO_2, has been assumed, except for the region at relatively elevated temperatures and fluid pressures where partial melting—an igneous or ultrametamorphic process—would ensue (e.g., see Figure 4.41b); in addition, neither strongly oxidizing nor strongly reducing conditions have been considered, but rather an intermediate f_{O_2} range has been selected, similar to that defined by, say, the magnetite + quartz = fayalite + O_2 equilibrium (see Figure 2.5, curve 4). The mineral assemblages diagnostic of the various metamorphic facies will be produced under the appropriate physical conditions only if several important protolith compositions are involved; so it is assumed that aluminous pelitic units, graywackes or feldspathic sandstones, and mafic igneous lithologies are present throughout to indicate the grade. And of course, the maintenance of chemical equilibrium along a (prograde) geothermal gradient has been postulated.

Although the influence of chemical complexities is not easy to assess quantitatively, the diagram seems topologically reasonable, with individual facies boundaries accurate to perhaps $\pm 50C°$. Note that for the multivariant equilibria ($n - 1$ degrees of freedom) which relate adjacent metamorphic facies assemblages (n degrees of freedom), intervening P–T zones exist which separate the distinctly different mineral assemblages for any bulk chemistry; at fixed rock bulk composition, the compositions of the individual minerals and the phase proportions change gradually with change in state variables. The

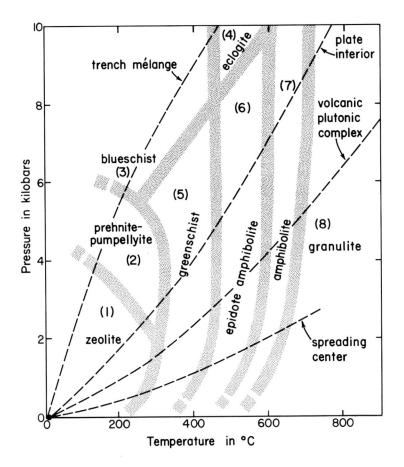

FIGURE 6.46.

Diagrammatic metamorphic petrogenetic grid for crustal rock types with specific mineral facies located in P–T space. An H_2O-rich fluid is assumed to be ubiquitous except at relatively high temperatures and pressures, where the activity of H_2O in nonmigmatitic terranes is thought to be low (otherwise, substantial partial melting would be expected). Characteristic prograde metamorphic P–T paths for rock sections located in contrasting plate-tectonic environments are also illustrated.

constant bandwidth shown diagrammatically is only $30C°$, but this is a purely arbitrary approximation which attempts to simplify the figure for clarity; in reality, the widths of the transition zones certainly must be variable functions of all the intensive parameters of the system. Moreover, metamorphic facies boundaries are complex functions of the bulk composition of the protolith and the chemical potentials of the mobile components (if any); for instance, as Zen (1961) and Albee and Zen (1969) have demonstrated, the zeolite-greenschist facies boundary is a strong function of μ_{CO_2}/μ_{H_2O}, larger values of

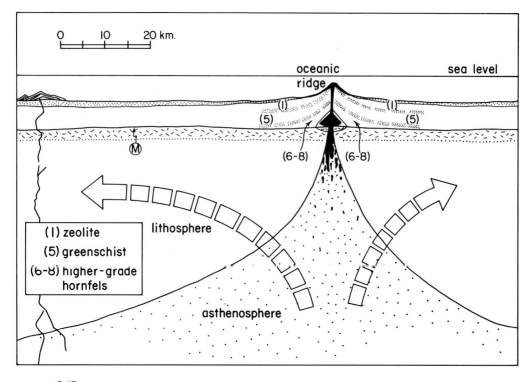

FIGURE 6.47.

Schematic distribution of metamorphic facies types near a divergent plate margin, assuming the over-all correctness of Figures 5.16 and 6.46. Relatively high-temperature, low-pressure recrystallization is confined to the vicinity of the rising axial plume. Metamorphic facies numbers correspond to those of Figure 6.46. The partial fusion of undepleted rising mantle material to yield oceanic tholeiitic melt (black) is also shown.

the ratio favoring greenschist assemblages at the expense of zeolites, even at very low temperatures and pressures.

Assuming the appropriateness of thermal structures illustrated previously as Figures 5.16 and 5.17, and of the schematic metamorphic petrogenetic grid of Figure 6.46, for crustal rocks, the spatial disposition of metamorphic facies in the neighborhood of divergent and convergent lithospheric plate boundaries can be approximated, as shown in Figures 6.47 and 6.48, respectively. Chemical equilibrium again has been assumed. Local departures from the inferred thermal regime, variations in rock bulk compositions, or the extent to which phase equilibrium has been attained undoubtedly account for some of the complicated natural occurrences, but the general relationship between recrystallization and sea-floor spreading seems to be reasonably well-understood.

It is apparent from Figure 6.47 that a relatively simple metamorphic zonation might be developed in the vicinity of an oceanic ridge. Rocks characteristic

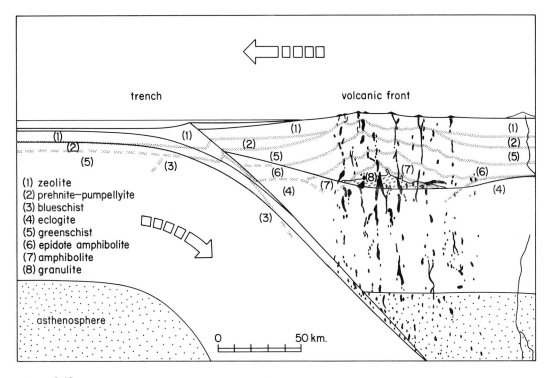

FIGURE 6.48.

Schematic distribution of metamorphic facies types near a convergent plate margin, assuming the over-all correctness of Figures 5.17 and 6.46. Relatively high-pressure, low-temperature recrystallization takes place in the narrow, asymmetric subduction zone; whereas a broad, bilaterally symmetric zone of high-temperature, low-pressure metamorphism typifies the magmatic arc. Metamorphic facies numbers correspond to those of Figure 6.46. The partial melting of descending, transformed oceanic crust and of relatively primitive mantle material to provide island arc or continental margin tholeiitic and calc-alkaline melts (black), and the anatexis of H_2O-rich basal portions of the continental crust (checks) are also illustrated. The width of the volcanic belt is only 20 to 30 km at any one time; so in the figure, magmatic activity shown has been integrated over a considerable time span.

of the several high-temperature, low-pressure metamorphic facies could be formed in the vicinity of the spreading center and subsequently transported laterally during the sea-floor spreading process. Concomitant with declining temperatures, retrograde reactions would be expected to occur to a greater or lesser degree depending on the chemical kinetics. Both prograde and retrograde reaction rates in turn depend on such factors as the accessibility of sea water and extent of granulation.

Near a convergent plate junction, similar to the one illustrated schematically in Figure 6.48, the nature of the thermal structure evidently produces an oceanward, relatively high-pressure, low-temperature metamorphic belt, and a landward, relatively high-temperature, low-pressure metamorphic terrane.

The former is a narrow, elongate belt of rocks which exhibits pronounced asymmetry in mineral assemblages, with highest grade, apparently most deeply subducted, blueschistic and eclogitic sections lying at the ancient plate junction, and successively more feebly recrystallized rocks extending seaward. In contrast, the landward terrane occupies a much broader region and seems to be characterized by telescoped, relatively high-temperature metamorphic mineral parageneses, and by a crude bilateral symmetry about the volcanic-plutonic axis.

Thus the disposition of metamorphic belts, and the paragenetic sequences observed within any specific terrane, are complicated functions of the original lithologies and of the attending P–T regime. The latter is obviously related to the large-scale tectonic processes to which the region has been subjected. As with igneous petrology and mantle equilibria, we may conclude that the various aspects of metamorphic petrology can be understood fully only by an integrated appreciation of the principles of classical thermodynamics and phase equilibrium on the one hand, and of plate tectonic-petrologic interactions on the other.

Appendixes

APPENDIX 1

Conversion data
from Bridgman (1963, pp. 28–35).

Pressure = force/area = (force · distance)/volume = energy/volume, etc.

	barye	bar	atmos.	kgwt/cm²	p.s.i.	mmHg
1 barye = 1 dyne/cm²	1	10^{-6}	$0.9869 \cdot 10^{-6}$	$1.020 \cdot 10^{-6}$	$14.504 \cdot 10^{-6}$	$7.50 \cdot 10^{-4}$
1 bar	10^6	1	0.98692	1.0197	14.504	750.1
1 atmosphere	$1.013 \cdot 10^6$	1.0133	1	1.0332	14.696	760
1 kgwt/cm²	$9.807 \cdot 10^5$	0.9807	0.9678	1	14.223	735.6
1 lbwt/in² (p.s.i.)	$6.895 \cdot 10^4$	0.6895	0.06805	0.0703	1	51.7
1 mmHg	1333.2	$1.333 \cdot 10^{-3}$	$1.316 \cdot 10^{-3}$	$1.3595 \cdot 10^{-3}$	0.01934	1

1 bar ~ 1 atm ~ 1 kg/cm² ~ 14.5 p.s.i. is the hydrostatic pressure equivalent to a load of 3.50 meters, or 11.5 feet, of crustal rocks (assuming crust density = 2.85).

1,000 bars is the hydrostatic pressure equivalent to a load of 3.5 km or 2.2 miles of crustal rocks.

1 km of crustal rocks is equivalent to a load of 285 bars.

1 mile of crustal rocks is equivalent to a load of 460 bars.

Energy = force · distance = volume · pressure

	erg	joule	calorie	cm³bar	cm³atm	kgwt cm
1 erg = 1 dyne cm	1	10^{-7}	$2.389 \cdot 10^{-8}$	10^{-6}	$9.869 \cdot 10^{-7}$	$1.0197 \cdot 10^{-6}$
1 joule	10^7	1	0.2389	10.00	9.869	10.197
1 calorie	$4.186 \cdot 10^7$	4.186	1	41.86	41.311	42.685
1 cm³ bar	10^6	0.1000	$2.389 \cdot 10^{-2}$	1	0.9869	1.0197
1 cm³ atmosphere	$1.013 \cdot 10^6$	0.1013	$2.421 \cdot 10^{-2}$	1.0132	1	1.0332
1 kgwt cm	$9.807 \cdot 10^5$	$9.807 \cdot 10^{-2}$	$2.343 \cdot 10^{-2}$	0.9807	0.9678	1

R (gas constant) = energy/(degree · mole) = (force · distance)/(degree · mole) = (volume · pressure)/(degree · mole)

$\quad$ = 1.987 cal/(°K · mole)

$\quad$ = 8.31470 joules/(°K · mole)

$\quad$ = 82.0597 cm³ atm/(°K · mole)

$\quad$ = 83.147 cm³ bar/(°K · mole)

$\quad$ = $8.3147 \cdot 10^7$ erg/(°K · mole)

$\quad$ = .9807 kgwt cm/(°K · mole)

Entropy = energy/°K = force · distance/°K = volume · pressure/°K

Volume = distance³ = energy/pressure = (energy · distance²)/force = (force · distance)/pressure, etc.

$\log n = 0.43429 \ln n$

$\ln n = 2.303 \log n$

N (Avogadro's number) = $6.0248 \cdot 10^{23}$/mole

One-atmosphere thermodynamic properties
of minerals and related substances at 298.15°K,
from Robie and Waldbaum (1968, pp. 11–25).

Mineral	Formula	Gram formula weight	S° (cal/deg/gfw)	Molar volume (cm³)	ΔH° (cal/gfw)	ΔG° (cal/fgw)
graphite	C	12.011	1.372 ± 0.005	5.2982 ± 0.0009	0	0
diamond	C	12.011	0.568 ± 0.003	3.4166 ± 0.0003	453 ± 10	693 ± 15
α-iron	Fe	55.847	6.52 ± 0.03	7.092 ± 0.004	0	0
oxygen	O_2	31.999	48.996 ± 0.010	24,465.0 ± 3.4	0	0
troilite	FeS	87.911	14.42 ± 0.04	18.20 ± 0.03	−24,130 ± 350	−24,219 ± 360
pyrite	FeS_2	119.975	12.65 ± 0.03	23.940 ± 0.007	−41,000 ± 400	−38,296 ± 410
corundum	Al_2O_3	101.961	12.18 ± 0.03	25.575 ± 0.007	−400,400 ± 300	−378,082 ± 310
carbon dioxide	CO_2	44.010	51.06 ± 0.01	24,465.0 ± 3.4	−94,051 ± 30	−94,257 ± 40
lime	CaO	56.079	9.5 ± 0.5	16.764 ± 0.005	−151,790 ± 300	−144,352 ± 340
wüstite	$Fe_{0.947}O$	68.887	13.76 ± 0.10	12.04 ± 0.04	−63,640 ± 200	−58,599 ± 210
hematite	Fe_2O_3	159.692	20.89 ± 0.05	30.274 ± 0.02	−197,300 ± 300	−177,728 ± 310
magnetite	Fe_3O_4	231.539	36.03 ± 0.10	44.524 ± 0.008	−267,400 ± 500	−243,094 ± 510

water	H_2O	18.015	16.71 ± 0.03	18.069 ± 0.003	−68,315 ± 10	−56,688 ± 20
steam	H_2O	18.015	45.104 ± 0.010	24,465.0 ± 3.4	−57,796 ± 10	−54,635 ± 20
periclase	MgO	40.311	6.44 ± 0.04	11.248 ± 0.004	−143,800 ± 100	−136,087 ± 110
brucite	$Mg(OH)_2$	58.327	15.09 ± 0.05	24.63 ± 0.07	−221,200 ± 700	−199,460 ± 730
α-quartz	SiO_2	60.085	9.88 ± 0.02	22.688 ± 0.001	−217,650 ± 400	−204,646 ± 410
rutile	TiO_2	79.899	12.04 ± 0.04	18.820 ± 0.008	−225,760 ± 100	−212,559 ± 110
spinel	$MgAl_2O_4$	142.273	19.26 ± 0.10	39.71 ± 0.03	−552,800 ± 500	−522,961 ± 510
aragonite	$CaCO_3$	100.089	21.18 ± 0.30	34.15 ± 0.05	−288,651 ± 340	−269,678 ± 350
calcite	$CaCO_3$	100.089	22.15 ± 0.20	36.934 ± 0.015	−288,592 ± 320	−269,908 ± 330
dolomite	$CaMg(CO_3)_2$	184.411	37.09 ± 0.07	64.34 ± 0.03	−557,613 ± 520	−518,734 ± 530
kyanite	Al_2SiO_5	162.046	20.02 ± 0.08	44.09 ± 0.07	−619,930 ± 540	−584,000 ± 550
andalusite	Al_2SiO_5	162.046	22.28 ± 0.10	51.53 ± 0.04	−619,390 ± 710	−584,134 ± 720
sillimanite	Al_2SiO_5	162.046	22.97 ± 0.10	49.90 ± 0.04	−618,650 ± 710	−583,600 ± 720
sphene	$CaTiSiO_5$	196.063	30.88 ± 0.20	55.65 ± 0.17	−622,050 ± 570	−588,246 ± 580
fayalite	Fe_2SiO_4	203.778	35.45 ± 0.40	46.39 ± 0.09	−353,544 ± 700	−329,668 ± 720
forsterite	Mg_2SiO_4	140.708	22.75 ± 0.20	43.79 ± 0.03	−520,370 ± 520	−491,938 ± 530
wollastonite	$CaSiO_3$	116.164	19.60 ± 0.20	39.93 ± 0.10	−390,640 ± 870	−370,313 ± 880
diopside	$CaMgSi_2O_6$	216.560	34.20 ± 0.20	66.09 ± 0.10	−767,390 ± 2180	−725,784 ± 2190
jadeite	$NaAlSi_2O_6$	202.140	31.90 ± 0.30	60.40 ± 0.10	−719,871 ± 1000	−677,206 ± 1010
tremolite	$Ca_2Mg_5Si_8O_{22}(OH)_2$	812.410	131.19 ± 0.30	272.92 ± 0.73	−2,952,935 ± 4140	−2,779,137 ± 4150
anorthite	$CaAl_2Si_2O_8$	278.210	48.45 ± 0.30	100.79 ± 0.05	−1,009,300 ± 1150	−955,626 ± 1160
high albite	$NaAlSi_3O_8$	262.224	54.67 ± 0.45	100.43 ± 0.09	−934,513 ± 770	−882,687 ± 790
low albite	$NaAlSi_3O_8$	262.224	50.20 ± 0.40	100.07 ± 0.13	−937,146 ± 740	−883,988 ± 760
nepheline	$NaAlSiO_4$	142.055	29.72 ± 0.30	54.16 ± 0.06	−497,029 ± 1000	−469,664 ± 1010
sanidine	$KAlSi_3O_8$	278.337	56.94 ± 1.00	109.05 ± 0.10	−944,378 ± 930	−892,263 ± 980
muscovite	$KAl_2Si_3AlO_{10}(OH)_2$	398.313	69.0 ± 0.7	140.71 ± 0.18	−1,421,180 ± 1300	−1,330,103 ± 1320
talc	$Mg_3Si_4O_{10}(OH)_2$	379.289	62.34 ± 0.15	136.25 ± 0.26	−1,415,205 ± 1710	−1,324,486 ± 1720

APPENDIX **3**

One-atmosphere thermodynamic properties
of minerals and related substances
at elevated temperatures, from Robie
and Waldbaum (1968, pp. 35–235).

The solid horizontal line across all columns at a specified temperature indicates that
a transition occurs in the phase of interest. The short horizontal line in the last three
columns denotes a change in the stability state of a constituting reference element at
some point in the 100°K temperature interval designated.

Carbon (reference state); gram formula weight, 12.011
Formula, C: Graphite 298.15° to 2,000°K.

| | | | | Formation from the elements | | |
Temp. (°K)	$H_T - H_{298}$ (Kcal)	S_T (cal/deg·gfw)	$-(G_T - H_{298})/T$	Enthalpy	Free energy (Kcal/gfw)	Log K
298.15	0.000	1.372	1.372	0.000	0.000	0.000
uncertainty		0.005	0.005			
400	0.250	2.088	1.463	.000	.000	.000
500	0.569	2.797	1.659	.000	.000	.000
600	0.947	3.484	1.906	.000	.000	.000
700	1.372	4.139	2.179	.000	.000	.000
800	1.831	4.752	2.463	.000	.000	.000
900	2.318	5.324	2.748	.000	.000	.000
1,000	2.824	5.857	3.033	.000	.000	.000
1,100	3.347	6.355	3.312	.000	.000	.000
1,200	3.883	6.822	3.586	.000	.000	.000
1,300	4.432	7.261	3.852	.000	.000	.000
1,400	4.986	7.674	4.111	.000	.000	.000
1,500	5.552	8.063	4.362	.000	.000	.000
1,600	6.122	8.430	4.604	.000	.000	.000
1,700	6.696	8.778	4.839	.000	.000	.000
1,800	7.275	9.109	5.067	.000	.000	.000
1,900	7.857	9.424	5.289	.000	.000	.000
2,000	8.442	9.724	5.503	.000	.000	.000

One-atmosphere thermodynamic properties
of minerals and related substances
at elevated temperatures, from Robie
and Waldbaum (1968, pp. 35–235).

Diamond; gram formula weight, 12.011
Formula, C: Diamond 298.15° to 1,200°K.

				Formation from the elements		
Temp. (°K)	$H_T - H_{298}$ (Kcal)	S_T	$-(G_T - H_{298})/T$ (cal/deg·gfw)	Enthalpy	Free energy (Kcal/gfw)	Log K
298.15	0.000	0.57	0.57	0.453	0.693	−0.508
uncertainty		0.003	0.003	0.010	0.020	0.015
400	0.200	1.14	0.64	0.403	0.783	−0.428
500	0.486	1.77	0.80	0.370	0.884	−0.387
600	0.842	2.42	1.01	0.348	0.988	−0.360
700	1.251	3.05	1.26	0.332	1.096	−0.342
800	1.700	3.65	1.52	0.322	1.205	−0.329
900	2.180	4.22	1.80	0.315	1.310	−0.318
1,000	2.685	4.75	2.06	0.314	1.423	−0.311
1,100	3.212	5.25	2.33	0.318	1.536	−0.305
1,200	3.760	5.73	2.59	0.330	1.643	−0.299

Iron (reference state); gram formula weight, 55.847

Formula, Fe: α crystals (body-centered cubic) 298.15° to 1,184°K. Curie point 1,033°K. γ crystals (face-centered cubic) 1,184° to 1,665°K. δ crystals (body-centered cubic) 1,665° to melting point 1,809°K. Liquid 1,809° to 2,000°K.

Temp. (°K)	$H_T - H_{298}$ (Kcal)	S_T (cal/deg·gfw)	$-(G_T - H_{298})/T$	Formation from the elements		
				Enthalpy (Kcal/gfw)	Free energy	Log K
298.15	0.000	6.520	6.520	0.000	0.000	0.000
uncertainty		0.030	0.030			
400	0.635	8.350	6.762	.000	.000	.000
500	1.310	9.860	7.240	.000	.000	.000
600	2.040	11.190	7.790	.000	.000	.000
700	2.830	12.400	8.357	.000	.000	.000
800	3.700	13.560	8.935	.000	.000	.000
900	4.680	14.720	9.520	.000	.000	.000
1,000	5.840	15.940	10.100	.000	.000	.000
1,100	7.220	17.260	10.696	.000	.000	.000
1,184	8.030	17.970	11.188	.000	.000	.000
1,184	8.255	18.160	11.188	.000	.000	.000
1,200	8.390	18.270	11.278	.000	.000	.000
1,300	9.210	18.920	11.835	.000	.000	.000
1,400	10.050	19.540	12.361	.000	.000	.000
1,500	10.900	20.130	12.863	.000	.000	.000
1,600	11.770	20.690	13.334	.000	.000	.000
1,665	12.340	21.040	13.630	.000	.000	.000
1,665	12.600	21.200	13.630	.000	.000	.000
1,700	12.940	21.400	13.788	.000	.000	.000
1,800	13.930	21.970	14.231	.000	.000	.000
1,809	14.020	22.020	14.270	.000	.000	.000
1,809	17.650	24.030	14.270	.000	.000	.000
1,900	18.610	24.540	14.745	.000	.000	.000
2,000	19.660	25.080	15.250	.000	.000	.000

One-atmosphere thermodynamic properties
of minerals and related substances
at elevated temperatures, from Robie
and Waldbaum (1968, pp. 35–235).

Oxygen (reference state); gram formula weight, 31.999
Formula, O_2: Ideal diatomic gas 298.15° to 2,000°K.

| Temp. (°K) | $H_T - H_{298}$ (Kcal) | S_T (cal/deg·gfw) | $-(G_T - H_{298})/T$ | Formation from the elements | | |
| | | | | Enthalpy | Free energy | |
				(Kcal/gfw)		Log K
298.15	0.000	48.996	48.996	0.000	0.000	0.000
uncertainty		0.010	0.010			
400	0.724	51.083	49.273	.000	.000	.000
500	1.455	52.714	49.804	.000	.000	.000
600	2.210	54.090	50.407	.000	.000	.000
700	2.988	55.289	51.020	.000	.000	.000
800	3.786	56.353	51.620	.000	.000	.000
900	4.600	57.312	52.201	.000	.000	.000
1,000	5.427	58.184	52.757	.000	.000	.000
1,100	6.266	58.983	53.287	.000	.000	.000
1,200	7.114	59.721	53.793	.000	.000	.000
1,300	7.971	60.407	54.275	.000	.000	.000
1,400	8.835	61.047	54.736	.000	.000	.000
1,500	9.706	61.648	55.177	.000	.000	.000
1,600	10.583	62.214	55.600	.000	.000	.000
1,700	11.465	62.749	56.005	.000	.000	.000
1,800	12.354	63.257	56.394	.000	.000	.000
1,900	13.249	63.741	56.768	.000	.000	.000
2,000	14.149	64.202	57.127	.000	.000	.000

Troilite; gram formula weight, 87.911
Formula, FeS: α crystals 298.15° to 411°K. β crystals 411° to Curie point 598°K.
  γ crystals 598° to melting point, 1,468°K.

Temp. (°K)	$H_T - H_{298}$ (Kcal)	S_T (cal/deg·gfw)	$-(G_T - H_{298})/T$	Formation from the elements		
				Enthalpy	Free energy (Kcal/gfw)	Log K
298.15	0.000	14.42	14.42	−24.130	−24.219	17.753
uncertainty		0.04	0.04	0.350	0.360	0.264
400	1.470	18.63	14.95	−24.404	−24.258	13.254
411	1.640	19.05	15.06	−24.391	−24.253	12.896
411	2.210	20.44	15.06	−23.821	−24.253	12.896
500	3.760	23.85	16.33	−23.728	−24.354	10.645
598	5.460	26.95	17.82	−23.580	−24.492	8.951
598	5.580	27.15	17.82	−23.460	−24.492	8.951
600	5.610	27.21	17.86	−23.464	−24.495	8.922
700	7.020	29.38	19.35	−23.644	−24.655	7.698
800	8.430	31.26	20.72	−36.929	−25.999	7.102
900	9.840	32.92	21.99	−36.937	−24.626	5.980
1,000	11.250	34.41	23.16	−37.128	−23.254	5.082
1,100	12.680	35.77	24.24	−37.521	−21.840	4.339
1,200	14.150	37.05	25.26	−37.665	−20.419	3.719
1,300	15.680	38.28	26.22	−37.400	−19.006	3.195
1,400	17.260	39.45	27.12	−37.107	−17.605	2.748
1,468	18.350	40.21	27.71	−36.897	−16.701	2.486
1,468	26.080	45.47	27.71	−29.167	−16.701	2.486
1,500	26.620	45.84	28.09	−29.045	−16.388	2.388
1,600	28.320	46.93	29.23	−28.663	−15.546	2.124
1,700	30.020	47.96	30.30	−28.582	−14.727	1.893
1,800	31.720	48.93	31.31	−28.322	−13.909	1.689
1,900	33.420	49.85	32.26	−31.752	−12.942	1.489
2,000	35.120	50.73	33.17	−31.553	−11.969	1.308

Pyrite; gram formula weight, 119.975
Formula, FeS_2: Crystals 298.15° to 1,000°K.

Temp. (°K)	$H_T - H_{298}$ (Kcal)	S_T (cal/deg·gfw)	$-(G_T - H_{298}(/T$ (cal/deg·gfw)	Formation from the elements		
				Enthalpy (Kcal/gfw)	Free energy (Kcal/gfw)	Log K
298.15	0.000	12.65	12.65	−41.000	−38.296	28.071
uncertainty		0.03	0.03	0.400	0.410	0.301
400	1.670	17.46	13.28	−42.183	−37.311	20.386
500	3.350	21.20	14.50	−43.056	−35.988	15.730
600	5.060	24.32	15.89	−43.788	−34.504	12.568
700	6.820	27.03	17.29	−44.418	−32.909	10.274
800	8.650	29.47	18.66	−71.108	−33.655	9.194
900	10.550	31.71	19.99	−71.064	−28.973	7.036
1,000	12.520	33.79	21.27	−71.136	−24.298	5.310

Corundum; gram formula weight, 101.961
Formula, Al_2O_3: Crystals 298.15° to melting point 2,345°K.

Temp. (°K)	$H_T - H_{298}$ (Kcal)	S_T (cal/deg·gfw)	$-(G_T - H_{298})/T$ (cal/deg·gfw)	Formation from the elements		
				Enthalpy (Kcal/gfw)	Free energy (Kcal/gfw)	Log K
298.15	0.000	12.18	12.18	−400.400	−378.082	277.141
uncertainty		0.03	0.03	0.300	0.310	0.227
400	2.147	18.34	12.98	−400.559	−370.640	202.508
500	4.577	23.76	14.60	−400.487	−362.896	158.621
600	7.193	28.52	16.53	−400.322	−355.393	129.451
700	9.940	32.75	18.55	−400.112	−347.920	108.625
800	12.778	36.54	20.57	−399.905	−340.479	93.014
900	15.685	39.97	22.54	−399.743	−333.061	80.878
1,000	18.644	43.08	24.44	−404.690	−325.290	71.092
1,100	21.644	45.94	26.27	−404.467	−317.364	63.054
1,200	24.677	48.58	28.02	−404.224	−309.456	56.359
1,300	27.735	51.03	29.70	−403.967	−301.568	50.698
1,400	30.814	53.31	31.30	−403.702	−293.701	45.849
1,500	33.910	55.45	32.84	−403.431	−285.855	41.649
1,600	37.020	57.45	34.32	−403.154	−278.023	37.976
1,700	40.140	59.35	35.73	−402.875	−270.212	34.738
1,800	43.269	61.13	37.10	−402.596	−262.413	31.861
1,900	46.405	62.83	38.41	−402.320	−254.634	29.290
2,000	49.547	64.44	39.67	−402.046	−246.871	26.977

One-atmosphere thermodynamic properties
of minerals and related substances
at elevated temperatures, from Robie
and Waldbaum (1968, pp. 35–235).

Carbon dioxide; gram formula weight, 44.010
Formula, CO_2: Ideal gas 298.15° to 2,000°K.

Temp. (°K)	$H_T - H_{298}$ (Kcal)	S_T (cal/deg·gfw)	$-(G_T - H_{298})/T$	Formation from the elements		
				Enthalpy	Free energy	
				(Kcal/gfw)		Log K
298.15	0.000	51.06	51.06	−94.051	−94.257	69.092
uncertainty		0.01	0.01	0.030	0.040	0.029
400	0.958	53.82	51.42	−94.067	−94.326	51.537
500	1.987	56.11	52.14	−94.088	−94.387	41.257
600	3.087	58.11	52.97	−94.121	−94.445	34.401
700	4.245	59.90	53.83	−94.166	−94.495	29.503
800	5.453	61.51	54.69	−94.215	−94.539	25.827
900	6.702	62.98	55.53	−94.267	−94.577	22.966
1,000	7.984	64.33	56.35	−94.318	−94.609	20.677
1,100	9.296	65.58	57.13	−94.368	−94.636	18.802
1,200	10.632	66.74	57.88	−94.416	−94.657	17.239
1,300	11.988	67.83	58.61	−94.466	−94.675	15.916
1,400	13.362	68.85	59.30	−94.512	−94.688	14.781
1,500	14.750	69.80	59.97	−94.559	−94.700	13.798
1,600	16.152	70.71	60.61	−94.604	−94.710	12.937
1,700	17.565	71.57	61.23	−94.647	−94.713	12.176
1,800	18.987	72.38	61.83	−94.693	−94.716	11.500
1,900	20.418	73.15	62.41	−94.739	−94.716	10.895
2,000	21.857	73.89	62.96	−94.785	−94.715	10.350

One-atmosphere thermodynamic properties
of minerals and related substances
at elevated temperatures, from Robie
and Waldbaum (1968, pp. 35–235).

Lime; gram formula weight, 56.079
Formula, CaO: Crystals 298.15° to 2,000°K.

Temp. (°K)	$H_T - H_{298}$ (Kcal)	S_T (cal/deg·gfw)	$-(G_T - H_{298})/T$	Formation from the elements		Log K
				Enthalpy	Free energy (Kcal/gfw)	
298.15	0.000	9.50	9.50	−151.790	−144.352	105.812
uncertainty		0.50	0.50	0.300	0.340	0.249
400	1.100	12.67	9.92	−151.702	−141.821	77.487
500	2.230	15.19	10.73	−151.622	−139.364	60.916
600	3.400	17.32	11.65	−151.555	−136.918	49.872
700	4.600	19.17	12.60	−151.509	−134.482	41.987
800	5.820	20.80	13.52	−151.523	−132.054	36.075
900	7.040	22.23	14.41	−151.540	−129.611	31.474
1,000	8.270	23.53	15.26	−151.623	−127.171	27.793
1,100	9.520	24.72	16.07	−151.753	−124.724	24.780
1,200	10.800	25.84	16.84	−153.657	−122.132	22.243
1,300	12.110	26.88	17.56	−153.515	−119.490	20.088
1,400	13.430	27.86	18.27	−153.367	−116.893	18.248
1,500	14.760	28.78	18.94	−153.213	−114.282	16.651
1,600	16.100	29.64	19.58	−153.051	−111.696	15.257
1,700	17.440	30.46	20.20	−152.892	−109.127	14.029
1,800	18.780	31.22	20.79	−189.016	−105.625	12.825
1,900	20.130	31.95	21.36	−188.613	−101.004	11.618
2,000	21.480	32.65	21.91	−188.213	−96.432	10.538

Wüstite; gram formula weight, 68.887
Formula, $Fe_{.947}O$: Crystals 298.15° to melting point 1,650°K. Liquid 1,650°
to 2,000°K.

Temp. (°K)	$H_T - H_{298}$ (Kcal)	S_T (cal/deg·gfw)	$-(G_T - H_{298})/T$	Formation from the elements		
				Enthalpy	Free Energy	
				(Kcal/gfw)		Log K
298.15	0.000	13.76	13.76	−63.640	−58.599	42.954
uncertainty		0.10	0.10	0.200	0.210	0.154
400	1.210	17.24	14.22	−63.393	−56.911	31.095
500	2.440	19.99	15.11	−63.168	−55.318	24.179
600	3.700	22.29	16.13	−62.977	−53.768	19.585
700	4.980	24.26	17.15	−62.834	−52.248	16.312
800	6.280	25.99	18.14	−62.757	−50.738	13.861
900	7.590	27.54	19.11	−62.782	−49.235	11.956
1,000	8.920	28.94	20.02	−62.964	−47.721	10.429
1,100	10.280	30.23	20.89	−63.330	−46.167	9.173
1,200	11.670	31.44	21.72	−63.472	−44.611	8.125
1,300	13.080	32.57	22.51	−63.267	−43.057	7.238
1,400	14.520	33.64	23.27	−63.055	−41.517	6.481
1,500	15.980	34.64	23.99	−62.835	−39.971	5.824
1,600	17.460	35.60	24.69	−62.618	−38.463	5.254
1,650	18.210	36.06	25.02	−62.511	−37.700	4.994
1,650	25.700	40.60	25.02	−55.021	−37.700	4.994
1,700	26.510	41.08	25.49	−55.117	−37.171	4.779
1,800	28.140	42.02	26.39	−54.869	−36.131	4.387
1,900	29.770	42.90	27.24	−58.118	−34.927	4.018
2,000	31.400	43.73	28.03	−57.933	−33.697	3.682

One-atmosphere thermodynamic properties
of minerals and related substances
at elevated temperatures, from Robie
and Waldbaum (1968, pp. 35–235).

Hematite; gram formula weight, 159.692
Formula, Fe_2O_3: α crystals 298.15° to 950°K. β crystals 950° to 1,050°K. γ crystals
1,050° to melting point, 1,895°K.

				Formation from the elements		
Temp. (°K)	$H_T - H_{298}$ (Kcal)	S_T (cal/deg·gfw)	$-(G_T - H_{298})/T$	Enthalpy	Free energy (Kcal/gfw)	Log K
298.15	0.000	20.89	20.89	−197.300	−177.728	130.278
uncertainty		0.05	0.05	0.300	0.310	0.227
400	2.750	28.80	21.92	−196.906	−171.096	93.482
500	5.770	35.53	23.99	−196.332	−164.701	71.991
600	9.010	41.43	26.41	−195.685	−158.433	57.709
700	12.460	46.74	28.94	−194.982	−152.286	47.546
800	16.130	51.64	31.48	−194.249	−146.241	39.951
900	20.030	56.23	33.97	−193.530	−140.269	34.062
950	22.060	58.43	35.21	−193.176	−137.300	31.586
950	22.220	58.60	35.21	−193.016	−137.300	31.586
1,000	24.020	60.44	36.42	−193.100	−134.383	29.369
1,050	25.820	62.20	37.61	−193.870	−131.430	27.356
1,100	27.500	63.76	38.76	−193.639	−128.480	25.527
1,200	30.870	66.69	40.96	−193.881	−122.562	22.322
1,300	34.250	69.40	43.05	−193.426	−116.660	19.612
1,400	37.650	71.92	45.03	−193.002	−110.778	17.293
1,500	41.070	74.28	46.90	−192.589	−104.910	15.285
1,600	44.540	76.52	48.68	−192.174	−99.083	13.534
1,700	48.100	78.68	50.38	−192.277	−93.262	11.990
1,800	51.880	80.84	52.02	−191.811	−87.435	10.616

Magnetite; gram formula weight, 231.539
Formula, Fe_3O_4: α crystals 298.15° to 900°K. β crystals 900° to melting
  point 1,870°K.

Temp. (°K)	$H_T - H_{298}$ (Kcal)	S_T (cal/deg·gfw)	$-(G_T - H_{298})/T$	Formation from the elements		
				Enthalpy (Kcal/gfw)	Free energy	Log K
298.15	0.000	36.03	36.03	−267.400	−243.094	178.192
uncertainty		0.10	0.10	0.500	0.510	0.374
400	3.990	47.51	37.53	−266.763	−234.881	128.332
500	8.320	57.15	40.51	−265.920	−226.991	99.217
600	13.060	65.78	44.01	−264.880	−219.298	79.879
700	18.340	73.91	47.71	−263.526	−211.818	66.132
800	24.260	81.80	51.47	−261.812	−204.543	55.878
900	30.550	89.21	55.27	−260.090	−197.473	47.953
900	30.550	89.21	55.27	−260.090	−197.473	47.953
1,000	35.350	94.27	58.92	−260.424	−190.506	41.635
1,100	40.150	98.84	62.34	−261.442	−183.445	36.447
1,200	44.950	103.02	65.56	−261.848	−176.370	32.121
1,300	49.750	106.86	68.59	−261.222	−169.294	28.461
1,400	54.550	110.42	71.46	−260.670	−162.258	25.330
1,500	59.350	113.73	74.16	−260.162	−155.228	22.617
1,600	64.150	116.83	76.74	−259.726	−148.257	20.251
1,700	68.950	119.74	79.18	−260.200	−141.271	18.162
1,800	73.750	122.48	81.51	−260.148	−134.249	16.300

One-atmosphere thermodynamic properties
of minerals and related substances
at elevated temperatures, from Robie
and Waldbaum (1968, pp. 35–235).

Water; gram formula weight, 18.015
Formula, H_2O: Liquid 298.15° to 373.15°K. Ideal gas 373.15° to 2,000°K.

| Temp. (°K) | $H_T - H_{298}$ (Kcal) | S_T (cal/deg·gfw) | $-(G_T - H_{298})/T$ | Formation from the elements | | |
| | | | | Enthalpy | Free energy | |
				(Kcal/gfw)		Log K
298.15	0.000	16.71	16.71	−68.315	−56.688	41.554
uncertainty		0.03	0.03	0.010	0.020	0.015
373.15	1.352	20.76	17.14	67.747	50.020	31.622
373.15	11.069	46.80	17.14	−58.030	−53.820	31.522
500	12.173	49.33	24.98	−58.275	−52.359	22.886
600	13.028	50.89	29.18	−58.498	−51.156	18.633
700	13.909	52.25	32.38	−58.708	−49.916	15.585
800	14.819	53.46	34.94	−58.903	−48.644	13.289
900	15.759	54.57	37.06	−59.082	−47.353	11.499
1,000	16.728	55.59	38.86	−59.244	−46.040	10.062
1,100	17.729	56.54	40.42	−59.389	−44.709	8.883
1,200	18.759	57.44	41.81	−59.517	−43.373	7.899
1,300	19.817	58.29	43.05	−59.631	−42.028	7.065
1,400	20.903	59.09	44.16	−59.731	−40.663	6.348
1,500	22.014	59.86	45.18	−59.822	−39.302	5.726
1,600	23.149	60.59	46.12	−59.903	−37.929	5.181
1,700	24.306	61.29	46.99	−59.974	−36.549	4.699
1,800	25.483	61.96	47.80	−60.039	−35.166	4.270
1,900	26.679	62.61	48.57	−60.096	−33.788	3.887
2,000	27.892	63.23	49.28	−60.148	−32.399	3.540

Steam; gram formula weight, 18.015
Formula, H_2O: Ideal gas 298.15° to 2,000°K.

Temp. (°K)	$H_T - H_{298}$ (Kcal)	S_T (cal/deg·gfw)	$-(G_T - H_{298})/T$	Formation from the elements		Log K
				Enthalpy	Free energy	
				(Kcal/gfw)		
298.15	0.000	45.10	45.10	−57.796	−54.635	40.048
uncertainty		0.01	0.01	0.010	0.020	0.015
400	0.825	47.48	45.42	−58.040	−53.517	29.240
500	1.654	49.33	46.02	−58.275	−52.360	22.886
600	2.509	50.89	46.71	−58.498	−51.155	18.633
700	3.390	52.25	47.40	−58.708	−49.914	15.584
800	4.300	53.46	48.09	−58.903	−48.646	13.289
900	5.240	54.57	48.75	−59.082	−47.351	11.498
1,000	6.209	55.59	49.38	−59.244	−46.040	10.062
1,100	7.210	56.54	49.99	−59.389	−44.712	8.883
1,200	8.240	57.44	50.57	−59.517	−43.372	7.899
1,300	9.298	58.29	51.13	−59.631	−42.022	7.065
1,400	10.384	59.09	51.67	−59.731	−40.663	6.348
1,500	11.495	59.86	52.19	−59.822	−39.298	5.726
1,600	12.630	60.59	52.70	−59.903	−37.927	5.181
1,700	13.787	61.29	53.18	−59.974	−36.551	4.699
1,800	14.964	61.96	53.65	−60.039	−35.171	4.270
1,900	16.160	62.61	54.10	−60.096	−33.788	3.887
2,000	17.373	63.23	54.55	−60.148	−32.403	3.541

One-atmosphere thermodynamic properties
of minerals and related substances
at elevated temperatures, from Robie
and Waldbaum (1968, pp. 35–235).

Periclase; gram formula weight, 40.311
Formula, MgO: Crystals 298.15° to 2,000°K.

Temp. (°K)	$H_T - H_{298}$ (Kcal)	S_T (cal/def·gfw)	$-(G_T - H_{298})/T$	Formation from the elements		Log K
				Enthalpy	Free energy	
				(Kcal/gfw)		
298.15	0.000	6.44	6.44	−143.800	−130.007	99.755
uncertainty		0.04	0.04	0.100	0.110	0.081
400	0.980	9.26	6.81	−143.803	−133.449	72.913
500	2.036	11.61	7.54	−143.751	−130.869	57.203
600	3.148	13.64	8.39	−143.683	−128.294	46.731
700	4.297	15.41	9.27	−143.617	−125.734	39.256
800	5.474	16.98	10.14	−143.561	−123.187	33.653
900	6.673	18.39	10.98	−143.520	−120.638	29.295
1,000	7.891	19.67	11.78	−145.632	−117.924	25.772
1,100	9.123	20.85	12.56	−145.600	−115.149	22.878
1,200	10.368	21.93	13.29	−145.559	−112.393	20.469
1,300	11.620	22.91	13.97	−145.515	−109.592	18.424
1,400	12.870	23.84	14.65	−175.822	−106.008	16.549
1,500	14.140	24.71	15.28	−175.485	−101.024	14.719
1,600	15.420	25.53	15.89	−175.140	−96.062	13.121
1,700	16.700	26.31	16.49	−174.798	−91.136	11.716
1,800	17.980	27.04	17.05	−174.460	−86.221	10.469
1,900	19.270	27.74	17.60	−174.114	−81.335	9.356
2,000	20.560	28.40	18.12	−173.771	−76.458	8.355

Brucite; gram formula weight, 58.327
Formula, Mg(OH)$_2$: Crystals 298.15° to 600°K. At approximately 540°K the
partial pressure of H$_2$O in equilibrium with brucite reaches one atmosphere.

Temp. (°K)	$H_T - H_{298}$ (Kcal)	S_T (cal/deg·gfw)	$-(G_T - H_{298})/T$	Formation from the elements		
				Enthalpy	Free energy (Kcal/gfw)	Log K
298.15	0.000	15.09	15.09	−221.200	−199.458	146.206
uncertainty		0.03	0.03	0.780	0.790	0.579
400	1.890	20.53	15.80	−221.362	−192.002	104.905
500	3.890	24.99	17.21	−221.431	−184.656	80.713
600	6.080	28.98	18.85	−221.362	−177.303	64.582

Quartz; gram formula weight, 60.085
Formula, SiO$_2$: α quartz 298.15 to 848°K. β quartz 848° to 2,000°K. β quartz is
metastable above 1,140°K.

Temp. (°K)	$H_T - H_{298}$ (Kcal)	S_T (cal/deg·gfw)	$-(G_T - H_{298})/T$	Formation from the elements		
				Enthalpy	Free Energy (Kcal/gfw)	Log K
298.15	0.000	9.88	9.88	−217.650	−204.646	150.009
uncertainty		0.02	0.02	0.400	0.410	0.301
400	1.200	13.33	10.33	−217.690	−200.197	109.382
500	2.560	16.36	11.24	−217.607	−195.830	85.597
600	4.040	19.05	12.32	−217.454	−191.486	69.749
700	5.630	21.50	13.46	−217.233	−187.176	58.439
800	7.320	23.76	14.61	−216.947	−182.905	49.967
848	8.170	24.79	15.16	−216.787	−180.870	46.615
848	8.460	25.13	15.16	−216.497	−180.870	46.615
900	9.300	26.09	15.76	−216.401	−178.680	43.389
1,000	10.920	27.80	16.88	−216.238	−174.494	38.136
1,100	12.570	29.37	17.94	−216.066	−170.325	33.840
1,140	13.247	29.97	18.35	−215.997	−168.662	32.334
1,200	14.250	30.83	18.95	−215.880	−166.175	30.264
1,300	15.940	32.18	19.92	−215.699	−162.039	27.241
1,400	17.640	33.44	20.84	−215.521	−157.915	24.652
1,500	19.360	34.63	21.72	−215.336	−153.824	22.412
1,600	21.100	35.76	22.57	−215.143	−149.745	20.454
1,700	22.860	36.82	23.37	−227.011	−145.549	18.712
1,800	24.630	37.84	24.16	−226.740	−140.777	17.093
1,900	26.420	38.81	24.90	−226.455	−136.013	15.645
2,000	28.220	39.73	25.62	−226.165	−131.261	14.343

One-atmosphere thermodynamic properties
of minerals and related substances
at elevated temperatures, from Robie
and Waldbaum (1968, pp. 35–235).

Rutile; gram formula weight, 79.899
Formula, TiO_2: Crystals 298.15° to melting point, 2,103°K.

Temp. (°Kfi	$H_T - H_{298}$ (Kcal)	S_T (cal/deg·gfw)	$-(G_T - H_{298})/T$	Formation from the elements		
				Enthalpy (Kcal/gfw)	Free energy	Log K
298.15	0.000	12.04	12.04	−225.760	−212.559	155.810
uncertainty		0.04	0.04	0.100	0.110	0.081
400	1.540	16.47	12.62	−225.571	−208.074	113.686
500	3.100	19.95	13.75	−225.385	−203.723	89.047
600	4.735	22.93	15.04	−225.170	−199.412	72.636
700	6.440	25.55	16.35	−224.932	−195.127	60.921
800	8.160	27.85	17.65	−224.723	−190.889	52.148
900	9.900	29.90	18.90	−224.535	−186.670	45.330
1,000	11.650	31.74	20.09	−224.373	−182.469	39.878
1,100	13.420	33.43	21.23	−224.227	−178.288	35.422
1,200	15.200	34.98	22.31	−225.071	−174.082	31.705
1,300	17.000	36.42	23.34	−224.851	−169.839	28.552
1,400	18.820	37.77	24.33	−224.642	−165.622	25.855
1,500	20.660	39.04	25.27	−224.445	−161.418	23.519
1,600	22.530	40.24	26.16	−224.249	−157.203	21.473
1,700	24.420	41.39	27.03	−224.063	−153.022	19.672
1,800	26.340	42.48	27.85	−223.879	−148.842	18.072
1,900	28.280	43.53	28.65	−223.705	−144.682	16.642
2,000	30.250	44.54	29.41	−227.222	−140.418	15.344

Spinel; gram formula weight, 142.273
Formula, MgAl$_2$O$_4$: Crystals 298.15° to 2,000°K.

Temp. (°K)	$H_T - H_{298}$ (Kcal)	S_T	$-(G_T - H_{298})/T$ (cal/deg·gfw)	Formation from the elements		
				Enthalpy	Free energy	
				(Kcal/gfw)		Log K
298.15	0.000	19.26	19.26	−552.800	−522.961	383.339
uncertainty		0.10	0.10	0.500	0.510	0.374
400	3.140	28.28	20.43	−552.949	−512.948	280.261
500	6.619	36.03	22.79	−552.833	−502.690	219.725
600	10.324	42.78	25.57	−552.622	−492.677	179.457
700	14.188	48.73	28.47	−552.378	−482.703	150.706
800	18.173	54.05	31.34	−552.145	−472.771	129.155
900	22.262	58.87	34.13	−551.959	−462.854	112.396
1,000	26.441	63.27	36.83	−559.017	−452.423	98.877
1,100	30.704	67.33	39.42	−558.730	−441.772	87.772
1,200	35.047	71.11	41.91	−558.381	−431.165	78.526
1,300	39.464	74.65	44.29	−557.974	−420.572	70.704
1,400	43.954	77.97	46.58	−587.855	−409.191	63.877
1,500	48.515	81.12	48.78	−587.051	−396.460	57.764
1,600	53.147	84.11	50.89	−586.188	−383.782	52.422
1,700	57.847	86.96	52.93	−585.267	−371.157	47.715
1,800	62.615	89.68	54.90	−584.290	−358.588	43.538
1,900	67.450	92.30	56.80	−583.260	−346.079	39.808
2,000	72.352	94.81	58.64	−582.173	−333.629	36.457

Aragonite; gram formula weight, 100.089
Formula, CaCO$_3$: Crystals 298.15° to 1,000°K.

Temp. (°K)	$H_T - H_{298}$ (Kcal)	S_T	$-(G_T - H_{298})/T$ (cal/deg·gfw)	Formation from the elements		
				Enthalpy	Free energy	
				(Kcal/gfw)		Log K
298.15	0.000	21.18	21.18	−288.651	−269.678	197.679
uncertainty		0.30	0.30	0.340	0.350	0.257
400	2.130	27.31	21.98	−288.507	−263.214	143.813
500	4.440	32.45	23.57	−288.297	−256.913	112.296
600	6.900	36.93	25.43	−288.073	−250.658	91.302
700	9.500	40.94	27.37	−287.830	−244.442	76.318
800	12.220	44.57	29.29	−287.601	−238.264	65.090
900	15.060	47.91	31.18	−287.299	−232.109	56.364
1,000	18.000	51.01	33.01	−287.005	−225.992	49.390

One-atmosphere thermodynamic properties
of minerals and related substances
at elevated temperatures, from Robie
and Waldbaum (1968, pp. 35–235).

Calcite; gram formula weight, 100.089
Formula, $CaCO_3$: Crystals 298.15° to 1,400°K.

Temp. (°K)	$H_T - H_{298}$ (Kcal)	S_T (cal/deg·gfw)	$-(G_T - H_{298})/T$ (cal/deg·gfw)	Formation from the elements		Log K
				Enthalpy (Kcal/gfw)	Free energy (Kcal/gfw)	
298.15	0.000	22.15	22.15	−288.592	−269.908	197.847
uncertainty		0.20	0.20	0.320	0.330	0.242
400	2.220	28.53	22.98	−288.358	−263.553	143.998
500	4.610	33.86	24.64	−288.068	−257.389	112.505
600	7.200	38.58	26.58	−287.714	−251.289	91.532
700	9.890	42.72	28.59	−287.381	−245.239	76.567
800	12.660	46.42	30.59	−287.102	−239.245	65.358
900	15.500	49.76	32.54	−286.800	−233.275	56.647
1,000	18.430	52.85	34.42	−286.516	−227.343	49.686
1,100	21.450	55.73	36.23	−286.238	−221.449	43.998
1,200	24.550	58.42	37.96	−287.706	−215.426	39.234
1,300	27.619	60.87	39.63	−287.211	−209.411	35.205
1,400	30.790	63.22	41.23	−286.632	−203.456	31.761

Dolomite; gram formula weight, 184.411
Formula, $CaMg(CO_3)_2$: Crystals 298.15° to 1,000°K.

Temp. (°K)	$H_T - H_{298}$ (Kcal)	S_T (cal/deg·gfw)	$-(G_T - H_{298})/T$	Formation from the elements		
				Enthalpy (Kcal/gfw)	Free energy	Log K
298.15	0.000	37.09	37.09	−557.613	−518.734	380.241
uncertainty		0.07	0.07	0.520	0.530	0.388
400	4.150	49.01	38.63	−557.406	−505.468	276.174
500	8.743	59.24	41.75	−556.968	−492.535	215.286
600	13.752	68.37	45.45	−556.371	−479.704	174.732
700	19.121	76.63	49.31	−555.645	−466.972	145.795
800	24.797	84.21	53.21	−554.838	−454.372	124.128
900	30.709	91.17	57.05	−553.923	−441.860	107.298
1,000	36.760	97.53	60.77	−555.182	−429.266	93.816

Kyanite; gram formula weight, 162.046
Formula, Al_2SiO_5: Crystals 298.15° to 1,800°K.

Temp. (°K)	$H_T - H_{298}$ (Kcal)	S_T (cal/deg·gfw)	$-(G_T - H_{298})/T$	Formation from the elements		
				Enthalpy (Kcal/gfw)	Free energy	Log K
298.15	0.000	20.02	20.02	−619.930	−584.000	428.082
uncertainty		0.08	0.08	0.540	0.550	0.403
400	3.320	29.55	21.25	−620.156	−571.894	312.467
500	7.080	37.93	23.77	−620.031	−559.569	244.587
600	11.170	45.38	26.76	−619.719	−547.506	199.428
700	15.460	51.99	29.90	−619.335	−535.500	167.190
800	19.900	57.91	33.03	−618.930	−523.547	143.026
900	24.450	63.27	36.10	−618.559	−511.648	124.245
1,000	29.090	68.16	39.07	−623.282	−499.414	109.147
1,100	33.810	72.66	41.92	−622.817	−487.053	96.768
1,200	38.620	76.84	44.66	−622.291	−474.731	86.460
1,300	43.510	80.76	47.29	−621.711	−462.467	77.748
1,400	48.480	84.44	49.81	−621.077	−450.233	70.284
1,500	53.530	87.92	52.23	−620.387	−438.062	63.825
1,600	58.660	91.23	54.57	−619.637	−425.934	58.180
1,700	63.870	94.39	56.82	−630.896	−413.751	53.191
1,800	69.160	97.41	58.99	−629.955	−400.994	48.687

One-atmosphere thermodynamic properties
of minerals and related substances
at elevated temperatures, from Robie
and Waldbaum (1968, pp. 35–235).

Andalusite; gram formula weight, 162.046
Formula, Al_2SiO_5: Crystals 298.15° to 1,800°K.

Temp. (°K)	$H_T - H_{298}$ (Kcal)	S_T (cal/deg·gfw)	$-(G_T - H_{298})/T$ (cal/deg·gfw)	Formation from the elements		
				Enthalpy (Kcal/gfw)	Free energy (Kcal/gfw)	Log K
298.15	0.000	22.28	22.28	−619.390	−584.134	428.180
uncertainty		0.10	0.10	0.710	0.720	0.528
400	3.340	31.87	23.52	−619.596	−572.262	312.668
500	7.120	40.29	26.05	−619.451	−560.169	244.849
600	11.170	47.67	29.05	−619.179	−548.340	199.732
700	15.420	54.22	32.19	−618.835	−536.561	167.521
800	19.810	60.08	35.32	−618.480	−524.833	143.377
900	24.320	65.39	38.37	−618.149	−513.146	124.609
1,000	28.920	70.23	41.31	−622.912	−501.114	109.518
1,100	33.600	74.69	44.14	−622.487	−488.956	97.146
1,200	38.350	78.82	46.86	−622.021	−476.837	86.844
1,300	43.170	82.68	49.47	−621.511	−464.763	78.134
1,400	48.060	86.31	51.98	−620.957	−452.731	70.674
1,500	53.020	89.73	54.38	−620.357	−440.747	64.217
1,600	58.040	92.97	56.69	−619.717	−428.798	58.571
1,700	63.120	96.05	58.92	−631.106	−416.783	53.581
1,800	68.260	98.89	60.97	−630.315	−404.018	49.054

Sillimanite; gram formula weight, 162.046
Formula, Al_2SiO_5: Crystals 298.15° to 1,800°K.

Temp. (°K)	$H_T - H_{298}$ (Kcal)	S_T (cal/deg·gfw)	$-(G_T - H_{298})/T$	Formation from the elements		
				Enthalpy	Free energy	
				(Kcal/gfw)		Log K
298.15	0.000	22.97	22.97	−618.650	−583.599	427.789
uncertainty		0.10	0.10	0.710	0.720	0.528
400	3.350	32.59	24.21	−618.846	−571.800	312.416
500	7.090	40.92	26.74	−618.741	−559.774	244.676
600	11.100	48.23	29.73	−618.509	−548.006	199.610
700	15.300	54.70	32.84	−618.215	−536.277	167.433
800	19.630	60.48	35.94	−617.920	−524.593	143.312
900	24.080	65.72	38.96	−617.649	−512.943	124.559
1,000	28.640	70.52	41.88	−622.452	−500.944	109.481
1,100	33.310	74.97	44.69	−622.037	−488.814	97.118
1,200	38.080	79.12	47.39	−621.551	−476.727	86.824
1,300	42.950	83.02	49.98	−620.991	−464.685	78.120
1,400	47.910	86.70	52.48	−620.367	−452.687	70.667
1,500	52.960	90.18	54.87	−619.677	−440.742	64.216
1,600	58.090	93.49	57.18	−618.927	−428.840	58.577
1,700	63.300	96.65	59.41	−630.186	−416.883	53.594
1,800	68.590	99.67	61.56	−629.245	−404.352	49.095

One-atmosphere thermodynamic properties
of minerals and related substances
at elevated temperatures, from Robie
and Waldbaum (1968, pp. 35–235).

Sphene; gram formula weight, 196.063
Formula, CaTiSiO$_5$: Crystals 298.15° to melting point 1,670°K. Liquid 1,670°
to 2,000°K.

Temp. (°K)	$H_T - H_{298}$ (Kcal)	S_T (cal/deg·gfw)	$-(G_T - H_{298})/T$	Formation from the elements		
				Enthalpy	Free energy	
				(Kcal/gfw)		Log K
298.15	0.000	30.88	30.88	−622.050	−588.246	431.195
uncertainty		0.20	0.20	0.570	0.580	0.425
400	3.750	41.66	32.28	−621.903	−576.708	315.098
500	7.690	50.44	35.06	−621.664	−565.437	247.152
600	11.860	58.04	38.27	−621.344	−554.225	201.876
700	16.230	64.77	41.58	−620.964	−543.059	169.550
800	20.750	70.81	44.87	−620.593	−531.967	145.326
900	25.380	76.26	48.06	−620.186	−520.907	126.493
1,000	30.070	81.20	51.13	−619.854	−509.885	111.435
1,100	34.800	85.71	54.07	−619.606	−498.906	99.123
1,200	39.580	89.87	56.89	−622.128	−487.773	88.835
1,300	44.430	93.75	59.57	−621.535	−476.589	80.122
1,400	49.350	97.39	62.14	−620.920	−465.468	72.663
1,500	54.340	100.84	64.61	−620.284	−454.399	66.206
1,600	59.400	104.10	66.97	−619.623	−443.360	60.560
1,670	62.980	106.29	68.58	−619.140	−435.134	56.945
1,670	92.570	124.01	68.58	−589.550	−435.134	56.945
1,700	94.570	125.20	69.57	−600.966	−432.798	55.640
1,800	101.250	129.02	72.77	−634.985	−422.059	51.245
1,900	107.930	132.63	75.82	−632.523	−410.295	47.195
2,000	114.610	136.06	78.75	−633.790	−398.581	43.555

Fayalite; gram formula weight, 203.778
Formula, Fe_2SiO_4: Crystals 298.15° to melting point 1,490°K. Liquid 1,490°
  to 2,000°K.

Temp. (°K)	$H_T - H_{298}$ (Kcal)	S_T (cal/deg·gfw)	$-(G_T - H_{298})/T$	Formation from the elements		
				Enthalpy	Free energy	
				(Kcal/gfw)		Log K
298.15	0.000	35.45	35.45	−353.544	−329.668	241.652
uncertainty		0.40	0.40	0.700	0.720	0.528
400	3.440	45.35	36.75	−353.338	−321.540	175.681
500	7.210	53.75	39.33	−352.926	−313.627	137.086
600	11.190	61.00	42.35	−352.488	−305.808	111.390
700	15.320	67.36	45.47	−352.085	−298.067	93.061
800	19.560	73.02	48.57	−351.787	−290.374	79.326
900	23.890	78.12	51.58	−351.665	−282.694	68.647
1,000	28.310	82.78	54.47	−351.849	−275.021	60.106
1,100	32.850	87.10	57.24	−352.386	−267.294	53.106
1,200	37.510	91.16	59.90	−352.408	−259.586	47.277
1,300	42.290	94.98	62.45	−351.634	−251.893	42.347
1,400	47.190	98.62	64.91	−350.800	−244.268	38.132
1,490	51.690	100.98	66.29	−349.985	−236.310	34.661
1,490	73.720	115.76	66.29	−327.955	−236.310	34.661
1,500	74.300	116.90	67.37	−327.796	−236.827	34.506
1,600	80.050	120.61	70.58	−326.210	−230.821	31.529
1,700	85.800	124.10	73.63	−337.310	−224.790	28.899
1,800	91.550	127.39	76.53	−335.928	−218.201	26.493
1,900	97.300	130.49	79.28	−341.938	−211.328	24.308
2,000	103.050	133.44	81.91	−340.698	−204.490	22.346

One-atmosphere thermodynamic properties
of minerals and related substances
at elevated temperatures, from Robie
and Waldbaum (1968, pp. 35–235).

Forsterite; gram formula weight, 140.708
Formula, Mg_2SiO_4: Crystals 298.15° to melting point 2,163°K.

Temp. (°K)	$H_T - H_{298}$ (Kcal)	S_T (cal/deg·gfw)	$-(G_T - H_{298})/T$	Formation from the elements		
				Enthalpy	Free energy	
				(Kcal/gfw)		Log K
298.15	0.000	22.75	22.75	−520.370	−491.938	360.599
uncertainty		0.20	0.20	0.520	0.530	0.388
400	3.100	31.66	23.91	−520.476	−482.202	263.462
500	6.520	39.38	26.34	−520.342	−472.698	206.616
600	10.180	45.95	28.98	−520.096	−463.126	168.693
700	14.010	51.85	31.84	−519.801	−453.652	141.636
800	17.960	57.12	34.67	−519.497	−444.228	121.357
900	22.000	61.88	37.44	−519.207	−434.828	105.590
1,000	26.130	66.23	40.10	−523.195	−425.117	92.909
1,100	30.340	70.24	42.66	−522.862	−415.308	82.514
1,200	34.630	73.97	45.11	−522.474	−405.568	73.864
1,300	39.000	77.47	47.47	−522.030	−395.834	66.546
1,400	43.450	80.77	49.73	−582.216	−384.491	60.022
1,500	47.950	83.87	51.90	−581.116	−370.411	53.969
1,600	52.470	86.79	54.00	−580.014	−356.411	48.683
1,700	57.000	89.55	56.02	−590.988	−342.387	44.017
1,800	61.540	92.14	57.95	−589.830	−327.786	39.799
1,900	66.090	94.60	59.82	−588.674	−313.261	36.033
2,000	70.650	96.94	61.61	−587.518	−298.806	32.652

Wollastonite; gram formula weight, 116.164
Formula, $CaSiO_3$: Crystals 298.15° to 1,400°K. Pseudowollastonite is the stable
  phase above 1,398°K.

Temp. (°K)	$H_T - H_{298}$ (Kcal)	S_T (cal/deg·gfw)	$-(G_T - H_{298})/T$	Formation from the elements		
				Enthalpy	Free energy	
				(Kcal/gfw)		Log K
298.15	0.000	19.60	19.60	−390.640	−370.263	271.410
uncertainty		0.20	0.20	0.870	0.880	0.645
400	2.300	26.21	20.46	−390.592	−363.302	198.499
500	4.780	31.74	22.18	−390.439	−356.499	155.825
600	7.390	36.49	24.17	−390.259	−349.726	127.387
700	10.140	40.72	26.23	−390.032	−342.983	107.084
800	13.000	44.54	28.29	−389.810	−336.282	91.868
900	15.890	47.94	30.28	−389.591	−329.599	80.037
1,000	18.810	51.02	32.21	−389.441	−322.935	70.577
1,100	21.770	53.84	34.05	−389.339	−316.294	62.842
1,200	24.800	56.48	35.81	−390.987	−309.529	56.373
1,300	27.880	58.94	37.49	−390.584	−302.743	50.896
1,400	31.000	61.25	39.11	−390.158	−296.008	46.209

Diopside; gram formula weight, 216.560
Formula, $CaMg(SiO_3)_2$: Crystals 298.15° to melting point, 1,664°K.

Temp. (°K)	$H_T - H_{298}$ (Kcal)	S_T (cal/deg·gfw)	$-(G_T - H_{298})/T$	Formation from the elements		
				Enthalpy	Free energy	
				(Kcal/gfw)		Log K
298.15	0.000	34.20	34.20	−767.390	−725.784	532.012
uncertainty		0.20	0.20	2.180	2.190	1.605
400	4.320	46.61	35.81	−767.545	−711.533	388.763
500	8.940	56.90	39.02	−767.534	−697.528	304.888
600	14.060	66.24	42.81	−767.214	−683.562	248.987
700	19.540	74.66	46.75	−766.709	−669.642	209.071
800	25.420	82.52	50.74	−765.992	−655.841	179.167
900	31.340	89.48	54.66	−765.335	−642.093	155.921
1,000	37.280	95.74	58.46	−766.953	−628.241	137.302
1,100	43.250	101.43	62.11	−766.518	−614.389	122.068
1,200	49.250	106.65	65.61	−767.894	−600.454	109.357
1,300	55.300	111.49	68.95	−767.239	−586.512	98.601
1,400	61.440	116.04	72.15	−796.872	−571.815	89.264
1,500*	67.660	120.34	75.23	−795.830	−555.799	80.980
1,600	73.980	124.41	78.17	−794.718	−539.840	73.738

One-atmosphere thermodynamic properties
of minerals and related substances
at elevated temperatures, from Robie
and Waldbaum (1968, pp. 35–235).

Jadeite; gram formula weight, 202.140
Formula, $NaAl(SiO_3)_2$: Crystals 298.15° to 1,200°K.

Temp. (°K)	$H_T - H_{298}$ (Kcal)	S_T (cal/deg·gfw)	$-(G_T - H_{298})/T$ (cal/deg·gfw)	Formation from the elements		
				Enthalpy (Kcal/gfw)	Free energy (Kcal/gfw)	Log K
298.15	0.000	31.90	31.90	−719.871	−677.206	496.404
uncertainty		0.30	0.30	1.000	1.010	0.740
400	4.250	44.10	33.47	−720.791	−662.633	362.045
500	8.970	54.62	36.68	−720.728	−647.958	283.221
600	14.040	63.86	40.46	−720.447	−633.433	230.727
700	19.360	72.05	44.39	−720.034	−618.959	193.247
800	24.860	79.39	48.31	−719.552	−604.551	165.155
900	30.490	86.02	52.14	−719.056	−590.205	143.321
1,000	36.240	92.08	55.84	−721.070	−575.720	125.823
1,100	42.120	97.68	59.39	−720.440	−561.213	111.502
1,200	48.160	102.94	62.81	−742.935	−546.323	99.499

Tremolite; gram formula weight, 812.409
Formula, $Ca_2Mg_5Si_8O_{22}(OH)_2$: Crystals 298.15° to 1,100°K.

Temp. (°K)	$H_T - H_{298}$ (Kcal)	S_T (cal/deg·gfw)	$-(G_T - H_{298})/T$ (cal/deg·gfw)	Formation from the elements		
				Enthalpy (Kcal/gfw)	Free energy (Kcal/gfw)	Log K
298.15	0.000	131.19	131.19	−2,952.935	−2,779.137	2,037.157
uncertainty		0.30	0.30	4.140	4.150	3.042
400	17.375	181.14	137.70	−2,953.488	−2,719.647	1,485.942
500	36.516	223.75	150.72	−2,952.751	−2,661.239	1,163.223
600	57.054	261.13	166.04	−2,951.329	−2,603.034	948.152
700	78.638	294.44	182.10	−2,949.511	−2,545.161	794.633
800	101.089	324.43	198.07	−2,947.470	−2,487.580	679.573
900	124.309	351.74	213.62	−2,945.105	−2,430.170	590.124
1,000	148.241	376.93	228.69	−2,953.240	−2,372.150	518.432
1,100	172.847	400.35	243.22	−2,950.360	−2,314.120	459.772

Anorthite; gram formula weight, 278.210
Formula, $CaAl_2Si_2O_8$: Crystals 298.15° to melting point 1,825°K.

Temp. (°K)	$H_T - H_{298}$ (Kcal)	S_T (cal/deg·gfw)	$-(G_T - H_{298})/T$	Formation from the elements		
				Enthalpy	Free energy	
				(Kcal/gfw)		Log K
298.15	0.000	48.45	48.45	−1,009.300	−955.626	700.491
uncertainty		0.30	0.30	1.150	1.160	0.850
400	5.570	64.47	50.54	−1,009.528	−937.460	512.203
500	11.750	78.23	54.73	−1,009.311	−919.188	401.776
600	18.450	90.44	59.69	−1,008.818	−901.214	328.266
700	25.410	101.16	64.86	−1,008.287	−883.317	275.783
800	32.570	110.72	70.01	−1,007.800	−865.506	236.444
900	39.910	119.35	75.01	−1,007.310	−847.733	205.857
1,000	47.430	127.28	79.85	−1,011.924	−829.650	181.319
1,100	55.130	134.62	84.50	−1,011.336	−811.459	161.222
1,200	62.970	141.44	88.96	−1,012.458	−793.184	144.458
1,300	70.930	147.80	93.24	−1,011.486	−774.930	130.277
1,400	79.050	153.82	97.36	−1,010.396	−756.783	118.139
1,500	87.450	159.62	101.32	−1,009.066	−738.733	107.633
1,600	96.170	165.24	105.13	−1,007.452	−720.770	98.452
1,700	105.230	170.73	108.83	−1,029.670	−702.699	90.338

High-albite; gram formula weight, 262.224
Formula, $NaAlSi_3O_8$: Crystals 298.15° to 1,400°K.

Temp. (°K)	$H_T - H_{298}$ (Kcal)	S_T (cal/deg·gfw	$-(G_T - H_{298})/T$	Formation from the elements		
				Enthalpy	Free energy	
				(Kcal/gfw)		Log K
298.15	0.000	54.67	54.67	−934.513	−882.687	647.025
uncertainty		0.45	0.45	0.770	0.790	0.579
400	5.530	70.56	56.74	−935.393	−864.995	472.610
500	11.627	84.14	60.89	−935.230	−847.263	370.338
600	18.254	96.21	65.78	−934.719	−829.715	302.223
700	25.162	106.85	70.90	−934.087	−812.264	253.600
800	32.279	116.33	75.99	−933.392	−794.897	217.155
900	39.576	124.91	80.94	−932.663	−777.614	188.830
1,000	47.044	132.78	85.73	−934.416	−760.220	166.145
1,100	54.661	140.02	90.33	−933.527	−742.824	147.585
1,200	62.398	146.74	94.74	−955.819	−725.067	132.052
1,300	70.236	153.01	98.98	−954.620	−705.882	118.669
1,400	78.153	158.86	103.04	−953.389	−686.773	107.210

One-atmosphere thermodynamic properties
of minerals and related substances
at elevated temperatures, from Robie
and Waldbaum (1968, pp. 35–235).

Low-albite; gram formula weight, 262.224
Formula, NaAlSi$_3$O$_8$: Crystals 298.15° to 1,400°K.

| Temp. (°K) | $H_T - H_{298}$ (Kcal) | S_T (cal/deg·gfw) | $-(G_T - H_{298})/T$ | Formation from the elements | | |
				Enthalpy	Free energy (Kcal/gfw)	Log K
298.15	0.000	50.20	50.20	−937.146	−883.988	647.979
uncertainty		0.40	0.40	0.740	0.760	0.557
400	5.410	65.75	52.22	−938.146	−865.822	473.062
500	11.390	79.07	56.29	−938.100	−847.598	370.484
600	17.900	90.93	61.10	−937.706	−829.536	302.158
700	24.690	101.40	66.13	−937.192	−811.555	253.378
800	31.690	110.74	71.13	−936.614	−793.643	216.813
900	38.870	119.19	76.00	−936.002	−775.802	188.390
1,000	46.220	126.94	80.72	−937.873	−757.839	165.625
1,100	53.720	134.08	85.24	−937.101	−739.866	146.997
1,200	61.340	140.71	89.59	−959.510	−721.521	131.407
1,300	69.060	146.89	93.77	−958.429	−701.740	117.973
1,400	76.860	152.67	97.77	−957.315	−682.027	106.469

Nepheline; gram formula weight, 142.055
Formula, $NaAlSiO_4$: Crystals 298.15° to 1,521°K. Carnegieite
  is the stable phase above 1,521°K.

Temp. (°K)	$H_T - H_{298}$ (Kcal)	S_T (cal/deg·gfw)	$-(G_T - H_{298})/T$	Formation from the elements		Log K
				Enthalpy	Free energy	
				(Kcal/gfw)		
298.15	0.000	29.72	29.72	−497.029	−469.664	344.272
uncertainty		0.30	0.30	1.000	1.010	0.740
400	3.095	38.62	30.88	−497.864	−460.339	251.517
500	6.280	46.93	34.37	−498.059	−451.401	197.307
600	10.420	53.49	36.12	−497.381	−441.543	160.832
700	14.150	59.23	39.02	−497.189	−432.247	134.953
800	18.000	64.37	41.87	−496.953	−422.987	115.554
900	21.970	69.05	44.64	−496.683	−413.761	100.475
1,000	26.050	73.34	47.29	−498.910	−404.364	88.373
1,100	30.370	77.46	49.85	−498.362	−394.941	78.467
1,200	35.050	81.53	52.32	−520.723	−385.121	70.140
1,300	39.330	84.96	54.71	−520.064	−373.854	62.850
1,400	43.620	88.14	56.98	−519.416	−362.630	56.609
1,500	47.920	91.11	59.16	−518.778	−351.471	51.209

High-sanidine; gram formula weight, 278.337
Formula, $KAlSi_3O_8$: Crystals 298.15° to melting point 1,473°K.

Temp. (°K)	$H_T - H_{298}$ (Kcal)	S_T (cal/deg·gfw)	$-(G_T - H_{298})/T$	Formation from the elements		Log K
				Enthalpy	Free energy	
				(Kcal/gfw)		
298.15	0.000	56.94	56.94	−944.378	−892.263	654.045
uncertainty		1.00	1.00	0.930	0.980	0.718
400	5.500	72.75	59.00	−945.259	−874.405	477.751
500	11.550	86.23	63.13	−945.145	−856.558	374.400
600	17.950	97.89	67.97	−944.866	−838.868	305.557
700	24.800	108.44	73.01	−944.302	−821.243	256.403
800	32.000	118.05	78.05	−943.541	−803.715	219.564
900	39.400	126.76	82.98	−942.731	−786.280	190.934
1,000	46.900	134.66	87.76	−944.483	−768.733	168.006
1,100	54.500	141.91	92.36	−962.552	−750.182	149.047
1,200	62.200	148.61	96.78	−961.438	−730.931	133.120
1,300	70.000	154.84	100.99	−960.277	−711.760	119.657
1,400	77.900	160.70	105.06	−959.063	−692.686	108.133

One-atmosphere thermodynamic properties
of minerals and related substances
at elevated temperatures, from Robie
and Waldbaum (1968, pp. 35–235).

Muscovite; gram formula weight, 398.313
Formula, $KAl_2 [AlSi_3O_{10}] (OH)_2$: Crystals 298.15° to 1,500°K.

Temp. (°K)	$H_T - H_{298}$ (Kcal)	S_T (cal/deg·gfw)	$-(G_T - H_{298})/T$	Formation from the elements		Log K
				Enthalpy	Free energy (Kcal/gfw)	
298.15	0.000	69.00	69.00	−1,421.180	−1,330.103	974.989
uncertainty		0.70	0.70	1.300	1.320	0.968
400	8.670	93.91	72.23	−1,422.266	−1,299.103	709.795
500	18.340	115.44	78.76	−1,421.955	−1,267.921	554.206
600	28.770	134.45	86.50	−1,421.174	−1,237.193	450.646
700	39.660	151.23	94.57	−1,420.198	−1,206.604	376.718
800	50.910	166.24	102.60	−1,419.123	−1,176.162	321.312
900	62.520	179.91	110.44	−1,417.967	−1,145.857	278.251
1,000	74.490	192.52	118.03	−1,424.287	−1,115.119	243.708
1,100	86.785	204.23	125.34	−1,441.583	−1,083.255	215.222
1,200	99.382	215.19	132.37	−1,439.520	−1,050.773	191.371
1,300	112.272	225.51	139.14	−1,437.243	−1,018.475	171.221
1,400	125.449	235.27	145.66	−1,434.752	− 986.342	153.974
1,500	138.909	244.55	151.95	−1,432.050	− 954.438	139.061

Talc; gram formula weight, 379.289
Formula, $Mg_3Si_4O_{10} (OH)_2$: Crystals 298.15° to 1,100°K.

Temp. (°K)	$H_T - H_{298}$ (Kcal)	S_T (cal/deg·gfw)	$-(G_T - H_{298})/T$	Formation from the elements		Log K
				Enthalpy	Free energy (Kcal/gfw)	
298.15	0.000	62.34	62.34	−1,415.205	1,324.486	970.871
uncertainty		0.15	0.15	1.710	1.720	1.261
400	8.561	86.94	65.54	−1,415.622	−1,293.412	706.686
500	17.997	107.97	71.98	−1,415.372	−1,262.882	552.004
600	28.148	126.46	79.55	−1,414.737	−1,232.432	448.912
700	38.889	143.01	87.45	−1,413.812	−1,202.119	375.317
800	50.152	158.03	95.34	−1,412.633	−1,171.957	320.163
900	61.903	171.87	103.09	−1,411.211	−1,141.942	277.301
1,000	74.121	184.75	110.63	−1,415.944	−1,111.578	242.935
1,100	86.793	195.81	116.91	−1,413.928	−1,080.098	214.595

Gibbs free energy of H_2O in calories per mole as a function of temperature and pressure, from Fisher and Zen (1971).

Temperature in °C	Pressure in bars								
	100	200	300	400	500	600	700	800	900
100	−53,755	−53,711	−53,668	−53,622	−53,578	−53,536	−53,493	−53,449	−53,406
120	−53,018	−52,972	−52,926	−52,882	−52,838	−52,792	−52,749	−52,704	−52,661
140	−52,286	−52,240	−52,194	−52,149	−52,103	−52,058	−52,012	−51,968	−51,924
160	−51,560	−51,514	−51,466	−51,421	−51,375	−51,329	−51,283	−51,236	−51,191
180	−50,841	−50,792	−50,745	−50,699	−50,650	−50,603	−50,557	−50,511	−50,465
200	−50,132	−50,082	−50,033	−49,985	−49,936	−49,889	−49,841	−49,794	−49,747
220	−49,423	−49,372	−49,332	−49,272	−49,224	−49,174	−49,125	−49,078	−49,030
240	−48,726	−48,674	−48,623	−48,570	−48,519	−48,470	−48,421	−48,371	−48,322
260	−48,037	−47,982	−47,928	−47,875	−47,824	−47,772	−47,721	−47,669	−47,620
280	−47,348	−47,292	−47,237	−47,183	−47,128	−47,077	−47,023	−46,971	−46,920
300	−46,678	−46,616	−46,560	−46,501	−46,446	−46,393	−46,335	−46,284	−46,233
320	−46,093	−45,947	−45,883	−45,829	−45,768	−45,713	−45,655	−45,596	−45,540
340	−45,626	−45,281	−45,217	−45,155	−45,090	−45,031	−44,973	−44,920	−44,861
360	−45,176	−44,637	−44,562	−44,498	−44,432	−44,369	−44,305	−44,244	−44,185
380	−44,723	−44,075	−43,920	−43,843	−43,771	−43,706	−43,645	−43,580	−43,518
400	−44,270	−43,563	−43,301	−43,206	−43,127	−43,058	−42,986	−42,919	−42,854
420	−43,819	−43,063	−42,726	−42,586	−42,496	−42,414	−42,339	−42,271	−42,205
440	−43,372	−42,562	−42,178	−41,986	−41,869	−41,778	−41,698	−41,626	−41,554
460	−42,919	−42,068	−41,647	−41,408	−41.267	−41,162	−41,069	−40,989	−40,912
480	−42,473	−41,578	−41,125	−40,854	−40,677	−40,556	−40,455	−40,365	−40,282
500	−42,025	−41,091	−40,603	−40,302	−40,102	−39,956	−39,841	−39,741	−39,650
520	−41′,578	−40,604	−40,087	−39,762	−39,539	−39,372	−39,241	−39,133	−39,035
540	−41,132	−40,118	−39,576	−39,224	−38,979	−38,793	−38,650	−38,528	−38,419
560	−40,684	−39,635	−39,070	−38,695	−38,428	−38,226	−38,064	−37,931	−37,817
580	−40,237	−39,155	−38,562	−38,166	−37,879	−37,664	−37,490	−37,344	−37,218
600	−39,789	−38,671	−38,050	−37,640	−37,341	−37,108	−36,917	−36,757	−36,621
620	−39,337	−38,187	−37,546	−37,118	−36,794	−36,550	−36,346	−36,181	−36,031
640	−38,887	−37,704	−37,042	−36,589	−36,254	−35,994	−35,783	−35,606	−35,449
660	−38,437	−37,220	−36,538	−36,071	−35,723	−35,447	−35,223	−35,029	−34,870
680	−37,994	−36,740	−36,036	−35,550	−35,191	−34,901	−34,668	−34,468	−34,295
700	−37,539	−36,256	−35,531	−35,032	−34,654	−34,356	−34,112	−33,900	−33,722
720	−37,093	−35,778	−35,032	−34,516	−34,129	−33,819	−33,561	−33,339	−33,149
740	−36,642	−35,295	−34,529	−33,999	−33,595	−33,277	−33,004	−32,783	−32,580
760	−36,187	−34,814	−34,029	−33,493	−33,067	−32,735	−32,457	−32,219	−32,017
780	−35,731	−34,324	−33,524	−32,960	−32,531	−32,195	−31,903	−31,657	−31,443
800	−35,284	−33,843	−33,020	−32,447	−32,008	−31,652	−31,356·	−31,103	−30,883

Gibbs free energy of H_2O in calories per mole as a function of temperature and pressure, from Fisher and Zen (1971).

Temperature in °C	Pressure in bars								
	1,000	1,100	1,200	1,300	1,400	1,500	1,600	1,700	1,800
100	−53,362	−53,319	−53,276	−53,233	−53,191	−53,148	−53,106	−53,064	−53,022
120	−52,619	−52,575	−52,532	−52,489	−52,446	−52,403	−52,360	−52,317	−52,275
140	−51,879	−51,835	−51,791	−51,748	−51,705	−51,661	−51,618	−51,574	−51,531
160	−51,147	−51,102	−51,057	−51,013	−50,969	−50,924	−50,881	−50,837	−50,794
180	−50,420	−50,374	−50,329	−50,284	−50,239	−50,194	−50,149	−50,104	−50,060
200	−49,701	−49,655	−49,608	−49,562	−49,516	−49,471	−49,425	−49,380	−49,335
220	−48,981	−48,934	−48,887	−48,840	−48,793	−48,746	−48,700	−48,654	−48,608
240	−48,273	−48,224	−48,176	−48,128	−48,081	−48,033	−47,986	−47,939	−47,892
260	−47,570	−47,521	−47,472	−47,423	−47,374	−47,326	−47,278	−47,230	−47,182
280	−46,868	−46,817	−46,767	−46,717	−46,667	−46,618	−46,569	−46,520	−46,472
300	−46,180	−46,128	−46,076	−46,025	−45,974	−45,924	−45,874	−45,824	−45,774
320	−45,489	−45,435	−45,382	−45,330	−45,278	−45,226	−45,175	−45,124	−45,074
340	−44,804	−44,749	−44,694	−44,641	−44,587	−44,534	−44,482	−44,430	−44,378
360	−44,130	−44,073	−44,016	−43,961	−43,906	−43,852	−43,798	−43,745	−43,692
380	−43,458	−43,399	−43,341	−43,283	−43,227	−43,171	−43,116	−43,061	−43,007
400	−42,792	−42,730	−42,670	−42,611	−42,552	−42,495	−42,438	−42,382	−42,326
420	−42,134	−42,070	−42,008	−41,946	−41,886	−41,826	−41,767	−41,709	−41,652
440	−41,483	−41,416	−41,350	−41,286	−41,224	−41,162	−41,102	−41,042	−40,983
460	−40,837	−40,766	−40,698	−40,631	−40,566	−40,502	−40,439	−40,378	−40,317
480	−40,201	−40,126	−40,054	−39,984	−39,917	−39,850	−39,785	−39,721	−39,658
500	−39,564	−39,484	−39,409	−39,335	−39,264	−39,195	−39,128	−39,062	−38,997
520	−38,941	−38,856	−38,776	−38,699	−38,624	−38,552	−38,482	−38,413	−38,346
540	−38,324	−38,234	−38,149	−38,068	−37,990	−37,914	−37,841	−37,770	−37,700
560	−37,709	−37,613	−37,523	−37,437	−37,355	−37,276	−37,200	−37,126	−37,053
580	−37,105	−37,002	−36,906	−36,816	−36,730	−36,647	−36,568	−36,491	−36,415
600	−36,505	−36,395	−36,293	−36,198	−36,108	−36,021	−35,938	−35,858	−35,780
620	−35,906	−35,789	−35,682	−35,581	−35,486	−35,396	−35,309	−35,225	−35,144
640	−35,312	−35,188	−35,074	−34,968	−34,869	−34,774	−34,683	−34,596	−34,512
660	−34,721	−34,590	−34,470	−34,358	−34,254	−34,155	−34,060	−33,969	−33,882
680	−34,141	−34,003	−33,877	−33,760	−33,650	−33,547	−33,448	−33,354	−33,263
700	−33,557	−33,412	−33,279	−33,157	−33,042	−32,934	−32,832	−32,734	−32,640
720	−32,984	−32,832	−32,693	−32,565	−32,446	−32,333	−32,227	−32,125	−32,027
740	−32,407	−32,249	−32,104	−31,970	−31,846	−31,729	−31,618	−31,513	−31,411
760	−31,828	−31,663	−31,512	−31,373	−31,244	−31,123	−31,008	−30,898	−30,794
780	−31,257	−31,086	−30,929	−30,785	−30,651	−30,525	−30,405	−30,292	−30,184
800	−30,686	−30,508	−30,345	−30,196	−30,057	−29,926	−29,803	−29,686	−29,574
820	−30,117	−29,933	−29,765	−29,610	−29,466	−29,331	−29,204	−29,083	−28,968
840	−29,550	−29,360	−29,186	−29,026	−28,878	−28,738	−28,607	−28,482	−28,363
860	−28,985	−28,789	−28,610	−28,445	−28,291	−28,148	−28,012	−27,884	−27,761
880	−28,420	−28,218	−28,033	−27,863	−27,705	−27,557	−27,417	−27,285	−27,159
900	−27,857	−27,648	−27,459	−27,284	−27,121	−26,969	−26,826	−26,690	−26,561
920	−27,296	−27,082	−26,887	−26,708	−26,541	−26,385	−26,238	−26,099	−25,967
940	−26,735	−26,516	−26,317	−26,133	−25,962	−25,802	−25,652	−25,509	−25,373
960	−26,174	−25,950	−25,746	−25,558	−25,383	−25,220	−25,065	−24,919	−24,780
980	−25,615	−25,387	−25,178	−24,986	−24,807	−24,639	−24,481	−24,332	−24,189
1,000	−25,057	−24,824	−24,611	−24,414	−24,231	−24,059	−23,898	−23,744	−23,599

Temperature in °C	Pressure in bars								
	1,900	2,000	2,100	2,200	2,300	2,400	2,500	2,600	2,700
100	−52,980	−52,938	−52,897	−52,855	−52,814	−52,773	−52,732	−52,691	−52,650
120	−52,233	−52,190	−52,148	−52,106	−52,064	−52,023	−51,981	−51,940	−51,898
140	−51,489	−51,446	−51,403	−51,361	−51,318	−51,276	−51,234	−51,192	−51,150
160	−50,750	−50,707	−50,664	−50,621	−50,578	−50,535	−50,492	−50,450	−50,407
180	−50,016	−49,973	−49,929	−49,885	−49,842	−49,798	−49,755	−49,712	−49,669
200	−49,290	−49,245	−49,200	−49,156	−49,112	−49,069	−49,025	−48,981	−48,938
220	−48,562	−48,517	−48,472	−48,426	−48,382	−48,337	−48,292	−48,248	−48,204
240	−47,846	−47,800	−47,753	−47,708	−47,662	−47,617	−47,571	−47,526	−47,481
260	−47,135	−47,088	−47,041	−46,994	−46,948	−46,901	−46,855	−46,810	−46,764
280	−46,423	−46,375	−46,328	−46,280	−46,233	−46,186	−46,139	−46,092	−46,046
300	−45,725	−45,676	−45,627	−45,579	−45,531	−45,483	−45,435	−45,388	−45,341
320	−45,023	−44,973	−44,924	−44,874	−44,825	−44,777	−44,728	−44,680	−44,632
340	−44,327	−44,276	−44,225	−44,175	−44,125	−44,075	−44,026	−43,976	−43,928
360	−43,639	−43,587	−43,535	−43,484	−43,433	−43,382	−43,332	−43,282	−43,232
380	−42,953	−42,899	−42,847	−42,794	−42,742	−42,690	−42,639	−42,588	−42,537
400	−42,271	−42,216	−42,162	−42,108	−42,055	−42,002	−41,949	−41,897	−41,845
420	−41,595	−41,539	−41,484	−41,429	−41,374	−41,320	−41,266	−41,213	−41,160
440	−40,925	−40,867	−40,809	−40,753	−40,697	−40,642	−40,587	−40,532	−40,479
460	−40,257	−40,198	−40,138	−40,080	−40,023	−39,966	−39,910	−39,855	−39,799
480	−39,597	−39,536	−39,475	−39,415	−39,357	−39,298	−39,241	−39,184	−39,128
500	−38,933	−38,870	−38,808	−38,747	−38,686	−38,627	−38,568	−38,509	−38,452
520	−38,280	−38,215	−38,152	−38,089	−38,027	−37,966	−37,906	−37,846	−37,787
540	−37,632	−37,565	−37,501	−37,436	−37,372	−37,309	−37,247	−37,186	−37,125
560	−36,983	−36,914	−36,849	−36,782	−36,717	−36,652	−36,588	−36,525	−36,463
580	−36,342	−36,271	−36,205	−36,137	−36,069	−36,002	−35,937	−35,872	−35,809
600	−35,704	−35,630	−35,564	−35,493	−35,423	−35,355	−35,287	−35,221	−35,156
620	−35,066	−34,989	−34,921	−34,848	−34,776	−34,706	−34,637	−34,569	−34,502
640	−34,430	−34,351	−34,282	−34,206	−34,132	−34,060	−33,989	−33,919	−33,850
660	−33,797	−33,715	−33,644	−33,566	−33,489	−33,415	−33,342	−33,270	−33,199
680	−33,176	−33,091	−33,016	−32,936	−32,857	−32,780	−32,705	−32,632	−32,559
700	−32,549	−32,461	−32,383	−32,300	−32,219	−32,140	−32,063	−31,987	−31,913
720	−31,933	−31,842	−31,760	−31,675	−31,591	−31,510	−31,431	−31,353	−31,277
740	−31,314	−31,220	−31,134	−31,046	−30,960	−30,877	−30,795	−30,715	−30,637
760	−30,693	−30,596	−30,505	−30,414	−30,326	−30,240	−30,156	−30,074	−29,994
780	−30,080	−29,980	−29,885	−29,791	−29,700	−29,612	−29,526	−29,442	−29,360
800	−29,467	−29,364	−29,264	−29,168	−29,074	−28,984	−28,895	−28,809	−28,725
820	−28,857	−28,751	−28,647	−28,548	−28,452	−28,359	−28,269	−28,180	−28,094
840	−28,250	−28,140	−28,033	−27,931	−27,833	−27,737	−27,644	−27,554	−27,465
860	−27,644	−27,532	−27,422	−27,318	−27,217	−27,119	−27,023	−26,930	−26,840
880	−27,039	−26,924	−26,812	−26,705	−26,601	−26,501	−26,403	−26,308	−26,215
900	−26,438	−26,320	−26,206	−26,096	−25,990	−25,887	−25,787	−25,689	−25,594
920	−25,840	−25,719	−25,602	−25,490	−25,381	−25,275	−25,173	−25,073	−24,976
940	−25,244	−25,119	−25,000	−24,884	−24,773	−24,665	−24,560	−24,459	−24,359
960	−24,647	−24,520	−24,398	−24,280	−24,166	−24,055	−23,948	−23,844	−23,743
980	−24,053	−23,923	−23,798	−23,677	−23,560	−23,448	−23,338	−23,232	−23,129
1,000	−23,459	−23,326	−23,198	−23,075	−22,955	−22,840	−22,729	−22,620	−22,515

Gibbs free energy of H_2O in calories
per mole as a function of temperature and
pressure, from Fisher and Zen (1971).

Temperature in °C	Pressure in bars								
	2,800	2,900	3,000	3,100	3,200	3,300	3,400	3,500	3,600
100	−52,609	−52,568	−52,528	−52,487	−52,447	−52,406	−52,366	−52,326	−52,286
120	−51,857	−51,816	−51,775	−51,734	−51,693	−51,652	−51,612	−51,571	−51,531
140	−51,109	−51,067	−51,025	−50,984	−50,943	−50,902	−50,860	−50,819	−50,779
160	−50,365	−50,323	−50,281	−50,239	−50,197	−50,156	−50,114	−50,073	−50,031
180	−49,626	−49,584	−49,541	−49,499	−49,456	−49,414	−49,372	−49,330	−49,288
200	−48,894	−48,851	−48,808	−48,765	−48,722	−48,679	−48,637	−48,594	−48,552
220	−48,161	−48,117	−48,073	−48,029	−47,986	−47,943	−47,900	−47,857	−47,814
240	−47,437	−47,392	−47,348	−47,303	−47,260	−47,216	−47,173	−47,129	−47,086
260	−46,719	−46,673	−46,628	−46,583	−46,539	−46,494	−46,450	−46,406	−46,361
280	−46,000	−45,954	−45,908	−45,863	−45,817	−45,772	−45,727	−45,682	−45,637
300	−45,294	−45,247	−45,201	−45,154	−45,108	−45,062	−45,017	−44,971	−44,926
320	−44,584	−44,537	−44,489	−44,442	−44,395	−44,349	−44,302	−44,256	−44,210
340	−43,879	−43,831	−43,783	−43,735	−43,687	−43,640	−43,592	−43,545	−43,499
360	−43,182	−43,133	−43,084	−43,035	−42,987	−42,939	−42,891	−42,843	−42,796
380	−42,486	−42,436	−42,386	−42,337	−42,288	−42,239	−42,190	−42,141	−42,093
400	−41,794	−41,743	−41,692	−41,642	−41,592	−41,542	−41,492	−41,443	−41,394
420	−41,108	−41,056	−41,004	−40,952	−40,901	−40,851	−40,800	−40,750	−40,700
440	−40,425	−40,372	−40,319	−40,267	−40,215	−40,163	−40,111	−40,060	−40,009
460	−39,745	−39,690	−39,637	−39,583	−39,530	−39,477	−39,425	−39,373	−39,321
480	−39,072	−39,016	−38,961	−38,907	−38,852	−38,799	−38,745	−38,692	−38,640
500	−38,395	−38,338	−38,282	−38,226	−38,171	−38,116	−38,062	−38,008	−37,954
520	−37,728	−37,670	−37,613	−37,556	−37,500	−37,444	−37,388	−37,333	−37,279
540	−37,066	−37,006	−36,948	−36,890	−36,832	−36,775	−36,718	−36,662	−36,607
560	−36,402	−36,341	−36,281	−36,222	−36,163	−36,105	−36,047	−35,990	−35,933
580	−35,746	−35,684	−35,622	−35,562	−35,502	−35,442	−35,383	−35,325	−35,267
600	−35,091	−35,028	−34,965	−34,903	−34,841	−34,781	−34,721	−34,661	−34,602
620	−34,436	−34,371	−34,306	−34,243	−34,180	−34,118	−34,057	−33,996	−33,936
640	−33,782	−33,716	−33,650	−33,585	−33,521	−33,458	−33,395	−33,333	−33,272
660	−33,130	−33,062	−32,994	−32,928	−32,863	−32,798	−32,734	−32,671	−32,608
680	−32,448	−32,418	−32,349	−32,281	−32,214	−32,148	−32,083	−32,018	−31,954
700	−31,840	−31,768	−31,698	−31,628	−31,560	−31,492	−31,425	−31,360	−31,294
720	−31,202	−31,129	−31,056	−30,985	−30,915	−30,846	−30,778	−30,711	−30,645
740	−30,560	−30,485	−30,411	−30,339	−30,267	−30,196	−30,127	−30,058	−29,990
760	−29,916	−29,839	−29,763	−29,689	−29,616	−29,543	−29,472	−29,402	−29,333
780	−29,279	−29,200	−29,123	−29,047	−28,972	−28,898	−28,826	−28,754	−28,684
800	−28,642	−28,562	−28,482	−28,405	−28,328	−28,253	−28,179	−28,106	−28,034
820	−28,009	−27,927	−27,846	−27,766	−27,688	−27,611	−27,536	−27,461	−27,388
840	−27,379	−27,294	−27,211	−27,130	−27,050	−26,972	−26,894	−26,818	−26,744
860	−26,751	−26,665	−26,580	−26,497	−26,415	−26,335	−26,256	−26,179	−26,103
880	−26,125	−26,036	−25,950	−25,865	−25,782	−25,700	−25,620	−25,540	−25,463
900	−25,502	−25,411	−25,323	−25,236	−25,151	−25,068	−24,986	−24,905	−24,826
920	−24,882	−24,789	−24,699	−24,611	−24,524	−24,439	−24,355	−24,273	−24,193
940	−24,263	−24,168	−24,076	−23,986	−23,898	−23,811	−23,726	−23,643	−23,561
960	−23,644	−23,548	−23,454	−23,362	−23,272	−23,184	−23,098	−23,013	−22,930
980	−23,028	−22,930	−22,835	−22,741	−22,649	−22,560	−22,472	−22,386	−22,301
1,000	−22,413	−22,313	−22,215	−22,120	−22,027	−21,936	−21,847	−21,760	−21,674

Temperature in °C	Pressure in bars								
	3,700	3,800	3,900	4,000	4,100	4,200	4,300	4,400	4,500
100	−52,246	−52,206	−52,166	−52,127	−52,087	−52,047	−52,008	−51,969	−51,929
120	−51,490	−51,450	−51,410	−51,370	−51,330	−51,290	−51,250	−51,210	−51,170
140	−50,738	−50,697	−50,656	−50,616	−50,575	−50,535	−50,445	−50,455	−50,415
160	−49,990	−49,949	−49,908	−49,867	−49,826	−49,785	−49,744	−49,704	−49,663
180	−49,246	−49,205	−49,163	−49,122	−49,081	−49,039	−48,998	−48,957	−48,916
200	−48,510	−48,468	−48,425	−48,384	−48,342	−48,300	−48,258	−48,217	−48,175
220	−47,771	−47,728	−47,686	−47,643	−47,601	−47,559	−47,516	−47,474	−47,433
240	−47,042	−46,999	−46,956	−46,913	−46,870	−46,827	−46,785	−46,742	−46,700
260	−46,319	−46,275	−46,231	−46,187	−46,144	−46,101	−46,058	−46,015	−45,972
280	−45,593	−45,548	−45,504	−45,460	−45,416	−45,373	−45,329	−45,286	−45,242
300	−44,881	−44,835	−44,791	−44,746	−44,701	−44,657	−44,613	−44,568	−44,524
320	−44,164	−44,118	−44,073	−44,028	−43,982	−43,937	−43,892	−43,848	−43,803
340	−43,462	−43,406	−43,360	−43,314	−43,268	−43,222	−43,177	−43,131	−43,086
360	−42,748	−42,701	−42,654	−42,608	−42,561	−42,515	−42,469	−42,423	−42,377
380	−42,045	−41,997	−41,950	−41,902	−41,855	−41,808	−41,761	−41,714	−41,668
400	−41,345	−41,296	−41,248	−41,200	−41,152	−41,104	−41,056	−41,009	−40,962
420	−40,650	−40,601	−40,552	−40,503	−40,454	−40,406	−40,357	−40,309	−40,262
440	−39,959	−39,909	−39,859	−39,809	−39,760	−39,710	−39,661	−39,613	−39,564
460	−39,270	−39,219	−39,168	−39,117	−39,067	−39,017	−38,967	−38,918	−38,868
480	−38,587	−38,535	−38,484	−38,432	−38,381	−38,330	−38,280	−38,230	−38,180
500	−37,901	−37,848	−37,795	−37,743	−37,691	−37,639	−37,588	−37,537	−37,486
520	−37,224	−37,170	−37,117	−37,064	−37,011	−36,958	−36,906	−36,854	−36,803
540	−36,551	−36,496	−36,442	−36,388	−36,334	−36,281	−36,228	−36,175	−36,122
560	−35,877	−35,821	−35,766	−35,711	−35,656	−35,602	−35,548	−35,494	−35,441
580	−35,210	−35,153	−35,096	−35,040	−34,985	−34,929	−34,875	−34,820	−34,766
600	−34,544	−34,486	−34,428	−34,371	−34,315	−34,258	−34,202	−34,147	−34,092
620	−33,876	−33,817	−33,759	−33,701	−33,643	−33,586	−33,529	−33,473	−33,417
640	−33,211	−33,151	−33,091	−33,032	−32,973	−32,915	−32,857	−32,800	−32,743
660	−32,546	−32,485	−32,424	−32,364	−32,304	−32,245	−32,186	−32,128	−32,070
680	−31,891	−31,829	−31,767	−31,706	−31,645	−31,585	−31,525	−31,465	−31,407
700	−31,230	−31,166	−31,103	−31,041	−30,979	−30,918	−30,857	−30,797	−30,737
720	−30,579	−30,514	−30,450	−30,386	−30,323	−30,261	−30,199	−30,138	−30,077
740	−29,923	−29,857	−29,792	−29,727	−29,663	−29,599	−29,536	−29,474	−29,412
760	−29,265	−29,197	−29,131	−29,065	−28,999	−28,935	−28,871	−28,807	−28,745
780	−28,614	−28,545	−28,477	−28,410	−28,344	−28,278	−28,213	−28,148	−28,084
800	−27,963	−27,893	−27,824	−27,755	−27,688	−27,621	−27,554	−27,489	−27,424
820	−27,315	−27,244	−27,173	−27,104	−27,035	−26,967	−26,899	−26,832	−26,766
840	−26,670	−26,597	−26,525	−26,454	−26,384	−26,315	−26,246	−26,178	−26,111
860	−26,027	−25,953	−25,880	−25,808	−25,736	−25,666	−25,596	−25,527	−25,459
880	−25,386	−25,311	−25,236	−25,163	−25,090	−25,018	−24,947	−24,877	−24,808
900	−24,748	−24,671	−24,595	−24,521	−24,447	−24,374	−24,302	−24,231	−24,160
920	−24,113	−24,035	−23,958	−23,882	−23,807	−23,733	−23,660	−23,588	−23,516
940	−23,480	−23,400	−23,322	−23,245	−23,169	−23,094	−23,020	−22,947	−22,874
960	−22,848	−22,767	−22,688	−22,609	−22,532	−22,456	−22,381	−22,307	−22,234
980	−22,218	−22,136	−22,056	−21,977	−21,899	−21,822	−21,746	−21,671	−21,597
1,000	−21,590	−21,507	−21,426	−21,346	−21,267	−21,189	−21,112	−21,037	−20,962

Gibbs free energy of H_2O in calories per mole as a function of temperature and pressure, from Fisher and Zen (1971).

Temperature in °C	Pressure in bars								
	4,600	4,700	4,800	4,900	5,000	5,100	5,200	5,300	5,400
100	−51,890	−51,851	−51,812	−51,773	−51,734	−51,695	−51,656	−51,617	−51,579
120	−51,131	−51,091	−51,052	−51,012	−50,973	−50,934	−50,895	−50,856	−50,817
140	−50,374	−50,335	−50,295	−50,255	−50,215	−50,176	−50,136	−50,097	−50,057
160	−49,623	−49,583	−49,542	−49,502	−49,462	−49,422	−49,382	−49,342	−49,302
180	−48,875	−48,834	−48,794	−48,753	−48,713	−48,672	−48,632	−48,591	−48,551
200	−48,134	−48,093	−48,052	−48,011	−47,970	−47,929	−47,888	−47,847	−47,807
220	−47,391	−47,349	−47,307	−47,266	−47,224	−47,183	−47,142	−47,101	−47,060
240	−46,657	−46,615	−46,573	−46,531	−46,489	−46,448	−46,406	−46,364	−46,323
260	−45,929	−45,886	−45,844	−45,801	−45,759	−45,716	−45,674	−45,632	−45,590
280	−45,199	−45,156	−45,113	−45,070	−45,027	−44,984	−44,941	−44,899	−44,856
300	−44,481	−44,438	−44,394	−44,351	−44,307	−44,264	−44,221	−44,178	−44,135
320	−43,759	−43,714	−43,670	−43,626	−43,582	−43,538	−43,495	−43,452	−43,409
340	−43,041	−42,996	−42,951	−42,907	−42,862	−42,818	−42,774	−42,729	−42,685
360	−42,331	−42,286	−42,240	−42,195	−42,150	−42,105	−42,060	−42,016	−41,971
380	−41,622	−41,576	−41,530	−41,484	−41,438	−41,393	−41,347	−41,302	−41,257
400	−40,915	−40,868	−40,822	−40,775	−40,729	−40,683	−40,637	−40,591	−40,545
420	−40,214	−40,167	−40,199	−40,072	−40,025	−39,978	−39,932	−39,885	−39,839
440	−39,516	−39,468	−39,420	−39,372	−39,324	−39,277	−39,230	−39,183	−39,136
460	−38,819	−38,771	−38,722	−38,673	−38,625	−38,577	−38,529	−38,482	−38,434
480	−38,130	−38,080	−38,031	−37,982	−37,933	−37,884	−37,835	−37,787	−37,739
500	−37,435	−37,385	−37,335	−37,285	−37,235	−37,186	−37,137	−37,088	−37,039
520	−36,751	−36,700	−36,649	−36,599	−36,548	−36,498	−36,448	−36,398	−36,349
540	−36,070	−36,018	−35,967	−35,915	−35,864	−35,813	−35,763	−35,712	−35,662
560	−35,388	−35,335	−35,282	−35,230	−35,178	−35,127	−35,075	−35,024	−34,973
580	−34,712	−34,658	−34,605	−34,552	−34,499	−34,447	−34,395	−34,343	−34,291
600	−34,037	−33,983	−33,929	−33,875	−33,821	−33,768	−33,715	−33,663	−33,610
620	−33,361	−33,306	−33,251	−33,196	−33,142	−33,088	−33,034	−32,981	−32,928
640	−32,686	−32,630	−32,574	−32,519	−32,464	−32,409	−32,355	−32,300	−32,247
660	−32,013	−31,955	−31,899	−31,842	−31,786	−31,731	−31,676	−31,621	−31,566
680	−31,348	−31,290	−31,233	−31,176	−31,119	−31,062	−31,006	−30,950	−30,895
700	−30,677	−30,619	−30,560	−30,502	−30,444	−30,387	−30,330	−30,273	−30,217
720	−30,016	−29,957	−29,897	−29,838	−29,780	−29,721	−29,663	−29,606	−29,549
740	−29,351	−29,290	−29,230	−29,170	−29,110	−29,051	−28,993	−28,934	−28,876
760	−28,682	−28,620	−28,559	−28,498	−28,438	−28,378	−28,318	−28,259	−28,200
780	−28,021	−27,958	−27,896	−27,834	−27,773	−27,712	−27,651	−27,591	−27,532
800	−27,359	−27,295	−27,232	−27,169	−27,107	−27,045	−26,984	−26,923	−26,862
820	−26,701	−26,636	−26,572	−26,508	−26,445	−26,382	−26,320	−26,258	−26,197
840	−26,045	−25,979	−25,913	−25,849	−25,785	−25,721	−25,658	−25,595	−25,533
860	−25,391	−25,325	−25,258	−25,192	−25,127	−25,063	−24,999	−24,935	−24,872
880	−24,739	−24,671	−24,604	−24,537	−24,471	−24,406	−24,341	−24,276	−24,212
900	−24,091	−24,022	−23,954	−23,886	−23,819	−23,752	−23,687	−23,621	−23,556
920	−23,446	−23,376	−23,307	−23,238	−23,170	−23,103	−23,036	−22,970	−22,904
940	−22,803	−22,732	−22,662	−22,592	−22,523	−22,455	−22,388	−22,321	−22,254
960	−22,161	−22,089	−22,018	−21,948	−21,879	−21,810	−21,741	−21,674	−21,607
980	−21,523	−21,451	−21,379	−21,308	−21,238	−21,168	−21,100	−21,031	−20,963
1,000	−20,888	−20,815	−20,743	−20,671	−20,600	−20,530	−20,461	−20,392	−20,323

Temperature in °C	Pressure in bars								
	5,500	5,600	5,700	5,800	5,900	6,000	6,100	6,200	6,300
100	−51,540	−51,502	−51,463	−51,425	−51,387	−51,349	−51,310	−51,272	−51,234
120	−50,778	−50,739	−50,700	−50,661	−50,623	−50,584	−50,546	−50,507	−50,469
140	−50,018	−49,979	−49,939	−49,900	−49,861	−49,822	−49,784	−49,745	−49,706
160	−49,263	−49,223	−49,183	−49,144	−49,105	−49,065	−49,026	−48,987	−48,948
180	−48,511	−48,471	−48,431	−48,391	−48,352	−48,312	−48,272	−48,233	−48,193
200	−47,766	−47,726	−47,685	−47,645	−47,605	−47,565	−47,525	−47,485	−47,445
220	−47,019	−46,978	−46,937	−46,896	−46,856	−46,815	−46,775	−46,734	−46,694
240	−46,281	−46,240	−46,199	−46,158	−46,117	−46,076	−46,035	−45,994	−45,953
260	−45,548	−45,507	−45,465	−45,423	−45,382	−45,340	−45,299	−45,258	−45,217
280	−44,814	−44,772	−44,730	−44,688	−44,646	−44,604	−44,562	−44,520	−44,479
300	−44,092	−44,049	−44,007	−43,964	−43,922	−43,879	−43,837	−43,795	−43,753
320	−43,365	−43,322	−43,279	−43,236	−43,193	−43,150	−43,108	−43,065	−43,022
340	−42,642	−42,598	−42,554	−42,512	−42,468	−42,425	−42,382	−42,339	−42,296
360	−41,927	−41,882	−41,838	−41,794	−41,750	−41,707	−41,663	−41,619	−41,576
380	−41,212	−41,167	−41,122	−41,078	−41,033	−40,989	−40,945	−40,901	−40,857
400	−40,500	−40,454	−40,409	−40,364	−40,319	−40,274	−40,230	−40,185	−40,141
420	−39,793	−39,747	−39,701	−39,656	−39,610	−39,565	−39,520	−39,474	−39,429
440	−39,089	−39,043	−38,996	−38,950	−38,904	−38,858	−38,812	−38,766	−38,721
460	−38,387	−38,340	−38,293	−38,246	−38,199	−38,153	−38,106	−38,060	−38,014
480	−37,691	−37,643	−37,595	−37,548	−37,501	−37,454	−37,407	−37,360	−37,313
500	−36,990	−36,942	−36,894	−36,846	−36,798	−36,750	−36,702	−36,655	−36,608
520	−36,300	−36,250	−36,202	−36,153	−36,104	−36,056	−36,008	−35,960	−35,912
540	−35,612	−35,562	−35,513	−35,463	−35,414	−35,365	−35,316	−35,268	−35,219
560	−34,923	−34,872	−34,822	−34,772	−34,722	−34,672	−34,623	−34,574	−34,525
580	−34,240	−34,189	−34,138	−34,087	−34,037	−33,986	−33,936	−33,886	−33,837
600	−33,558	−33,506	−33,455	−33,403	−33,352	−33,301	−33,250	−33,200	−33,150
620	−32,875	−32,822	−32,770	−32,718	−32,666	−32,614	−32,563	−32,512	−32,461
640	−32,193	−32,140	−32,087	−32,034	−31,981	−31,929	−31,877	−31,825	−31,773
660	−31,512	−31,457	−31,404	−31,350	−31,297	−31,244	−31,191	−31,139	−31,086
680	−30,840	−30,785	−30,730	−30,676	−30,622	−30,568	−30,515	−30,462	−30,409
700	−30,161	−30,106	−30,050	−29,995	−29,940	−29,886	−29,832	−29,778	−29,724
720	−29,492	−29,436	−29,380	−29,324	−29,268	−29,213	−29,158	−29,104	−29,049
740	−28,819	−28,762	−28,705	−28,648	−28,592	−28,536	−28,480	−28,425	−28,370
760	−28,142	−28,084	−28,026	−27,969	−27,912	−27,855	−27,799	−27,743	−27,687
780	−27,472	−27,414	−27,355	−27,297	−27,239	−27,182	−27,125	−27,068	−27,011
800	−26,802	−26,743	−26,683	−26,625	−26,566	−26,508	−26,450	−26,392	−26,335
820	−26,136	−26,075	−26,015	−25,955	−25,896	−25,837	−25,778	−25,720	−25,662
840	−25,471	−25,410	−25,349	−25,288	−25,228	−25,168	−25,109	−25,050	−24,991
860	−24,809	−24,747	−24,685	−24,624	−24,563	−24,502	−24,442	−24,382	−24,323
880	−24,149	−24,086	−24,023	−23,961	−23,899	−23,838	−23,777	−23,717	−23,656
900	−23,492	−23,428	−23,365	−23,302	−23,239	−23,177	−23,116	−23,054	−22,993
920	−22,839	−22,774	−22,710	−22,647	−22,583	−22,520	−22,458	−22,396	−22,334
940	−22,188	−22,123	−22,058	−21,994	−21,930	−21,866	−21,803	−21,740	−21,678
960	−21,540	−21,474	−21,408	−21,343	−21,278	−21,214	−21,150	−21,087	−21,024
980	−20,896	−20,829	−20,763	−20,697	−20,632	−20,567	−20,503	−20,439	−20,375
1,000	−20,256	−20,188	−20,121	−20,055	−19,989	−19,924	−19,859	−19,794	−19,730

Gibbs free energy of H_2O in calories per mole as a function of temperature and pressure, from Fisher and Zen (1971).

Temperature in °C	Pressure in bars								
	6,400	6,500	6,600	6,700	6,800	6,900	7,000	7,100	7,200
100	−51,196	−51,159	−51,121	−51,083	−51,045	−51,008	−50,970	−50,932	−50,895
120	−50,431	−50,392	−50,354	−50,316	−50,278	−50,240	−50,202	−50,164	−50,126
140	−49,667	−49,629	−49,590	−49,552	−49,513	−49,475	−49,437	−49,398	−49,360
160	−48,909	−48,870	−48,831	−48,792	−48,753	−48,714	−48,676	−48,637	−48,599
180	−48,154	−48,114	−48,075	−48,036	−47,997	−47,958	−47,919	−47,880	−47,841
200	−47,405	−47,365	−47,326	−47,286	−47,246	−47,207	−47,168	−47,128	−47,089
220	−46,654	−46,614	−46,574	−46,534	−46,494	−46,454	−46,414	−46,374	−46,335
240	−45,913	−45,872	−45,832	−45,791	−45,751	−45,711	−45,670	−45,630	−45,590
260	−45,176	−45,135	−45,094	−45,053	−45,012	−44,972	−44,931	−44,891	−44,850
280	−44,437	−44,396	−44,355	−44,313	−44,272	−44,231	−44,190	−44,149	−44,108
300	−43,711	−43,669	−43,627	−43,586	−43,544	−43,503	−43,461	−43,420	−43,379
320	−42,980	−42,938	−42,896	−42,853	−42,811	−42,770	−42,728	−42,686	−42,644
340	−42,253	−42,210	−42,168	−42,125	−42,083	−42,040	−41,998	−41,956	−41,914
360	−41,534	−41,490	−41,447	−41,404	−41,361	−41,318	−41,276	−41,233	−41,191
380	−40,813	−40,769	−40,726	−40,682	−40,639	−40,596	−40,553	−40,510	−40,467
400	−40,096	−40,052	−40,008	−39,964	−39,920	−39,876	−39,833	−39,789	−39,746
420	−39,385	−39,340	−39,295	−39,251	−39,206	−39,162	−39,118	−39,074	−39,030
440	−38,675	−38,630	−38,585	−38,540	−38,495	−38,450	−38,406	−38,361	−38,317
460	−37,968	−37,922	−37,876	−37,831	−37,786	−37,740	−37,695	−37,650	−37,605
480	−37,267	−37,220	−37,174	−37,128	−37,082	−37,036	−36,991	−36,945	−36,900
500	−36,561	−36,514	−36,467	−36,420	−36,374	−36,328	−36,281	−36,235	−36,189
520	−35,864	−35,817	−35,770	−35,722	−35,675	−35,628	−35,582	−35,535	−35,489
540	−35,171	−35,123	−35,075	−35,027	−34,980	−34,932	−34,885	−34,838	−34,791
560	−34,476	−34,427	−34,379	−34,330	−34,282	−34,234	−34,186	−34,139	−34,091
580	−33,787	−33,738	−33,689	−33,640	−33,591	−33,543	−33,494	−33,446	−33,398
600	−33,100	−33,050	−33,000	−32,950	−32,901	−32,852	−32,803	−32,754	−32,706
620	−32,410	−32,360	−32,309	−32,259	−32,209	−32,159	−32,110	−32,061	−32,011
640	−31,722	−31,671	−31,620	−31,569	−31,519	−31,468	−31,418	−31,368	−31,318
660	−31,034	−30,982	−30,931	−30,879	−30,828	−30,777	−30,727	−30,676	−30,626
680	−30,356	−30,304	−30,251	−30,199	−30,148	−30,096	−30,045	−29,994	−29,943
700	−29,671	−29,618	−29,565	−29,512	−29,460	−29,408	−29,356	−29,304	−29,253
720	−28,995	−28,942	−28,888	−28,835	−28,782	−28,729	−28,676	−28,624	−28,572
740	−28,315	−28,261	−28,207	−28,153	−28,099	−28,046	−27,992	−27,939	−27,887
760	−27,632	−27,576	−27,522	−27,467	−27,413	−27,359	−27,305	−27,251	−27,198
780	−26,955	−26,899	−26,844	−26,789	−26,734	−26,679	−26,624	−26,570	−26,516
800	−26,278	−26,222	−26,165	−26,109	−26,054	−25,998	−25,943	−25,888	−25,834
820	−25,604	−25,547	−25,490	−25,433	−25,377	−25,321	−25,265	−25,210	−25,155
040	24,933	24,875	24,817	−24,760	−24,703	−24,646	−24,589	−24,533	−24,477
860	−24,264	−24,205	−24,147	−24,089	−24,031	−23,974	−23,917	−23,860	−23,803
880	−23,597	−23,537	−23,478	−23,419	−23,361	−23,303	−23,245	−23,187	−23,130
900	−22,933	−22,873	−22,813	−22,753	−22,694	−22,635	−22,577	−22,519	−22,461
920	−22,273	−22,212	−22,152	−22,092	−22,032	−21,972	−21,913	−21,854	−21,796
940	−21,616	−21,554	−21,493	−21,432	−21,372	−21,312	−21,252	−21,192	−21,133
960	−20,961	−20,899	−20,837	−20,776	−20,714	−20,654	−20,593	−20,533	−20,473
980	−20,312	−20,249	−20,186	−20,124	−20,062	−20,000	−19,939	−19,878	−19,818
1,000	−19,666	−19,602	−19,539	−19,476	−19,414	−19,351	−19,289	−19,228	−19,166

Temperature in °C	Pressure in bars								
	7,300	7,400	7,500	7,600	7,700	7,800	7,900	8,000	8,100
100	−50,858	−50,820	−50,783	−50,746	−50,708	−50,671	−50,634	−50,597	−50,560
120	−50,089	−50,051	−50,013	−49,976	−49,938	−49,901	−49,863	−49,826	−49,789
140	−49,322	−49,284	−49,246	−49,208	−49,170	−49,133	−49,095	−49,057	−49,019
160	−48,560	−48,522	−48,483	−48,445	−48,407	−48,369	−48,331	−48,293	−48,255
180	−47,802	−47,763	−47,724	−47,686	−47,647	−47,609	−47,570	−47,532	−47,493
200	−47,050	−47,011	−46,972	−46,933	−46,894	−46,855	−46,816	−46,777	−46,739
220	−46,295	−46,256	−46,216	−46,177	−46,138	−46,098	−46,059	−46,020	−45,981
240	−45,550	−45,510	−45,471	−45,431	−45,391	−45,352	−45,312	−45,273	−45,233
260	−44,810	−44,769	−44,729	−44,689	−44,649	−44,609	−44,569	−44,529	−44,490
280	−44,068	−44,027	−43,986	−43,946	−43,905	−43,865	−43,825	−43,784	−43,744
300	−43,338	−43,296	−43,255	−43,215	−43,174	−43,133	−43,092	−43,052	−43,011
320	−42,603	−42,561	−42,520	−42,478	−42,437	−42,396	−42,355	−42,314	−42,273
340	−41,872	−41,830	−41,788	−41,746	−41,705	−41,663	−41,621	−41,580	−41,539
360	−41,148	−41,106	−41,064	−41,021	−40,979	−40,937	−40,895	−40,854	−40,812
380	−40,424	−40,382	−40,339	−40,296	−40,254	−40,211	−40,169	−40,127	−40,085
400	−39,702	−39,659	−39,616	−39,574	−39,531	−39,488	−39,445	−39,403	−39,360
420	−38,986	−38,942	−38,899	−38,855	−38,812	−38,769	−38,726	−38,682	−38,639
440	−38,273	−38,228	−38,184	−38,140	−38,097	−38,053	−38,009	−37,966	−37,922
460	−37,560	−37,516	−37,471	−37,427	−37,382	−37,338	−37,294	−37,250	−37,206
480	−36,854	−36,809	−36,764	−36,719	−36,675	−36,630	−36,585	−36,541	−36,497
500	−36,144	−36,098	−36,052	−36,007	−35,962	−35,917	−35,872	−35,827	−35,782
520	−35,442	−35,396	−35,350	−35,304	−35,259	−35,213	−35,167	−35,122	−35,077
540	−34,744	−34,697	−34,651	−34,605	−34,558	−34,512	−34,466	−34,420	−34,375
560	−34,044	−33,997	−33,950	−33,903	−33,856	−33,809	−33,763	−33,717	−33,670
580	−33,350	−33,302	−33,255	−33,207	−33,160	−33,113	−33,066	−33,019	−32,973
600	−32,657	−32,609	−32,561	−32,513	−32,465	−32,417	−32,370	−32,323	−32,275
620	−31,962	−31,914	−31,865	−31,816	−31,768	−31,720	−31,672	−31,624	−31,577
640	−31,269	−31,219	−31,170	−31,121	−31,072	−31,024	−30,975	−30,927	−30,879
660	−30,576	−30,526	−30,476	−30,426	−30,377	−30,328	−30,279	−30,230	−30,181
680	−29,892	−29,841	−29,791	−29,741	−29,691	−29,641	−29,592	−29,542	−29,493
700	−29,201	−29,150	−29,099	−29,049	−28,998	−28,948	−28,898	−28,848	−28,798
720	−28,520	−28,468	−28,417	−28,366	−28,315	−28,264	−28,213	−28,163	−28,113
740	−27,834	−27,782	−27,730	−27,678	−27,626	−27,575	−27,524	−27,473	−27,422
760	−27,145	−27,092	−27,039	−26,987	−26,935	−26,883	−26,831	−26,779	−26,728
780	−26,462	−26,409	−26,356	−26,303	−26,250	−26,198	−26,145	−26,093	−26,041
800	−25,779	−25,725	−25,672	−25,618	−25,565	−25,512	−25,459	−25,406	−25,354
820	−25,100	−25,045	−24,990	−24,936	−24,882	−24,829	−24,775	−24,722	−24,669
840	−24,422	−24,367	−24,311	−24,257	−24,202	−24,148	−24,094	−24,040	−23,987
860	−23,747	−23,691	−23,635	−23,580	−23,525	−23,470	−23,415	−23,361	−23,307
880	−23,073	−23,017	−22,960	−22,904	−22,849	−22,793	−22,738	−22,683	−22,628
900	−22,403	−22,346	−22,289	−22,233	−22,176	−22,120	−22,064	−22,009	−21,953
920	−21,738	−21,680	−21,622	−21,565	−21,508	−21,451	−21,395	−21,338	−21,282
940	−21,074	−21,016	−20,957	−20,899	−20,842	−20,784	−20,727	−20,670	−20,614
960	−20,413	−20,354	−20,295	−20,236	−20,178	−20,120	−20,062	−20,005	−19,947
980	−19,757	−19,697	−19,638	−19,578	−19,519	−19,460	−19,402	−19,343	−19,285
1,000	−19,105	−19,045	−18,984	−18,924	−18,864	−18,804	−18,745	−18,686	−18,627

Gibbs free energy of H_2O in calories per mole as a function of temperature and pressure, from Fisher and Zen (1971).

Temperature in °C	Pressure in bars								
	8,200	8,300	8,400	8,500	8,600	8,700	8,800	8,900	9,000
100	−50,523	−50,486	−50,449	−50,413	−50,376	−50,339	−50,302	−50,266	−50,229
120	−49,751	−49,714	−49,677	−49,640	−49,603	−49,566	−49,529	−49,492	−49,455
140	−48,982	−48,944	−48,907	−48,869	−48,832	−48,795	−48,757	−48,720	−48,683
160	−48,217	−48,179	−48,141	−48,103	−48,065	−48,028	−47,990	−47,953	−47,915
180	−47,455	−47,417	−47,379	−47,341	−47,303	−47,265	−47,227	−47,189	−47,151
200	−46,700	−46,661	−46,623	−46,584	−46,546	−46,508	−46,469	−46,431	−46,393
220	−45,942	−45,903	−45,864	−45,825	−45,787	−45,748	−45,709	−45,671	−45,632
240	−45,194	−45,154	−45,115	−45,076	−45,037	−44,998	−44,959	−44,920	−44,881
260	−44,450	−44,410	−44,371	−44,331	−44,292	−44,252	−44,213	−44,174	−44,134
280	−43,704	−43,664	−43,624	−43,584	−43,545	−43,505	−43,465	−43,425	−43,386
300	−42,971	−42,930	−42,890	−42,850	−42,809	−42,769	−42,729	−42,689	−42,649
320	−42,232	−42,191	−42,151	−42,110	−42,069	−42,029	−41,989	−41,948	−41,908
340	−41,497	−41,456	−41,415	−41,374	−41,333	−41,292	−41,252	−41,211	−41,170
360	−40,770	−40,729	−40,687	−40,646	−40,604	−40,563	−40,522	−40,481	−40,440
380	−40,043	−40,001	−39,959	−39,917	−39,875	−39,834	−39,792	−39,751	−39,709
400	−39,318	−39,275	−39,233	−39,191	−39,149	−39,107	−39,065	−39,023	−38,981
420	−38,598	−38,555	−38,512	−38,469	−38,427	−38,384	−38,342	−38,300	−38,258
440	−37,879	−37,836	−37,792	−37,749	−37,706	−37,664	−37,621	−37,579	−37,537
460	−37,163	−37,119	−37,075	−37,032	−36,989	−36,945	−36,902	−36,859	−36,816
480	−36,452	−36,408	−36,364	−36,320	−36,277	−36,233	−36,189	−36,146	−36,103
500	−35,737	−35,693	−35,648	−35,604	−35,560	−35,516	−35,472	−35,428	−35,384
520	−35,032	−34,987	−34,942	−34,897	−34,852	−34,808	−34,764	−34,719	−34,675
540	−34,329	−34,284	−34,238	−34,193	−34,148	−34,103	−34,058	−34,013	−33,969
560	−33,624	−33,578	−33,533	−33,487	−33,441	−33,396	−33,351	−33,306	−33,261
580	−32,926	−32,880	−32,833	−32,787	−32,741	−32,695	−32,650	−32,604	−32,559
600	−32,228	−32,182	−32,135	−32,088	−32,042	−31,995	−31,949	−31,903	−31,857
620	−31,529	−31,482	−31,434	−31,387	−31,340	−31,294	−31,247	−31,201	−31,154
640	−30,831	−30,783	−30,735	−30,688	−30,640	−30,953	−30,546	−30,499	−30,452
660	−30,133	−30,084	−30,036	−29,988	−29,940	−29,893	−29,845	−29,798	−29,750
680	−29,444	−29,395	−29,347	−29,298	−29,250	−29,202	−29,154	−29,106	−29,058
700	−28,749	−28,699	−28,650	−28,601	−28,552	−28,504	−28,455	−28,407	−28,359
720	−28,062	−28,013	−27,963	−27,913	−27,864	−27,815	−27,766	−27,717	−27,669
740	−27,372	−27,321	−27,271	−27,221	−27,171	−27,122	−27,072	−27,023	−26,974
760	−26,677	−26,626	−26,575	−26,525	−26,475	−26,424	−26,374	−26,325	−26,275
780	−25,990	−25,938	−25,887	−25,836	−25,785	−25,735	−25,684	−25,634	−25,584
800	−25,302	−25,250	−25,198	−25,146	−25,096	−25,044	−24,993	−24,942	−24,892
820	−24,616	−24,564	−24,512	−24,460	−24,408	−24,356	−24,305	−24,253	−24,202
840	−23,933	−23,880	−23,827	−23,775	−23,722	−23,670	−23,618	−23,567	−23,515
860	−23,253	−23,199	−23,146	−23,093	−23,040	−22,987	−22,935	−22,883	−22,831
880	−22,574	−22,520	−22,466	−22,412	−22,359	−22,305	−22,252	−22,200	−22,147
900	−21,898	−21,844	−21,789	−21,735	−21,681	−21,627	−21,574	−21,520	−21,467
920	−21,227	−21,171	−21,116	−21,061	−21,007	−20,952	−20,898	−20,844	−20,791
940	−20,557	−20,501	−20,446	−20,390	−20,335	−20,280	−20,225	−20,170	−20,116
960	−19,890	−19,834	−19,777	−19,721	−19,665	−19,609	−19,554	−19,499	−19,444
980	−19,228	−19,170	−19,113	−19,056	−18,999	−18,943	−18,887	−18,831	−18,775
1,000	−18,568	−18,510	−18,452	−18,394	−18,337	−18,279	−18,222	−18,165	−18,109

Temperature in °C	Pressure in bars									
	9,100	9,200	9,300	9,400	9,500	9,600	9,700	9,800	9,900	10,000
100	−50,193	−50,156	−50,120	−50,083	−50,047	−50,011	−49,975	−49,938	−49,902	−49,866
120	−49,418	−49,381	−49,345	−49,308	−49,271	−49,235	−49,198	−49,162	−49,125	−49,089
140	−48,646	−48,609	−48,572	−48,535	−48,498	−48,461	−48,424	−48,387	−48,350	−48,314
160	−47,878	−47,840	−47,803	−47,766	−47,728	−47,691	−47,654	−47,617	−47,580	−47,543
180	−47,113	−47,075	−47,038	−47,000	−46,963	−46,925	−46,888	−46,850	−46,813	−46,776
200	−46,355	−46,317	−46,279	−46,241	−46,203	−46,165	−46,127	−46,090	−46,052	−46,014
220	−45,594	−45,555	−45,517	−45,479	−45,441	−45,402	−45,364	−45,326	−45,288	−45,250
240	−44,842	−44,804	−44,765	−44,726	−44,688	−44,649	−44,611	−44,573	−44,534	−44,496
260	−44,095	−44,056	−44,017	−43,978	−43,939	−43,901	−43,862	−43,823	−43,784	−43,746
280	−43,346	−43,307	−43,268	−43,228	−43,189	−43,150	−43,111	−43,072	−43,033	−42,994
300	−42,610	−42,570	−42,530	−42,490	−42,451	−42,411	−42,372	−42,333	−42,293	−42,254
320	−41,868	−41,828	−41,788	−41,748	−41,708	−41,668	−41,628	−41,588	−41,549	−41,509
340	−41,130	−41,089	−41,049	−41,008	−40,968	−40,928	−40,888	−40,848	−40,808	−40,768
360	−40,399	−40,358	−40,317	−40,277	−40,236	−40,195	−40,155	−40,114	−40,074	−40,034
380	−39,668	−39,627	−39,585	−39,544	−39,503	−39,462	−39,422	−39,381	−39,340	−39,300
400	−38,939	−38,898	−38,856	−38,815	−38,773	−38,732	−38,691	−38,650	−38,609	−38,568
420	−38,216	−38,174	−38,132	−38,090	−38,048	−38,007	−37,965	−37,924	−37,882	−37,841
440	−37,494	−37,452	−37,410	−37,367	−37,325	−37,283	−37,241	−37,200	−37,158	−37,116
460	−36,773	−36,731	−36,688	−36,645	−36,604	−36,561	−36,519	−36,477	−36,435	−36,393
480	−36,059	−36,016	−35,973	−35,930	−35,887	−35,845	−35,802	−35,759	−35,717	−35,675
500	−35,340	−35,297	−35,253	−35,210	−35,167	−35,124	−35,081	−35,038	−34,995	−34,952
520	−34,631	−34,587	−24,543	−34,499	−34,456	−34,412	−34,369	−34,326	−34,283	−34,239
540	−33,924	−33,880	−33,836	−33,792	−33,748	−33,704	−33,660	−33,616	−33,573	−33,529
560	−33,216	−33,171	−33,126	−33,082	−33,037	−32,993	−32,949	−32,905	−32,861	−32,817
580	−32,513	−32,468	−32,423	−32,378	−32,333	−32,289	−32,244	−32,200	−32,155	−32,111
600	−31,812	−31,766	−31,721	−31,675	−31,630	−31,585	−31,540	−31,495	−31,450	−31,406
620	−31,108	−31,062	−31,016	−30,970	−30,925	−30,879	−30,834	−30,789	−30,744	−30,699
640	−30,406	−30,359	−30,313	−30,267	−30,221	−30,175	−30,129	−30,083	−30,038	−29,992
660	−29,703	−29,656	−29,610	−29,563	−29,517	−29,470	−29,424	−29,378	−29,332	−29,286
680	−29,010	−28,963	−28,916	−28,869	−28,822	−28,775	−28,729	−28,682	−28,636	−28,590
700	−28,311	−28,263	−28,215	−28,168	−28,120	−28,073	−28,026	−27,979	−27,932	−27,886
720	−27,620	−27,572	−27,524	−27,476	−27,428	−27,380	−27,333	−27,286	−27,238	−27,191
740	−26,925	−26,876	−26,827	−26,779	−26,731	−26,683	−26,635	−26,587	−26,540	−26,492
760	−26,226	−26,177	−26,128	−26,079	−26,030	−25,981	−25,933	−25,885	−25,837	−25,789
780	−25,534	−25,484	−25,435	−25,385	−25,336	−25,287	−25,239	−25,190	−25,142	−25,093
800	−24,841	−24,791	−24,741	−24,691	−24,642	−24,592	−24,543	−24,494	−24,445	−24,397
820	−24,152	−24,101	−24,050	−24,000	−23,950	−23,900	−23,851	−23,801	−23,752	−23,703
840	−23,464	−23,413	−23,362	−23,311	−23,261	−23,210	−23,160	−23,110	−23,060	−23,011
860	−22,779	−22,727	−22,676	−22,625	−22,574	−22,523	−22,472	−22,422	−22,372	−22,322
880	−22,095	−22,043	−21,991	−21,939	−21,888	−21,836	−21,785	−21,735	−21,684	−21,634
900	−21,414	−21,362	−21,309	−21,257	−21,205	−21,153	−21,102	−21,050	−20,999	−20,948
920	−20,737	−20,684	−20,631	−20,578	−20,526	−20,474	−20,422	−20,370	−20,318	−20,267
940	−20,062	−20,008	−19,955	−19,902	−19,848	−19,796	−19,743	−19,691	−19,638	−19,587
960	−19,389	−19,334	−19,280	−19,226	−19,173	−19,119	−19,066	−19,013	−18,960	−18,908
980	−18,719	−18,664	−18,609	−18,555	−18,500	−18,446	−18,392	−18,338	−18,285	−18,232
1,000	−18,052	−17,996	−17,941	−17,885	−17,830	−17,775	−17,720	−17,665	−17,611	−17,557

APPENDIX **5**

Specific volume of H_2O (in cm³/gram) as a function of temperature and pressure, from Kennedy and Holser (1966, pp. 374–379).

Pressure in bars	Temperature in °C										
	−10	0	10	20	30	40	50	60	70	80	90
1	—	1.00013	1.00027	1.00177	1.00434	1.00781	1.01208	1.01706	1.02271	1.02900	1.0359
10	—	.9996	.9998	1.0013	1.0039	1.0074	1.0116	1.0166	1.0223	1.0285	1.0354
50	—	.9975	.9980	.9986	1.0021	1.0056	1.0099	1.0148	1.0204	1.0266	1.0334
100	—	.9951	.9958	.9974	1.0000	1.0035	1.0077	1.0126	1.0181	1.0243	1.0310
150	—	.9927	.9935	.9952	.9978	1.0013	1.0055	1.0104	1.0159	1.0220	1.0287
200	—	.9903	.9913	.9930	.9956	.9991	1.0033	1.0082	1.0137	1.0198	1.0264
250	—	.9879	.9890	.9908	.9935	.9970	1.0012	1.0061	1.0115	1.0176	1.0241
300	—	.9856	.9868	.9887	.9914	.9949	.9991	1.0040	1.0094	1.0154	1.0219
350	—	.9833	.9846	.9866	.9894	.9929	.9971	1.0019	1.0073	1.0132	1.0197
400	—	.9810	.9825	.9845	.9874	.9909	.9951	.9999	1.0052	1.0111	1.0175
450	—	.9788	.9804	.9824	.9854	.9889	.9931	.9979	1.0032	1.0090	1.0154
500	—	.9766	.9783	.9804	.9834	.9870	.9911	.9959	1.0012	1.0070	1.0133
600	—	.9723	.9742	.9765	.9795	.9831	.9873	.9920	.9973	1.0030	1.0092
700	—	.9682	.9703	.9726	.9756	.9793	.9835	.9882	.9934	.9990	1.0051
800	—	.9641	.9664	.9688	.9719	.9756	.9798	.9845	.9896	.9952	1.0012
900	—	.9602	.9626	.9652	.9684	.9721	.9763	.9809	.9860	.9915	.9974
1,000	—	.9564	.9589	.9616	.9649	.9687	.9729	.9774	.9824	.9878	.9937
1,200	.945	.9491	.9519	.9548	.9582	.9620	.9662	.9707	.9756	.9809	.9866
1,400	.939	.9422	.9452	.9483	.9518	.9556	.9598	.9643	.9691	.9742	.9798
1,600	.934	.936	.939	.942	.946	.950	.954	.958	.963	.968	.974
1,800	.929	.930	.933	.937	.941	.945	.949	.953	.958	.963	.968
2,000	.922	.925	.928	.932	.936	.940	.944	.948	.953	.958	.963
2,500	.909	.912	.915	.919	.923	.927	.931	.935	.940	.945	.950
3,000	.897	.900	.904	.907	.911	.915	.919	.924	.928	.933	.938
3,500	.886	.889	.893	.897	.900	.904	.908	.913	.917	.921	.926
4,000	.876	.879	.883	.887	.890	.894	.898	.902	.906	.911	.916
5,000	.859	.862	.865	.869	.872	.876	.880	.884	.888	.893	.897
6,000	—	.847	.849	.853	.857	.861	.865	.869	.873	.877	.881
7,000	—	—	.835	.839	.843	.847	.851	.855	.859	.862	.866
8,000	—	—	—	.826	.830	.834	.838	.842	.846	.849	.853
9,000	—	—	—	—	.819	.823	.827	.831	.835	.838	.842
10,000	—	—	—	—	.809	.813	.817	.821	.825	.829	.833
15,000	—	—	—	—	—	—	—	.780	.784	.787	.791
20,000	—	—	—	—	—	—	—	—	—	.757	.760
25,000	—	—	—	—	—	—	—	—	—	—	—
30,000	—	—	—	—	—	—	—	—	—	—	—
40,000	—	—	—	—	—	—	—	—	—	—	—
50,000	—	—	—	—	—	—	—	—	—	—	—
100,000	—	—	—	—	—	—	—	—	—	—	—
150,000	—	—	—	—	—	—	—	—	—	—	—
200,000	—	—	—	—	—	—	—	—	—	—	—
250,000	—	—	—	—	—	—	—	—	—	—	—

Specific volume of H_2O (in cm³/gram) as a function of temperature and pressure, from Kennedy and Holser (1966, pp. 374–379).

Pressure in bars	Temperature, in °C								
	100	120	140	160	180	200	220	240	260
1	1.0434	1794.	1890.	1,984.	2,079.	2,171.	2,264.7	2,358.6	2,452.1
10	1.0429	1.0598	1.0794	1.1018	194.4	206.8	216.3	227.1	237.6
50	1.0408	1.0577	1.0769	1.0990	1.1242	1.1531	1.1868	1.2265	1.2747
100	1.0384	1.0550	1.0739	1.0955	1.1202	1.1483	1.1810	1.2189	1.2647
150	1.0359	1.0523	1.0709	1.0922	1.1163	1.1437	1.1753	1.2118	1.2553
200	1.0336	1.0497	1.0680	1.0889	1.1125	1.1392	1.1698	1.2050	1.2466
250	1.0313	1.0471	1.0652	1.0857	1.1088	1.1349	1.1645	1.1986	1.2384
300	1.0290	1.0445	1.0625	1.0826	1.1052	1.1307	1.1596	1.1926	1.2308
350	1.0267	1.0421	1.0598	1.0796	1.1018	1.1267	1.1548	1.1868	1.2236
400	1.0245	1.0397	1.0570	1.0765	1.0983	1.1227	1.1501	1.1812	1.2168
450	1.0223	1.0373	1.0543	1.0735	1.0949	1.1187	1.1455	1.1758	1.2100
500	1.0201	1.0349	1.0516	1.0704	1.0914	1.1147	1.1410	1.1705	1.2036
600	1.0158	1.030	1.046	1.064	1.085	1.1072	1.1325	1.1604	1.1916
700	1.0117	1.025	1.041	1.059	1.079	1.1003	1.1245	1.1509	1.1806
800	1.0076	1.021	1.037	1.054	1.073	1.0936	1.1169	1.1421	1.1704
900	1.0037	1.017	1.032	1.049	1.067	1.0872	1.1096	1.1339	1.1609
1,000	.9999	1.013	1.028	1.044	1.061	1.0811	1.1027	1.1261	1.1520
1,200	.9927	1.005	1.019	1.035	1.051	1.0696	1.0900	1.1117	1.1357
1,400	.9858	.998	1.011	1.026	1.041	1.0591	1.0783	1.0988	1.1211
1,600	.979	.991	1.004	1.018	1.033	1.050	1.068	1.087	1.108
1,800	.973	.985	.997	1.010	1.025	1.040	1.058	1.076	1.096
2,000	.968	.979	.991	1.003	1.017	1.032	1.048	1.065	1.084
2,500	.955	.965	.975	.985	.997	1.010	1.024	1.040	1.056
3,000	.943	—	—	—	—	—	—	—	—
3,500	.931	—	—	—	—	—	—	—	—
4,000	.921	—	—	—	—	—	—	—	—
5,000	.902	.911	.919	.927	.936	.944	—	—	—
6,000	.885	—	—	—	—	—	—	—	—
7,000	.870	—	—	—	—	—	—	—	—
8,000	.857	—	—	—	—	—	—	—	—
9,000	.846	—	—	—	—	—	—	—	—
10,000	.837	.845	.852	.859	.866	.872	—	—	—
15,000	.794	.800	.805	.810	.815	.821	—	—	—
20,000	.762	.766	.770	.774	.778	.783	—	—	—
25,000	—	.739	.742	.746	.750	.754	—	—	—
30,000	—	—	—	.724	.728	.732	—	—	—
40,000	—	—	—	—	—	.700	—	—	—
50,000	—	—	—	—	—	—	—	—	—
100,000	—	—	—	—	—	—	—	—	—
150,000	—	—	—	—	—	—	—	—	—
200,000	—	—	—	—	—	—	—	—	—
250,000	—	—	—	—	—	—	—	—	—

Pressure in bars	Temperature in °C								
	280	300	320	340	360	380	400	420	440
1	2,552.9	2,638.6	2,731.6	2,824.4	2,922.1	3,009.9	3,102.6	3,195.3	3,288.0
10	247.9	257.9	267.9	277.6	287.5	296.9	306.6	316.2	325.8
50	42.30	45.34	48.13	50.71	53.18	55.55	57.88	60.11	62.24
100	1.3217	1.3970	19.27	21.52	23.32	24.94	26.42	27.80	29.13
150	1.3086	1.3771	1.4725	1.6300	12.58	14.25	15.69	16.90	17.99
200	1.2968	1.3599	1.4431	1.5694	1.8235	8.267	9.965	11.20	12.25
250	1.2860	1.3448	1.4207	1.5265	1.6965	2.254	6.015	7.553	8.673
300	1.2762	1.3313	1.4008	1.4930	1.6272	1.883	2.799	4.900	6.224
350	1.2670	1.3191	1.3832	1.4653	1.5792	1.767	2.127	3.062	4.396
400	1.2582	1.3077	1.3676	1.4420	1.5448	1.695	1.916	2.363	3.209
450	1.2498	1.2970	1.3534	1.4222	1.5145	1.643	1.812	2.091	2.594
500	1.2418	1.2869	1.3404	1.4046	1.4869	1.596	1.732	1.944	2.276
600	1.2273	1.2688	1.3171	1.3735	1.4433	1.532	1.634	1.772	1.963
700	1.2142	1.2528	1.2971	1.3477	1.4089	1.484	1.571	1.677	1.812
800	1.2027	1.2384	1.2795	1.3257	1.3803	1.446	1.516	1.606	1.712
900	1.1911	1.2253	1.2636	1.3061	1.3558	1.414	1.477	1.552	1.641
1,000	1.1807	1.2131	1.2491	1.2885	1.3342	1.387	1.442	1.506	1.582
1,200	1.1620	1.1913	1.2234	1.2581	1.2978	1.342	1.389	1.440	1.498
1,400	1.1454	1.1724	1.2016	1.2326	1.2677	1.306	1.348	1.391	1.438
1,600	1.131	1.1559	1.183	1.211	1.242	1.277	1.315	1.353	1.393
1,800	1.117	1.142	1.166	1.192	1.221	1.252	1.287	1.321	1.357
2,000	1.105	1.127	1.151	1.174	1.203	1.229	1.260	1.293	1.326
2,500	1.075	1.095	1.116	1.137	1.161	1.183	1.209	1.236	1.265
3,000	—	—	—	—	—	—	—	—	—
3,500	—	—	—	—	—	—	—	—	—
4,000	—	—	—	—	—	—	—	—	—
5,000	—	.995	—	—	—	—	1.065	—	—
6,000	—	—	—	—	—	—	—	—	—
7,000	—	—	—	—	—	—	—	—	—
8,000	—	—	—	—	—	—	—	—	—
9,000	—	—	—	—	—	—	—	—	—
10,000	—	.907	—	—	—	—	.937	—	—
15,000	—	.846	—	—	—	—	.873	—	—
20,000	—	.806	—	—	—	—	.826	—	—
25,000	—	.774	—	—	—	—	.794	—	—
30,000	—	.751	—	—	—	—	.770	—	—
40,000	—	.713	—	—	—	—	.731	—	—
50,000	—	.681*	—	—	—	—	.698	—	—
100,000	—	.575*	—	—	—	—	.587	—	—
150,000	—	.514*	—	—	—	—	.523*	—	—
200,000	—	.478*	—	—	—	—	.483*	—	—
250,000	—	.454*	—	—	—	—	.459*	—	—

*Probably supercooled with respect to ice.

Specific volume of H_2O (in cm³/gram) as a function of temperature and pressure, from Kennedy and Holser (1966, pp. 374–3790.

Pressure in bars	Temperature in °C								
	460	480	500	550	600	650	700	750	800
1	3,380.6	3,473.3	3,565.7	3,797.	4,028.	4,261.	4,491.	4,721.	4,951.
10	335.3	345.0	354.3	377.8	401.2	424.3	447.6	470.8	494.0
50	64.39	66.46	68.58	73.66	78.58	83.55	88.39	93.21	98.06
100	30.40	31.63	32.81	35.56	38.28	40.99	43.56	46.06	48.54
150	18.99	19.92	20.81	22.89	24.84	26.75	28.59	30.35	32.06
200	13.18	14.01	14.78	16.53	18.15	19.70	21.14	22.49	23.83
250	9.611	10.44	11.17	12.73	14.12	15.43	16.64	17.78	18.89
300	7.207	8.012	8.709	10.17	11.43	12.60	13.66	14.65	15.61
350	5.417	6.259	6.954	8.340	9.507	10.56	11.53	12.43	13.28
400	4.114	4.943	5.640	6.978	8.077	9.057	9.948	10.76	11.53
450	3.260	4.001	4.645	5.929	6.971	7.888	8.717	9.465	10.18
500	2.724	3.294	3.901	5.112	6.095	6.960	7.735	8.440	9.104
600	2.225	2.570	2.959	3.949	4.823	5.593	6.279	6.909	7.498
700	1.986	2.207	2.468	3.219	3.975	4.657	5.269	5.840	6.371
800	1.844	2.004	2.190	2.757	3.390	3.978	4.524	5.053	5.542
900	1.744	1.867	2.012	2.453	2.969	3.480	3.968	4.450	4.906
1,000	1.672	1.774	1.892	2.246	2.670	3.111	3.547	3.988	4.406
1,200	1.562	1.636	1.720	1.973	2.281	2.618	2.966	3.328	3.682
1,400	1.490	1.549	1.616	1.806	2.037	2.303	2.590	2.885	3.182
1,600	1.437	1.486	1.538	1.688	1.869	2.085	2.322	2.571	2.823
1,800	1.395	1.436	1.481	1.607	1.753	1.927	2.124	2.338	2.555
2,000	1.361	1.397	1.435	1.540	1.666	1.814	1.980	2.161	2.348
2,500	1.296	1.325	1.356	1.436	1.525	1.626	1.735	1.857	1.993
3,000	—	—	—	—	—	—	—	—	—
3,500	—	—	—	—	—	—	—	—	—
4,000	—	—	—	—	—	—	—	—	—
5,000	—	—	1.139	—	—	—	—	1.350	—
6,000	—	—	—	—	—	—	—	—	—
7,000	—	—	—	—	—	—	—	—	—
8,000	—	—	—	—	—	—	—	1.078	—
9,000	—	—	—	—	—	—	—	.969	—
10,000	—	—	.978	—	—	—	—	.903	—
15,000	—	—	.900	—	—	—	—	.864	—
20,000	—	—	.848	—	—	—	—	.838	—
25,000	—	—	.814	—	—	—	—	.794	—
30,000	—	—	.789	—	—	—	—	.757	—
40,000	—	—	.749	—	—	—	—	.628	—
50,000	—	—	.715	—	—	—	—	.552	—
100,000	—	—	.599	—	—	—	—	.504	—
150,000	—	—	.531*	—	—	—	—	.473*	—
200,000	—	—	.489*	—	—	—	—	—	—
250,000	—	—	.463*	—	—	—	—	—	—

*Probably supercooled with respect to ice.

Pressure in bars	Temperature, in °C			
	850	900	950	1000
1	5,183.	5,413.	5,644.	5,875.
10	517.6	540.6	563.9	587.3
50	102.89	107.64	112.39	115.1
100	51.01	53.47	55.91	58.39
150	33.77	35.45	37.14	38.85
200	25.15	26.46	27.77	29.08
250	19.99	21.08	22.16	23.24
300	16.56	17.50	18.43	19.35
350	14.12	14.95	15.76	16.57
400	12.29	13.04	13.77	14.50
450	10.87	11.56	12.23	12.88
500	9.750	10.38	10.99	11.59
600	8.070	8.621	9.157	9.675
700	6.884	7.377	7.849	8.307
800	6.006	6.451	6.874	7.285
900	5.329	5.735	6.119	6.495
1,000	4.792	5.163	5.518	5.863
1,200	4.009	4.320	4.624	4.927
1,400	3.458	3.730	3.997	4.264
1,600	3.065	3.301	3.536	3.772
1,800	2.768	2.976	3.186	3.400
2,000	2.536	2.722	2.909	3.097
2,500	2.138	2.287	2.434	2.576
3,000	—	—	—	—
3,500	—	—	—	—
4,000	—	—	—	—
5,000	—	—	—	1.616
6,000	—	—	—	—
7,000	—	—	—	—
8,000	—	—	—	1.189
9,000	—	—	—	1.040
10,000	—	—	—	.959
15,000	—	—	—	.914
20,000	—	—	—	.886
25,000	—	—	—	.839
30,000	—	—	—	.799
40,000	—	—	—	.658
50,000	—	—	—	.573
100,000	—	—	—	.519
150,000	—	—	—	.484
200,000	—	—	—	—
250,000	—	—	—	—

Density of CO_2 (in grams/cm³) as a function
of temperature and pressure, from
Kennedy and Holser (1966, pp. 382–383).

Temperature in °C	Pressure in bars										
	25	50	75	100	150	200	250	300	350	400	450
0	.0601	.947	.954	.969	.997	1.0170	1.0350	1.0530	1.0670	1.0792	1.0900
10	.0561	.864	.891	.914	.950	.9770	1.0000	1.0190	1.0350	1.0502	1.0635
20	.0527	.1423	.810	.855	.901	.9335	.9600	.9832	1.0030	1.0200	1.0351
30	.0499	.1251	.655	.782	.850	.8887	.9190	.9460	.9685	.9882	1.0054
40	.0476	.1135	.2305	.638	.785	.8415	.8771	.9077	.9339	.9559	.9755
50	.0456	.1052	.1932	.3901	.705	.7855	.8347	.8687	.8990	.9233	.9451
60	.0437	.0984	.1726	.2868	.604	.7240	.7889	.8292	.8634	.8905	.9139
70	.0421	.0930	.1584	.2478	.504	.6605	.7379	.7882	.8270	.8575	.8821
80	.0406	.0883	.1469	.2215	.430	.5935	.6872	.7466	.7898	.8243	.8516
90	.0391	.0845	.1381	.2019	.373	.5325	.6359	.7040	.7522	.7909	.8212
100	.0378	.0810	.1305	.1877	.333	.4815	.5880	.6630	.7160	.7571	.7911
150	.0325	.0674	.1054	.1461	.2337	.3267	.4151	.4925	.5549	.6079	.6501
200	.0288	.0586	.0898	.1220	.1900	.2591	.3271	.3907	.4491	.5006	.5443
250	.0257	.0518	.0788	.1065	.1629	.2192	.2743	.3274	.3773	.4237	.4672
300	.0233	.0468	.0707	.0951	.1434	.1923	.2388	.2850	.3259	.3691	.4072
350	.0213	.0427	.0643	.0857	.1292	.1725	.2137	.2540	.2928	.3284	.3637
400	.0197	.0393	.0591	.0788	.1178	.1565	.1942	.2308	.2650	.2979	.3293
450	.0183	.0365	.0547	.0726	.1086	.1441	.1786	.2117	.2431	.2738	.3019
500	.0171	.0340	.0509	.0677	.1009	.1339	.1658	.1962	.2253	.2536	.2802
550	.0160	.0319	.0477	.0635	.0945	.1250	.1546	.1833	.2104	.2370	.2614
600	.0151	.0301	.0449	.0597	.0887	.1174	.1450	.1722	.1979	.2227	.2457
650	.0143	.0284	.0424	.0563	.0837	.1107	.1368	.1626	.1872	.2102	.2321
700	.0135	.0269	.0402	.0534	.0794	.1048	.1296	.1538	.1767	.1992	.2205
750	.0128	.0256	.0382	.0508	.0754	.0995	.1233	.1460	.1682	.1895	.2101
800	.0122	.0244	.0364	.0484	.0718	.0948	.1173	.1391	.1603	.1806	.2009
850	.0117	.0233	.0348	.0462	.0686	.0906	.1123	.1328	.1532	.1729	.1924
900	.0112	.0223	.0333	.0442	.0657	.0868	.1073	.1272	.1468	.1657	.1841
950	.0107	.0213	.0319	.0422	.0630	.0832	.1026	.1222	.1404	.1589	.1764
1,000	.0103	.0205	.0307	.0407	.0604	.0797	.0986	.1174	.1350	.1527	.1697

Temperature in °C	Pressure in bars									
	500	600	700	800	900	1,000	1,100	1,200	1,300	1,400
0	1.1024	1.1227	1.1405	1.1569	1.1698	1.1841	1.1990	1.2101	1.2211	1.2305
10	1.0759	1.0974	1.1167	1.1340	1.1485	1.1636	1.1783	1.1908	1.2022	1.2127
20	1.0482	1.0719	1.0928	1.1112	1.1272	1.1431	1.1586	1.1716	1.1835	1.1949
30	1.0200	1.0462	1.0689	1.0883	1.1060	1.1226	1.1386	1.1525	1.1650	1.1772
40	.9916	1.0204	1.0450	1.0659	1.0848	1.1022	1.1187	1.1335	1.1466	1.1595
50	.9630	.9945	1.0211	1.0434	1.0637	1.0819	1.0990	1.1146	1.1284	1.1419
60	.9340	.9686	.9973	1.0210	1.0427	1.0617	1.0795	1.0958	1.1104	1.1244
70	.9050	.9427	.9735	.9988	1.0218	1.0417	1.0603	1.0772	1.0926	1.1071
80	.8763	.9168	.9498	.9769	1.0010	1.0219	1.0413	1.0589	1.0750	1.0900
90	.8478	.8911	.9262	.9552	.9804	1.0023	1.0225	1.0409	1.0576	1.0731
100	.8195	.8655	.9038	.9337	.9600	.9830	1.0041	1.0231	1.0405	1.0565
150	.6889	.7469	.7928	.8309	.8628	.8912	.9171	.9386	.9590	.9781
200	.5833	.6485	.6998	.7423	.7789	.8111	.8391	.8634	.8860	.9072
250	.5064	.5709	.6244	.6687	.7065	.7411	.7707	.7971	.8217	.8441
300	.4444	.5088	.5626	.6076	.6457	.6805	.7115	.7390	.7651	.7890
350	.3967	.4583	.5117	.5563	.5945	.6291	.6607	.6884	.7155	.7400
400	.3601	.4172	.4684	.5130	.5514	.5859	.6173	.6450	.6722	.6969
450	.3311	.3840	.4325	.4758	.5147	.5487	.5798	.6081	.6350	.6591
500	.3068	.3565	.4020	.4438	.4824	.5165	.5470	.5757	.6021	.6264
550	.2866	.3335	.3768	.4167	.4539	.4877	.5182	.5467	.5728	.5975
600	.2693	.3139	.3553	.3932	.4289	.4620	.4929	.5208	.5467	.5715
650	.2546	.2969	.3366	.3729	.4066	.4392	.4699	.4978	.5232	.5477
700	.2420	.2819	.3201	.3553	.3876	.4192	.4489	.4770	.5019	.5262
750	.2303	.2685	.3052	.3393	.3711	.4015	.4301	.4578	.4826	.5066
800	.2197	.2567	.2916	.3249	.3560	.3856	.4134	.4403	.4651	.4888
850	.2105	.2461	.2796	.3118	.3422	.3710	.3981	.4245	.4491	.4724
900	.2020	.2365	.2689	.2999	.3296	.3579	.3841	.4102	.4345	.4573
950	.1941	.2276	.2592	.2893	.3180	.3458	.3717	.3971	.4208	.4434
1,000	.1868	.2194	.2503	.2796	.3079	.3346	.3603	.3850	.4078	.4305

References

Abelson, P. H., ed. 1959. Researches in Geochemistry. Wiley, New York. 511 pp.

———, ed. 1967. Researches in Geochemistry, Volume 2. Wiley, New York. 663 pp.

Ahrens, T. J., and G. Schubert. 1975. "Gabbro-eclogite reaction rate and its geophysical significance," *Rev. Geophysics and Space Physics,* 13, 383–400.

Akimoto, S., and H. Fujisawa. 1968. "Olivine-spinel solid solution equilibria in the system Mg_2SiO_4–Fe_2SiO_4," *Jour. Geophys. Research,* 73, 1467–1479.

Albee, A. L. 1965. "Phase equilibria in three assemblages of kyanite-zone pelitic schists, Lincoln Mountain Quadrangle, Central Vermont," *Jour. Petrology,* 6, 246–301.

Albee, A. L., and E-an Zen. 1969. "Dependence of the zeolitic facies on the chemical potentials of CO_2 and H_2O," in Contributions to Physico-chemical Petrology (Nauka, Moscow), pp. 249–260.

Allen, J. C., P. J. Modreski, C. Haygood, and A. L. Boettcher. 1972. "The role of water in the mantle of the Earth: the stability of amphiboles and micas," *24th Intl. Geol. Congress (Montreal),* sec. 2, pp. 231–240.

Althaus, E., E. Karotke, K. H. Nitsch, and H. G. F. Winkler. 1970. "An experimental reexamination of the upper stability limit of muscovite plus quartz," *Neues Jahrbuch Miner. Mh.,* 7, 325–336.

Anderson, D. L., and R. L. Kovach. 1969. "Universal dispersion tables, III: Free oscillation variational parameters," *Bull. Seismological Soc. America,* 59, 1667–1693.

Anderson, O. 1915. "The system anorthite–forsterite–silica," *Amer. Jour. Sci.,* 4th series, 39, 407–454.

Bailey, E. B., C. T. Clough, W. B. Wright, J. E. Richey, and G. V. Wilson. 1924. Tertiary and post-Tertiary geology of Mull, Loch Aline, and Oban. Mem. Geol. Survey, Scotland. 445 pp.

Barnes, H. L., and W. G. Ernst. 1963. "Ideality and ionization in hydrothermal fluids: the system MgO–H_2O–$NaOH$," *Amer. Jour. Sci.,* 261, 129–150.

Barrow, G. 1893. "On an intrusion of muscovite–biotite–gneiss in the southeastern Highlands of Scotland, and its accompanying metamorphism," *Quart. Jour. Geol. Soc. London,* 49, 330–358.

———. 1912. "On the geology of Lower Dee-side and the southern Highland border," *Proc. Geol. Assoc. London,* 23, 274–290.

Barth, T. F. W. 1961. "Ideas on the interrelation of igneous and sedimentary rock," *Finland Comm. géol. Bull.*, no. 196, pp. 321–326.

Bartholemé, P. 1962. "Iron-magnesium ratio in associated pyroxenes and olivines," in A. E. Engel, H. L. James, and B. F. Leonard, eds., Petrologic Studies: A Volume in Honor of A. F. Buddington (Geol. Soc. America), pp. 1–20.

Berman, R. 1962. "Graphite–diamond equilibrium boundary," in First International Congress on Diamonds in Industry (Ditchling Press, Sussex, England), pp. 291–295.

Best, M. G. 1975. "Migration of hydrous fluids in the upper mantle and potassium variation in calc-alkaline rocks." *Geology*, 3, 429–432.

Birch, F., and P. LeComte. 1960. "Temperature-pressure plane for albite composition," *Amer. Jour. Sci.*, 258, 209–217.

Boettcher, A. L. 1973. "Volcanism and orogenic belts—the origin of andesites," *Tectonophysics*, 17, 223–240.

Boettcher, A. L., and P. J. Wyllie. 1968a. "The calcite–aragonite transition measured in the system $CaO–CO_2–H_2O$," *Jour. Geology*, 76, 314–330.

———. 1968b. "Jadeite stability measured in the presence of silicate liquids in the system $NaAlSiO_4–SiO_2–H_2O$," *Geochim. et Cosmochim. Acta*, 32, 999–1012.

Boettcher, A. L., B. O. Mysen, and P. J. Modreski. 1975. "Melting in the mantle: phase relationships in natural and synthetic peridotite-H_2O and peridotite-H_2O-CO_2 systems at high pressures," *Phys. Chem. Earth*, 9, 855–867.

Bowen, N. L. 1913. "The melting phenomena of the plagioclase feldspars," *Amer. Jour. Sci.*, 4th series, 34, 577–599.

———. 1914. "The ternary system diopside–forsterite–silica," *Amer. Jour. Sci.*, 4th series, 38, 207–264.

———. 1915. "The crystallization of haplobasaltic, haplodioritic, and related magma," *Amer. Jour. Sci.*, 4th series, 40, 161–185.

———. 1928. The Evolution of the Igneous Rocks. Dover, New York. 334 pp.

———. 1940. "Progressive metamorphism of siliceous limestone and dolomites," *Jour. Geol.*, 48, 225–274.

Bowen, N. L., and O. Anderson. 1914. "The binary system $MgO–SiO_2$," *Amer. Jour. Sci.*, 4th series, 37, 487–500.

Bowen, N. L., and J. F. Schairer. 1935. "The system $MgO–FeO–SiO_2$," *Amer. Jour. Sci.*, 29, 151–217.

Boyd, F. R. 1959. "Hydrothermal investigations of amphiboles," in Abelson (1959), pp. 377–396.

———. 1973. "The pyroxene geotherm," *Geochim. et Cosmochim. Acta*, 37, 2533–2546.

———. 1974. "Ultramafic nodules from the Frank Smith kimberlite pipe, South Africa," *Carnegie Inst. Wash. Yearbook*, 73, 285–294.

Boyd, F. R., and J. L. England. 1960a. "Apparatus for phase-equilibrium measurements at pressures up to 50 kilobars and temperatures up to 1,750°C," *Jour. Geophys. Research*, 65, 741–748.

———. 1960b. "The quartz–coesite transition," *Jour. Geophys. Research*, 65, 749–756.

Boyd, F. R., J. L. England, and B. T. C. Davis. 1964. "Effects of pressure on the melting and polymorphism of enstatite, $MgSiO_3$," *Jour. Geophys. Research*, 69, 2101–2110.

Boyd, F. R., and P. H. Nixon. 1973. "Structure of the upper mantle beneath Lesotho," *Carnegie Inst. Wash. Yearbook*, 72, 431–445.

Boyd, F. R., and J. F. Schairer. 1964. "The system $MgSiO_3$–$CaMgSi_2O_6$," *Jour. Petrology,* 5, 275–309.

Bridgman, P. W. 1963. Dimensional Analysis. Yale Univ. Press, New Haven, Conn. 113 pp.

Broecker, W. S., and V. M. Oversby. 1971. Chemical Equilibria in the Earth. McGraw-Hill, New York. 318 pp.

Bundy, F. P., H. P. Bovenkerk, H. M. Strong, and R. H. Wentorf, Jr. 1961. "Diamond–graphite equilibrium line from growth and graphitization of diamond," *Jour. Chem. Physics,* 35, 383.

Cameron, K. L. 1975. "An experimental study of actinolite–cummingtonite phase relations with notes on the synthesis of Fe-rich anthophyllite," *Amer. Mineralogist,* 60, 375–390.

Carmichael, I. S. E., F. J. Turner, and J. Verhoogen. 1974. Igneous Petrology. McGraw-Hill, New York. 739 pp.

Clark, S. P., Jr., ed. 1966. Handbook of Physical Constants. Geol. Soc. America, Memoir 97. 587 pp.

Clark, S. P., and A. E. Ringwood. 1964. "Density distribution and constitution of the mantle," *Rev. Geophysics,* 2, 35–88.

Cohen, L. H., K. Ito, and G. C. Kennedy. 1967. "Melting and phase relations in an anhydrous basalt to 40 kilobars," *Amer. Jour. Sci.,* 265, 475–518.

Coombs, D. S. 1960. "Lower-grade mineral facies in New Zealand," *Proceedings 21st Internatl. Geol. Cong. (Copenhagen, 1960),* sec. 13, pp. 339–351.

———. 1961. "Some recent work on the lower grades of metamorphism," *Austral. Jour. Sci.,* 24, 203–215.

Coombs, D. S., A. J. Ellis, W. S. Fyfe, and A. M. Taylor. 1959. "The zeolite facies, with comments on the interpretation of hydrothermal syntheses," *Geochim. et Cosmochim. Acta,* 17, 53–107.

Crawford, W. A., and W. S. Fyfe. 1964. "Calcite–aragonite equilibrium at 100°C," *Science,* 144, 1569–1570.

Daly, R. A. 1933. Igneous Rocks and the Depths of the Earth. McGraw-Hill, New York. 598 pp.

Danielsson, A. 1950. "Das Calcit–Wollastonitgleichgewicht," *Geochim. et Cosmochim. Acta,* 1, 55–69.

Darken, L. S., and R. W. Gurry. 1953. Physical Chemistry of Metals. McGraw-Hill, New York, 535 pp.

Davis, B. T. C., and F. R. Boyd. 1966. "The join $Mg_2Si_2O_6$–$CaMgSi_2O_6$ at 30 kilobars pressure and its application to pyroxenes from kimberlites," *Jour. Geophys. Research,* 71, 3567–3576.

Deer, W. A., R. A. Howie, and J. Zussman. 1963. Rock-forming Minerals, Vol. 4: Framework Silicates. Wiley, New York. 435 pp.

Dickinson, W. R. 1970. "Relations of andesites, granites, and derivative sandstones to arc-trench tectonics," *Rev. Geophysics and Space Physics,* 8, 813–860.

Eggler, D. H. 1974. "Effect of CO_2 on the melting of peridotite," *Carnegie Inst. Wash. Yearbook,* 73, 215–224.

Elsasser, W. M. 1971. "Sea-floor spreading as thermal convection," *Jour. Geophys. Research,* 76, 1101–1112.

Engel, A. E. J., and C. G. Engel. 1964. "Composition of basalts from the mid-Atlantic ridge," *Science,* 144, 1330–1333.

Engel, A. E. J., C. G. Engel, and R. G. Havens. 1965. "Chemical characteristics of oceanic basalts and the upper mantle," *Geol. Soc. America Bull.,* 76, 719–734.

Epstein, S., and H. P. Taylor. 1967. "Variation of O^{18}/O^{16} in minerals and rocks," in Abelson (1967), pp. 29–62.

Ernst, W. G. 1962. "Synthesis, stability relations, and occurrence of riebeckite and riebeckite–arfvedsonite solid solutions," *Jour. Geol.,* 70, 689–736.

———. 1963a. "Petrogenesis of glaucophane schists," *Jour. Petrology,* 4, 1–30.

———. 1963b. "Significance of phengitic micas from low-grade schists," *Amer. Mineralogist,* 48, 1357–1373.

———. 1964. "Petrochemical study of coexisting minerals from low-grade schists, Eastern Shikoku, Japan," *Geochim. et Cosmochim. Acta,* 28, 1631–1668.

———. 1966. "Synthesis and stability relations of ferrotremolite," *Amer. Jour. Sci.,* 264, 37–65.

———. 1968. Amphiboles, Crystal Chemistry, Phase Relations, and Occurrence. Springer-Verlag, New York. 125 pp.

———. 1969. Earth Materials. Prentice-Hall, Englewood Cliffs, N. J. 150 pp.

Eskola, P. 1914. On the Petrology of the Orijärvi Region in Southwestern Finland. *Bull. Comm. géol. Finlande,* no. 40, 279 pp.

———. 1939. "Die metamorphen Gesteine," in T. F. W. Barth, C. W. Correns, and P. Eskola, Die Entstehung der Gesteine (Springer-Verlag, Berlin), pp. 263–407.

Essene, E., B. J. Henson, and D. H. Green, 1970. "Experimental study of amphibolite and eclogite stability," *Physics Earth and Planetary Interiors,* 3, 378–384.

Eugster, H. P. 1957. "Heterogeneous reactions involving oxidation and reduction at high pressures and temperatures," *Jour. Chem. Physics,* 26, 1760–1761.

Eugster, H. P., and D. R. Wones. 1962. "Stability relations of the ferruginous biotite, annite," *Jour. Petrology,* 3, 82–125.

Eugster, H. P., and G. B. Skippen. 1967. "Igneous and metamorphic reactions involving gas equilibria," in Abelson (1967), pp. 492–520.

Evans, B. W. 1965. "Application of a reaction-rate method to the breakdown equilibria of muscovite and muscovite plus quartz," *Amer. Jour. Sci.,* 263, 647–667.

Evans, B. W., and V. Trommsdorff. 1974. "Stability of enstatite + talc, and CO_2-metasomatism of metaperidotite, Val d'Efra, Lepontine Alps," *Amer. Jour. Sci.,* 274, 274–296.

Fisher, J. R., and E-an Zen. 1971. "Thermochemical calculations from hydrothermal phase-equilibrium data and the free energy of H_2O," *Amer. Jour. Sci.,* 270, 297–314.

Fletcher, R. C., and R. H. McCallister. 1974. "Spinodal decomposition as a possible mechanism in the exsolution of clinopyroxene," *Carnegie Inst. Wash. Yearbook,* 73, 396–399.

French, B. M. 1966. "Some geological implications of equilibrium between graphite and a C–H–O gas phase at high temperatures and pressures," *Rev. Geophysics,* 4, 223–253.

Fudali, R. F. 1963. "Experimental studies bearing on the origin of pseudoleucite and associated problems of alkalic rock systems," *Geol. Soc. America Bull.,* 74, 1101–1126.

Fyfe, W. S., F. J. Turner, and J. Verhoogen. 1958. Metamorphic Reactions and Metamorphic Facies. *Geol. Soc. America,* Memoir 73, 260 pp.

Ganguly, J. 1972. "Staurolite stability and related parageneses: theory, experiments,
and applications," *Jour. Petrology,* 13, 335–365.

Gilbert, M. C. 1966. "Synthesis and stability relations of the hornblende ferropargasite," *Amer. Jour. Sci.,* 264, 698–742.

Glasstone, S. 1947. Thermodynamics for Chemists. Van Nostrand, New York. 522 pp.

Goldschmidt, V. M. 1911. Die Kontaktmetamorphose in Kristianiagebiet. *Oslo
Vidensk. Skr.,* vol. I, Math.-Natur. Kl., no. 11. 405 pp.

Goldsmith, J. R. 1959. "Some aspects of the geochemistry of carbonates," in Abelson
(1959), pp. 336–358.

Goldsmith, J. R., and H. C. Heard. 1961. "Subsolidus relations in the system $CaCO_3$–
$MgCO_3$, *Jour. Geol.,* 69, 45–74.

Goldsmith, J. R., and R. C. Newton. 1969. *P-T-x* relations in the system $CaCO_3$–
$MgCO_3$ at high temperatures and pressures," *Amer. Jour. Sci.,* 267-A, 160–190.

Gordon, T. M. 1973. "Determination of internally consistent thermodynamic data from
phase-equilibrium experiments," *Jour. Geol.,* 81, 199–208.

Gordon, T. M., and H. J. Greenwood. 1971. "The stability of grossularite in H_2O–CO_2
mixtures," *Amer. Mineralogist,* 56, 1674–1688.

Green, D. H. 1972. "Magmatic activity as the major process in the chemical evolution
of the Earth's crust and mantle," *Tectonophysics,* 13, 47–71.

Green, D. H., and A. E. Ringwood. 1967a. "An experimental investigation of the
gabbro-to-eclogite transformation and its petrological applications," *Geochim. et
Cosmochim. Acta,* 31, 767–833.

———. 1967b. "The genesis of basalt magmas," *Contr. Mineral. and Petrology,* 15,
103–190.

Green, T. H., and A. E. Ringwood. 1968. "Genesis of the calc-alkaline igneous suite."
Contr. Mineral, and Petrology, 18. 105–162.

Greenwood, H. J. 1961. "The system $NaAlSi_2O_6$–H_2O–Argon: total pressure and
water pressure in metamorphism," *Jour. Geophys. Research.* 66, 3923–3946.

———. 1963. "The synthesis and stability of anthophyllite," *Jour. Petrology,* 4, 317–351.

———. 1967a. "Wollastonite: stability in H_2O–CO_2 mixtures and occurrence in a
contact-metamorphic aureole near Salmo, British Columbia, Canada," *Amer.
Mineralogist,* 52, 1669–1680.

———. 1967b. "The *n*-dimensional tie-line problem," *Geochim. et Cosmochim. Acta,*
31, 465–490.

———. 1967c. "Mineral equilibria in the system MgO–SiO_2–H_2O–CO_2," in Abelson,
(1967), pp. 542–567.

———. 1968. "Matrix methods and the phase rule in petrology," *Proc. 23rd Intl. Geol.
Congress (Prague),* 6, 267–279.

———. 1970. "Anthophyllite: corrections and comments on its stability," *Amer. Jour.
Sci.,* 270, 151–154.

———. 1972. "Al^{IV}-Si^{IV} disorder in sillimanite and its effect on phase relations of
the aluminum silicate minerals," *Geol. Soc. America Memoir,* 132, 553–571.

———. 1975. "Buffering of pore fluids by metamorphic reactions," *Amer. Jour. Sci.,*
275, 573–593.

Greig, J. W. 1927. "Immiscibility in silicate melts," *Amer. Jour. Sci.,* fifth series, 13, 1–44.

Griggs, D. T. 1972. "The sinking lithosphere and the focal mechanism of deep earthquakes," in E. C. Robertson, J. F. Hays, and L. Knopoff, eds., The Nature of the Solid Earth (McGraw-Hill, New York), pp. 361–384.

Grover, J. E., and P. M. Orville. 1969. "The partitioning of cations between coexisting single- and multi-site phases with application to the assemblages orthopyroxene-clinopyroxene and orthopyroxene–olivine," *Geochim. et Cosmochim. Acta,* 33, 205–226.

Harker, R. I., and O. F. Tuttle. 1956. "Experimental data on the P_{CO_2}–T curve for the reaction calcite + quartz = wollastonite + carbon dioxide," *Amer. Jour. Sci.,* 254, 239–256.

Hasebe, K., N. Fujii, and S. Uyeda. 1970. "Thermal processes under island arcs, *Tectonophysics,* 10, 335–355.

Hatherton, T., and W. R. Dickinson. 1969. "The relationship between andesitic volcanism and seismicity in Indonesia, the Lesser Antilles, and other island arcs," *Jour. Geophys. Research,* 74, 5301–5310.

Helz, R. T. 1973. "Phase relations of basalts in their melting range at P_{H_2O} = 5 kb as a function of oxygen fugacity," *Jour. Petrology,* 14, 249–302.

Hlabse, T., and O. J. Kleppa. 1968. "The thermochemistry of jadeite," *Amer. Mineralogist,* 53, 1281–1292.

Holloway, J. R. 1973. "The system pargasite–H_2O–CO_2: a model for melting of a hydrous mineral with a mixed-volatile fluid—I. Experimental results to 8 kbar," *Geochim. et Cosmochim. Acta,* 37, 651–666.

Holloway, J. R., and C. W. Burnham. 1972. "Melting relations of basalt with equilibrium water pressure less than total pressure," *Jour. Petrology,* 13, 1–29.

Hoschek, G. 1969. "The stability of staurolite and chloritoid and their significance in metamorphism of pelitic rocks," *Contr. Mineral. and Petrology,* 22, 208–232.

———. 1973. "Die Reaktion Phlogopit + Calcit + Quarz = Tremolit + Kalifeldspat + H_2O + CO_2," *Contr. Mineral. and Petrology,* 39, 231–237.

Hsu, L. C. 1968. "Selected phase relationships in the system Al–Mn–Fe–Si–O–H: a model for garnet equilibria," *Jour. Petrology,* 9, 40–83.

Ito, K., and G. C. Kennedy. 1967. "Melting and phase relations in a natural peridotite to 40 kilobars," *Amer. Jour. Sci.,* 265, 519–538.

———. 1971. "An experimental study of the basalt–garnet granulite–eclogite transition," *Geophys. Monograph Ser.,* 14, 303–314.

———. 1974. "The composition of liquids formed by partial melting of eclogites at high temperatures and pressures," *Jour. Geol.,* 82, 383–392.

Jamieson, J. C. 1953. "Phase equilibria in the system calcite–aragonite," *Jour. Chem. Physics,* 21, 1385–1390.

Johannes, W., P. M. Bell, H. K. Mao, A. L. Boettcher, D. W. Chipman, J. F. Hays, R. C. Newton, and F. Seifert. 1971. "An interlaboratory comparison of piston-cylinder pressure calibration using the albite breakdown reaction," *Contr. Mineral. and Petrology,* 32, 24–38.

Johannes, W., and D. Puhan. 1971. "The calcite–aragonite transition reinvestigated," *Contr. Mineral. and Petrology,* 31, 28–38.

Kennedy, G. C., and W. T. Holser. 1966. "Pressure-volume-temperature and phase relations of water and carbon dioxide," *Geol. Soc. America Memoir,* 97, 371–384.

Kern, R., and A. M. Weisbrod, 1967. Thermodynamics for Geologists. Freeman, Cooper, San Francisco. 304 pp.

Kerrick, D. M. 1968. "Experiments on the upper stability limit of pyrophyllite," *Amer. Jour. Sci.,* 266, 204–214.

———. 1974. "Review of metamorphic mixed-volatile (H_2O–CO_2) equilibria," *Amer. Mineralogist,* 59, 729–762.

Korzhinskii, D. S. 1959. Physicochemical Basis of the Analysis of the Paragenesis of Minerals. Consultants Bur., New York. 142 pp.

Kretz, R. 1959. "Chemical study of garnet, biotite, and hornblende from gneisses of southwestern Quebec, with emphasis on distribution of elements in coexisting minerals," *Jour. Geol.,* 67, 371–402.

———. 1963. "Distribution of magnesium and iron between orthopyroxene and calcic pyroxene in natural mineral assemblages," *Jour. Geol.,* 71, 773–785.

Kuno, H. 1966. "Lateral variation of basaltic magma type across continental margins and island arcs," *Bull. Volcan.,* 29, 195–202.

———. 1968. "Differentiation of basalt magmas," in H. H. Hess and A. Poldervaart, eds., Basalts (Wiley, New York), pp. 623–688.

Kushiro, I. 1972a. "Determination of liquidus relations in synthetic silicate systems with electron-probe analysis: the system forsterite–diopside–silica at 1 atmosphere," *Amer. Mineralogist,* 57, 1260–1271.

———. 1972b. "Effect of water on the composition of magmas formed at high pressures," *Jour. Petrology,* 13, 311–334.

———. 1973a. "The system diopside–anorthite–albite, determination of compositions of coexisting phases," *Carnegie Inst. Wash. Yearbook,* 72, 502–507.

———. 1973b. "Origin of some magmas in oceanic and circum-oceanic regions," *Tectonophysics,* 17, 211–222.

Kushiro, I., H. S. Yoder, Jr., and M. Nishikawa. 1968. "Effect of water on the melting of enstatite." *Geol. Soc. America Bull.,* 79, 1685–1692.

Lambert, I. B., and P. J. Wyllie. 1972. "Melting of gabbro (quartz eclogite) with excess water to 35 kilobars, with geological applications," *Jour. Geol.,* 80, 693–708.

Lewis, G. N., and M. Randall. 1961. Thermodynamics. Revised by K. S. Pitzer and L. Brewer. McGraw-Hill, New York. 723 pp.

Liou, J. G. 1971a. "*P–T* stabilities of laumontite, wairakite, lawsonite, and related minerals in the system $CaO \cdot Al_2O_3 \cdot 2SiO_2$–$SiO_2$–$H_2O$," *Jour. Petrology,* 12, 379–411.

———. 1971b. "Synthesis and stability relations of prehnite, $Ca_2Al_2Si_3O_{10}(OH)_2$" *Amer. Mineralogist,* 56, 507–531.

———. 1971c. "Analcime equilibria," *Lithos,* 389–402.

———. 1971d. "Stilbite–laumontite equilibrium," *Contr. Mineral. and Petrology,* 31, 171–177.

Lovering, J. F. 1958. "The nature of the Mohorovicic discontinuity, *Trans. Amer. Geophys. Union,* 39, 947–955.

Luth, W. C., R. H. Jahns, and O. F. Tuttle. 1964. "The granite system at pressures of 4 to 10 kilobars," *Jour. Geophys. Research,* 69, 759–773.

Luth, W. C., and O. F. Tuttle. 1966. "The alkali feldspar solvus in the system Na_2O–K_2O–Al_2O_3–SiO_2–H_2O," *Amer. Mineralogist,* 51, 1359–1373.

MacGregor, I. D. 1968. "Mafic and ultramafic inclusions as indicators of the depth of origin of basaltic magmas," *Jour. Geophys. Research,* 73, 3737–3745.

———. 1974. "The system MgO–Al$_2$O$_3$–SiO$_2$: solubility of Al$_2$O$_3$ in enstatite for spinel and garnet peridotite compositions," *Amer. Mineralogist,* 59, 110–119.

———. 1975. "Petrologic and geophysical significance of mantle xenoliths in basalts." *Rev. Geophysics and Space Physics,* 13, 90–93.

MacGregor, I. D., and A. R. Basu. 1974. "Thermal structure of the lithosphere: a petrologic model," *Science,* 185, 1007–1011.

Marsh, B. D., and I. S. E. Carmichael. 1974. "Benioff zone magmatism," *Jour. Geophys. Research,* 79, 1196–1206.

Mason, B. 1966. Principles of Geochemistry. Wiley, New York. 329 pp.

Matsuda, T., and S. Uyeda. 1970. "On the Pacific-type orogeny and its model-extension of the paired-belts concept and possible origin of marginal seas," *Tectonophysics,* 11, 5–27.

Matsui, Y., and S. Banno. 1965. "Intracrystalline exchange equilibria in silicate solid solutions," *Proc. Japan Acad.,* 41, 461–466.

Miyashiro, A. 1961. "Evolution of metamorphic belts," *Jour. Petrology,* 2, 277–311.

———. 1973. Metamorphism and Metamorphic Belts. George Allen & Unwin, London. 492 pp.

———. 1974. "Volcanic rock series in island arcs and active continental margins," *Amer. Jour. Sci.,* 274, 321–355.

Modreski, P. J., and A. L. Boettcher. 1973. "Phase relationships of phlogopite in the system K$_2$O–MgO–CaO–Al$_2$O$_3$–SiO$_2$–H$_2$O to 35 kilobars: a better model for micas in the interior of the Earth," *Amer. Jour. Sci.,* 273, 385–414.

Moore, W. J. 1972. Physical Chemistry. Prentice-Hall, New York. 977 pp.

Morse, S. A. 1970. "Alkali feldspars with water at 5 kb pressure," *Jour. Petrology,* 11, 221–253.

Muan, A., and E. F. Osborn. 1956. "Phase equilibria at liquidus temperatures in the system MgO–FeO–Fe$_2$O$_3$–SiO$_2$," *Jour. Amer. Ceram. Soc.,* 39, 121–140.

Mueller, R. F. 1960. "Compositional characteristics and equilibrium relations in mineral assemblages of a metamorphosed iron formation," *Amer. Jour. Sci.,* 258, 449–497.

———. 1962. "Energetics of certain silicate solid solutions," *Geochim. et Cosmochim. Acta,* 26, 581–598.

———. 1969. "Kinetics and thermodynamics of intracrystalline distributions," *Mineral. Soc. America,* special paper no. 2, pp. 83–94.

Mysen, B. O., and A. L. Boettcher. 1975. "Melting in a hydrous mantle, II: Geochemistry of crystals and liquids formed by anatexis of mantle peridotite with controlled activities of H$_2$O, CO$_2$, and O$_2$," *Jour. Petrology,* 16, 549–593.

Mysen, B. O., I. Kushiro, I. A. Nicholls, and A. E. Ringwood. 1974. "A possible mantle origin for andesitic magmas: discussion of a paper by Nicholls and Ringwood," *Earth and Planetary Sci. Lett.,* 21, 221–229.

Newton, R. C. 1966a. "Kyanite–andalusite equilibrium from 700° to 800°C," *Science,* 153, 170–172.

———. 1966b. "Some calc-silicate equilibrium relations," *Amer. Jour. Sci.,* 264, 204–222.

Newton, R. C., and G. C. Kennedy. 1963. "Some equilibrium reactions in the join CaAl$_2$Si$_2$O$_8$–H$_2$O," *Jour. Geophys. Research,* 68, 2967–2984.

Newton, R. C., and J. V. Smith. 1967. "Investigations concerning the breakdown of albite at depth in the earth," *Jour. Geol.,* 75, 268–286.

Newton, R. C., J. R. Goldsmith, and J. V. Smith. 1969. "Aragonite crystallization from strained calcite at reduced pressures and its bearing on aragonite in low-grade metamorphism," *Contr. Mineral. and Petrology,* 22, 335–348.

Nicholls, I. A. 1974. "Liquids in equilibrium with peridotitic mineral assemblages at high water pressures," *Contr. Mineral. and Petrology,* 45, 289–316.

Nicolas, A., and E. D. Jackson. 1972. "Répartition en deux provinces des péridotites des chaînes alpines longeant la Méditerranée: implications géotectoniques," *Schwiez. Mineral. Petrograph. Mitt.,* 52, 479–495.

Nitsch, K. H. 1968. "Die Stabilität von Lawsonit," *Naturwissenschaften,* 8, 388–389.

Nockolds, S. R. 1954. "Average chemical compositions of some igneous rocks," *Geol. Soc. America Bull.,* 65, 1007–1032.

O'Hara, M. J. 1965. "Primary magmas and the origin of basalts," *Scot. Jour. Geol.,* 1, 19–40.

———. 1967. "Mineral facies in ultrabasic rocks," in P. J. Wyllie, ed., Ultramafic and Related Rocks (Wiley, New York), pp. 7–18.

Onuki, H., and W. G. Ernst. 1969. "Coexisting sodic amphiboles and sodic pyroxenes from blueschist facies metamorphic rocks," *Mineral. Soc. America Special Paper,* no. 2, pp. 241–250.

Orville, P. M. 1963. "Alkali ion exchange between vapor and feldspar phases," *Amer. Jour. Sci.,* 261, 201–237.

———. 1967. "Unit cell parameters of the microcline–low albite and the sanidine–high albite solid-solution series," *Amer. Mineralogist,* 52, 55–86.

Orville, P. M., and H. J. Greenwood. 1965. "Determination of ΔH of reaction from experimental pressure-temperature curves," *Amer. Jour. Sci.,* 263, 678–683.

Osborn, E. F. 1942. "The system $CaSiO_3$–diopside–anorthite," *Amer. Jour. Sci.,* 240, 751–788.

———. 1959. "Role of oxygen pressure in the crystallization and differentiation of basaltic magma," *Amer. Jour. Sci.,* 257, 609–647.

———. 1962. "Reaction series for subalkaline igneous rocks based on different oxygen-pressure conditions," *Amer. Mineralogist,* 47, 211–226.

Oxburgh, E. R., and D. L. Turcotte. 1968. "Mid-ocean ridges and geothermal distribution during mantle convection," *Jour. Geophys. Research,* 73, 2643–2661.

———. 1970. "Thermal structure of island arcs," *Geol. Soc. America Bull.,* 81, 1665–1668.

Poldervaart, A. 1955. "Chemistry of the earth's crust," *Geol. Soc. America Special Paper,* no. 62, pp. 119–144.

Poty, B. P., H. A. Stalder, and A. M. Weisbrod. 1974. "Fluid inclusions studies in quartz from fissures of western and central Alps," *Schweiz. Min. Petr. Mitt.,* 54, 717–752.

Presnall, D. C. 1969. "The geometrical analysis of partial fusion," *Amer. Jour. Sci.,* 267, 1178–1194.

Presnall, D. C., and P. C. Bateman. 1973. "Fusion relations in the system $NaAlSi_3O_8$–$CaAl_2Si_2O_8$–$KAlSi_3O_8$–SiO_2–H_2O and generation of granitic magmas in the Sierra Nevada batholith," *Geol. Soc. America Bull.,* 84, 3181–3202.

Ramberg, H., and G. W. deVore. 1951. "The distribution of Fe^{++} and Mg^{++} in co-existing olivines and pyroxenes," *Jour. Geol.,* 59, 193–210.

Ricci, J. E. 1951. The Phase Rule and Heterogeneous Equilibrium. Van Nostrand, New York. 505 pp.

Richardson, S. W. 1968. "Staurolite stability in a part of the system Fe–Al–Si–O–H," *Jour. Petrology,* 9, 467–488.

Richardson, S. W., P. M. Bell, and M. C. Gilbert. 1968. "Kyanite–sillimanite equilibrium between 700° and 1500°C," *Amer. Jour. Sci.,* 266, 513–541.

Richardson, S. W., M. C. Gilbert, and P. M. Bell. 1969. "Experimental determination of kyanite–andalusite and andalusite–sillimanite equilibria: the aluminum silicate triple point," *Amer. Jour. Sci.,* 267, 259–272.

Richardson, W. A., and G. Sneesby. 1922. "The frequency distribution of igneous rocks, I: Frequency distribution of the major oxides in analyses of igneous rocks," *Mineral. Mag.,* 19, 303–313.

Ringwood, A. E. 1966. "The chemical composition and origin of the earth," in P. M. Hurley, ed., Advances in Earth Science (M.I.T. Press, Cambridge, Mass.), pp. 287–356.

———. 1974. "The petrological evolution of island arc systems, *Quart. Jour. Geol. Soc. London,* 130, 183–204.

———. 1975. Composition and Petrology of the Earth's Mantle. McGraw-Hill, New York. 618 pp.

Ringwood, A. E., and D. H. Green. 1966. "An experimental investigation of the gabbro-eclogite transformation and some geophysical implications," *Tectonophysics,* 3, 383–427.

Ringwood, A. E., and A. Major. 1970. "The system Mg_2SiO_4–Fe_2SiO_4 at high pressures and temperatures," *Physics Earth and Planetary Interiors,* 3, 89–108.

Robertson, E. C., F. Birch, and G. J. F. MacDonald. 1957. "Experimental determination of jadeite stability relations to 25,000 bars," *Amer. Jour. Sci.,* 255, 115–137.

Robie, R. A., and D. R. Waldbaum. 1968. "Thermodynamic properties of minerals and related substances at 298.15°K (25.0°C) and one atmosphere (1.013 bars) pressure and at higher temperatures," *U.S. Geol. Survey Bull.* no. 1259. 256 pp.

Robin, P.-Y. 1974. "Thermodynamic equilibrium across a coherent interface in a stressed crystal," *Amer. Mineralogist,* 59, 1286–1298.

Ronov, A. B., and A. A. Yaroshevsky. 1969. "Chemical composition of the earth's crust," *Amer. Geophys. Union Geophysical Monograph,* no. 13, pp. 37–57.

Saxena, S. K. 1973. Thermodynamics of Rock-Forming Crystalline Solutions. Springer-Verlag, New York. 188 pp.

Schairer, J. F. 1955. "The ternary systems leucite–corundum–spinel and leucite–forsterite–spinel," *Jour. Amer. Ceram. Soc.,* 38, 153–158.

———. 1957. "Melting relations of the common rock-forming silicates," *Jour. Amer. Ceram. Soc.,* 40, 215–235.

Schairer, J. F., and N. L. Bowen. 1947a. "The system anorthite–leucite–silica," *Comm. géol. Finlande Bull.,* No. 140, pp. 67–87.

———. 1947b. "Melting relations in the systems Na_2O–Al_2O_3–SiO_2 and K_2O–Al_2O_3–SiO_2," *Amer. Jour. Sci.,* 245, 193–204.

———. 1955. "The system K_2O–Al_2O_3–SiO_2," *Amer. Jour. Sci.,* 253, 681–746.

Schairer, J. F., and H. S. Yoder, Jr. 1960. "The nature of residual liquids from crystallization, with data on the system nepheline–diopside–silica," *Amer. Jour. Sci.,* 258A, 273–283.

Schuiling, R. D., and B. W. Vink. 1967. "Stability relations of some titanium-minerals (sphene, perovskite, rutile, anatase)," *Geochim. et Cosmochim. Acta,* 31, 2399–2411.

Shaw, H. R. 1967. "Hydrogen osmosis in hydrothermal experiments," in Abelson (1967), pp. 521–541.

Smyth, F. H., and L. H. Adams. 1923. "The system calcium oxide–carbon dioxide," *Jour. Amer. Chem. Soc.*, 45, 1167–1184.

Steiner, J. C., R. H. Jahns, and W. C. Luth. 1975. "Crystallization of alkali feldspar and quartz in the haplogranite system $NaAlSi_3O_8$–$KAlSi_3O_8$–SiO_2–H_2O at 4 kb," *Geol. Soc. America Bull.*, 86, 83–98.

Stewart, D. B., and E. H. Roseboom, Jr. 1962. "Lower-temperature terminations of the three-phase region plagioclase–alkali feldspar–liquid," *Jour. Petrology*, 3, 280–315.

Storre, B. 1970. "Stabilitätsbedingungen Grossular-führender Paragenesen im System CaO–Al_2O_3–SiO_2–CO_2–H_2O," *Contr. Mineral. and Petrology*, 29, 145–162.

Storre, B., and K. H. Nitsch. 1972. "Die Reaktion 2 Zoisit + 1 CO_2 = 3 Anorthit + 1 Calcit + 1 H_2O," *Contr. Mineral. and Petrology*, 35, 1–10.

Streckeisen, A. L. 1967. "Classification and nomenclature of igneous rocks," *Neues Jahrb. Mineral. Abh.*, 107, 144–240.

Thompson, A. B. 1970. "Laumontite equilibria and the zeolite facies," *Amer. Jour. Sci.*, 269, 267–275.

———. 1971. "Analcite–albite equilibria at low temperatures," *Amer. Jour. Sci.*, 271, 79–92.

Thompson, J. B., Jr. 1955. "The thermodynamic basis for the mineral facies concept," *Amer. Jour. Sci.*, 253, 65–103.

———. 1959. "Local equilibrium in metasomatic processes," in Abelson (1959), pp. 427–457.

———. 1969. "Chemical reactions in crystals," *Amer. Mineralogist*, 54, 341–375.

Tilley, C. E. 1925. "A preliminary study of metamorphic zones in the southern Highlands of Scotland," *Quart. Jour. Geol. Soc. London*, 81, 100–112.

Touray, J. C. 1968. "Recherches géochimiques sur les inclusions à CO_2 liquide," *Bull. Soc. franç. Minèral. Crist.*, 91, 367–382.

Trommsdorff, V., and B. W. Evans. 1974. "Alpine metamorphism of peridotitic rocks," *Schweiz. Mineral. und Petrograph. Mitt.*, 54, 333–352.

Turcotte, D. L., and E. R. Oxburgh. 1972. "Mantle convection and the new global tectonics," *Ann. Rev. Fluid Mech.*, 4, 33–68.

Turner, F. J. 1968. Metamorphic Petrology. McGraw-Hill, New York. 403 pp.

Tuttle, O. F., and N. L. Bowen. 1958. "Origin of granite in the light of experimental studies in the system $NaAlSi_3O_8$–$KAlSi_3O_8$–SiO_2–H_2O," *Geol. Soc. America Memoir*, no. 74. 153 pp.

Tuttle, O. F., and J. L. England. 1955. "Preliminary report on the system SiO_2–H_2O," *Geol. Soc. America Bull.*, 66, 149–152.

Ulmer, G. C., ed. 1971. Research Techniques for High Pressure and High Temperature. Springer-Verlag, New York. 367 pp.

Urey, H. C., and H. Craig. 1953. "The composition of the stone meteorites and the origin of the meteorites," *Geochim. et Cosmochim. Acta*, 4, 36–82.

Velde, B. 1965. "Phengite micas: synthesis, stability, and natural occurrence," *Amer. Jour. Sci.*, 263, 886–913.

von Platen, H. 1965. "Kristallisation granitscher Schmelzen," *Contr. Mineral. and Petrology*, 11, 334–381.

Wager, L. R., and W. A. Deer. 1939. "Geological investigations in east Greenland, part III: The petrology of the Skaergaard intrusion, Kangerdlugssuaq, East Greenland," *Meddelelser om Grønland*, vol. 105, no. 4. 352 pp.

Warner, R. D., and W. C. Luth. 1974. "The diopside–orthoenstatite two-phase region in the system $CaMgSi_2O_6$–$Mg_2Si_2O_6$," *Amer. Mineralogist*, 59, 98–109.

Washington, H. S. 1917. "Chemical analyses of igneous rocks, published from 1884 to 1913," *U.S. Geol. Survey Prof. Paper*, no. 99. 1201 pp.

Weisbrod, A. M., and B. P. Poty. 1975. "Thermodynamics and geochemistry of the hydrothermal evolution of the Mayres pegmatite (southeastern Massif Central, France)," *Petrologie*, 1, 1–16 and 89–102.

Winkler, H. G. F. 1974. Petrogenesis of Metamorphic Rocks. Springer-Verlag, New York. 320 pp.

Wones, D. R., and H. P. Eugster. 1965. "Stability of biotite: experiment, theory, and application," *Amer. Mineralogist*, 50, 1228–1272.

Wones, D. R., and M. C. Gilbert. 1969. "The fayalite–magnetite–quartz assemblage between 600° and 800°C," *Amer. Jour. Sci.*, 267A, 480–488.

Wyllie, P. J. 1970. "Ultramafic rocks and the upper mantle," *Mineral. Soc. America Special Paper*, no. 3, pp. 3–32.

———. 1971. The Dynamic Earth. Wiley, New York. 416 pp.

Yoder, H. S., Jr. 1969. "Calc-alkaline andesites: experimental data bearing on the origin of their assumed characteristics," *Oregon Dept. Geol. Mineral Indust. Bull.*, 65, 77–89.

Yoder, H. S., Jr., D. B. Stewart, and J. R. Smith. 1957. "Ternary feldspars," *Carnegie Inst. Wash. Yearbook*, 56, 206–214.

Yoder, H. S., Jr., and C. E. Tilley. 1962. "Origin of basalt magmas: an experimental study of natural and synthetic rock systems," *Jour. Petrology*, 3, 342–532.

Yund, R. A., and R. H. McCallister. 1970. "Kinetics and mechanisms of exsolution," *Chem. Geol.*, 6, 5–30.

Zen, E-an. 1961. "The zeolite facies: an interpretation," *Amer. Jour. Sci.*, 259, 401–409.

———. 1966. "Construction of pressure-temperature diagrams for multicomponent systems after the method of Schreinemakers—a geometric approach," *U.S. Geol. Survey Bull.*, no. 1225, 56 pp.

———. 1969. "The stability relations of the polymorphs of aluminum silicate: a survey and some comments," *Amer. Jour. Sci.*, 267, 297–309.

———. 1970. "Comments on the thermodynamic constants and hydrothermal stability relations of anthophyllite," *Amer. Jour. Sci.*, 270, 136–150.

Zen, E-an, and E. H. Roseboom, Jr. 1972. "Some topological relationships in multi-systems of $n + 3$ phases, III: Ternary systems," *Amer. Jour. Sci.*, 272, 677–710.

Author Index

Italicized page numbers indicate that the indexed author is cited in a figure (including legend), or in both text and figure.

Subject Index

Italicized page numbers indicate that the indexed subject will be found in a figure (including legend), or in both text and figure. Notice the major groupings of subentries under: definitions; igneous rock and magma types; metamorphic; minerals; systems, chemical; thermodynamic relationships

anatexis. *See* igneous differentiation
    processes, partial fusion
anhydrous melt, 44, Chapters 4, 5
apparatus, experimental synthesis, 25–31
    cold-seal apparatus, 27, *28*
    furnace windings (or core), 26, 27, 29
    hydrothermal pressure apparatus, 25, 27, *28*
    insulation, 26
    internally heated gas apparatus, 27, *28,* 29
    piston-cylinder device, 25, 29, *30, 31*
    pressure correction, piston-cylinder
      device, 30
    pressure vessels, 27, *28–31*
    quenching furnace, 25, *26*
    thermocouples, 26, 27, 29
asthenosphere, 149–150, 162, 170, 171, *173, 174, 252, 253*

binary and pseudobinary systems, 34, 43–45, 48–50, 75–96, *156, 158, 162*
binary eutectic
    limited solid solution, 94–96
    no solid solution, 43–45, 75–82
binary minimum, 90, *91–93*
binary peritectic, 82–86
binary solid solution, 86–92
binary solvus, 92–94, 129, *156,* 198, *199*

carbon dioxide
    thermodynamic properties, 258, 267, 308–309
    See also reactions
chemical potential of mixing, *79*
continental crust
    composition, *64,* 67, 142, 172

differentiation, 142–143
generation, 171, 172, *174*
continental margin, 171, 172, *174,* 253
conventions
    distinction between component and
      phase, 3
    fluid, usage of the term, 69
    heat and work, signs, 7
    reactants and products, signs, 5–6
    standard-state, 6
cotectic curve, 97
critical end point, 129, 134, *135–138*
critical point, *69, 91, 92,* 130, *134–138*
crust-mantle differentiation and plate
    tectonics, 170–177
crust, origin, 146
crystallization paths, curved, 101–102, 107–109, *120, 136*

definitions (*see also* thermodynamic
    relationships)
    closed system, 3
    continuous-reaction relation, 88
    degrees of freedom, 21
    discontinuous-reaction relation, 83
    disequilibrium, 5
    equilibrium, 4
    eutectic point, 74
    extensive parameter, 5
    fluid, 2, 69
    fractional crystallization, 65
    fractional fusion, 77
    gas, 2
    grade of metamorphism, 180
    heterogeneous reaction, 4